Additive Friction Stir Deposition

Additive Manufacturing Materials and Technologies

Additive Friction Stir Deposition

HANG Z. YU
Virginia Tech, United States

Series Editor

MA QIAN

ELSEVIER

Elsevier
Radarweg 29, PO Box 211, 1000 AE Amsterdam, Netherlands
The Boulevard, Langford Lane, Kidlington, Oxford OX5 1GB, United Kingdom
50 Hampshire Street, 5th Floor, Cambridge, MA 02139, United States

Notices
Knowledge and best practice in this field are constantly changing. As new research and experience broaden our understanding, changes in research methods, professional practices, or medical treatment may become necessary.

Practitioners and researchers must always rely on their own experience and knowledge in evaluating and using any information, methods, compounds, or experiments described herein. In using such information or methods they should be mindful of their own safety and the safety of others, including parties for whom they have a professional responsibility.

To the fullest extent of the law, neither the Publisher nor the authors, contributors, or editors, assume any liability for any injury and/or damage to persons or property as a matter of products liability, negligence or otherwise, or from any use or operation of any methods, products, instructions, or ideas contained in the material herein.

ISBN: 978-0-12-824374-9

Publisher: Matthew Deans
Acquisitions Editor: Dennis McGonagle
Editorial Project Manager: Joshua Mearns
Production Project Manager: Kamesh Ramajogi
Cover Designer: Matthew Limbert

Typeset by MPS Limited, Chennai, India

Contents

Preface

The ability to manufacture new products that give better performance and incur less cost is a measure of the future competencies of a nation. Whether they are bulk-scale metals and ceramics used for the aerospace and automotive industries or thin films and nanoscale devices for the semiconductor industry, we always want them to be manufactured in an efficient, a more reliable, and a scalable manner while possessing an unprecedented supply of excellent properties. In this regard, the importance of advanced manufacturing cannot be overstated—a common sense already shared by scientists, engineers, entrepreneurs, industry leaders, and policymakers. As an example, numerous advanced manufacturing programs or initiatives have been witnessed over the last few years; America Makes, Manufacturing USA, and the Advanced Manufacturing Initiative by European Space Agency, to name just a few. In recent TMS and MS&T conferences, more than 1/3 of the symposia were related to advanced manufacturing.

Among subfields of advanced manufacturing, bulk-scale metal additive manufacturing represents a frontier of innovation rapidly changing how engineers design and fabricate critical components, such as fuel nozzles in engines and combustor liners. At the time of writing, the mainstream metal additive manufacturing technologies are known to be based on selective melting and rapid solidification. As fusion-based additive processes, these technologies share the same liquid bonding mechanism with casting and therefore are faced with managing the same challenges, such as porosity, dendritic microstructure, hot cracking, and high residual stress. Because of such issues, casting is not widely used for creating critical metal components in traditional metal manufacturing; instead, these components are usually processed via thermomechanical routes, like forging, rolling, and extrusion. This naturally raises the question: to fundamentally avoid the quality issues in fusion-based metal additive manufacturing, should we adopt a metal additive process that is thermomechanical processing-based rather than liquid processing-based?

During recent years, this question has motivated me to research an emerging metal additive manufacturing technology with a thermomechanical processing nature. The process is termed additive friction stir deposition, which leverages the friction stir principle and a material feeding mechanism to enable location-specific printing and free forming in 3D space. Viewed as a forging-based additive process, additive friction stir deposition inherits the advantages of forging in porosity removal and equiaxed microstructure formation, allowing for excellent mechanical performance in the as-printed state. It has shown great promise for large-scale additive manufacturing, structural repair, materials recycling, as well as operation under extreme conditions. Despite its relatively young age, extensive research and development efforts during the last few years have propelled this technology to the forefront of innovation with focused attention from the aerospace and defense sectors.

The goal of this book is to offer a timely overview of this vigorously growing field, in the hope that more talents from academia and industry will join the R&D efforts of additive friction stir deposition. New R&D advances will improve our fundamental understanding of the process, thereby unlocking its full potential for widespread applications. The study on additive friction stir deposition is interdisciplinary, necessitating joint efforts from experts in physics, metallurgy, materials science, fluid mechanics, mechanical engineering, and component design. As a materials scientist by education and training, I attempted to include knowledge from all relevant disciplines in this book, which may have been a stretch at times. Fortunately, the colleagues in this research field come from a variety of academic backgrounds; learning from them and incorporating their work into this book has hopefully compensated for any negative effects from my stretch.

This book aims to provide a comprehensive picture of additive friction stir deposition, from process fundamentals and materials flow to microstructure evolution and mechanical properties. Some special topics like printing of composites, tooling, and niche applications are also included. However, this book is not a conclusion by any means; given how rapidly this field is growing these days, extensive updates are expected in the next edition. Some of the ideas discussed in the last chapter ("Future Perspectives"), such as the incorporation of artificial intelligence and printing of hierarchical materials, may have matured into a discipline with a large number of research results by that time.

I am grateful to many people without whom writing this book would not be possible. First I would like to thank Dr. Jack Lesko, who initiated the Advanced Manufacturing Team at Virginia Tech and connected me with the folks at Aeroprobe Inc. back in 2016. This was the first time I learned of a commercially available additive friction stir deposition process, which is now known as the MELD technology. I would also like to thank Ms. Nanci Hardwick, the CEO of MELD Manufacturing Corporation, and Dr. Chase Cox, the Director of Technology at MELD Manufacturing Corporation. Through interactions with them, I was able to realize how additive friction stir deposition may have a big impact in real life, especially in the realm of large-scale additive manufacturing.

I am particularly grateful to Prof. Rajiv Mishra from the University of North Texas and Dr. Zhili Feng from Oak Ridge National Laboratory, who, as world-renowned experts in friction stir research, never hesitated to offer me invaluable career advice. I am grateful to Mr. Kyu Cho from the Army Research Laboratory as well as Mr. Nam Phan from NAVAIR, who not only provided support for my research in this area but also provided vital guidance regarding the research direction for critical applications. Helpful discussions and guidance from the industry folks, including from Dr. George Luckey, Dr. Joy Forsmark, and Mr. Eric Poczatek from Ford Motor Company, Dr. Nanzhu Zhao from Nissan Motor Corporation, Dr. Changjie Sun from GE Research, Dr. Michael Eller from Lockheed Martin Corporation, and Dr. Jim Lua from GEM Inc. are also greatly appreciated.

I cannot say "thank you" enough to the great students at Virginia Tech who have worked and who are working with me, for their groundbreaking research on additive friction stir deposition as well as their tremendous help in writing this book. I would like to acknowledge my previous Ph.D. students: Dr. Mackenzie Perry, Dr. Joey Griffiths, Dr. Hunter Rauch, and Dr. David Garcia, as well as current graduate students and postdocs: Doug Hartley, Jake Yoder, Greg Hahn, Ryan Gottwald, Kendall Knight, Robert Callaham, Robert Palisin, Donnie Erb, Hannah Glaser, and Dr. Nikhil Gotawala. Most of them read part of the book draft and provided critical comments. I would also like to acknowledge a few undergraduate students for their input: Justin He, Adam Seigler, and Gideon Crawford.

I am grateful to my colleagues in the research field of additive friction stir deposition, including Prof. Paul Allison, Prof. Brian Jordon, and Prof. Luke Brewer from the University of Alabama, Prof. Rajiv Mishra from the University of North Texas, as well as Prof. Daniel Fabijanic from Deakin University. Their work has been instrumental in enriching my knowledge and understanding of additive friction stir deposition; incorporating their work has improved the book quality. I would like to thank the editors from Elsevier for their patience in helping me prepare the book: Ms. Christina Gifford, Mr. Dennis McGonagle, and Mr. Joshua Mearns.

A book project like this would not have been possible without strong support from my family. I would like to express my heartfelt gratitude to my wife, Yunhui, for her enduring encouragement of my career and unwavering efforts in caring for our toddler, Chris—who also deserves credit for being able to sleep through the night as a young baby during the time I worked on the book draft.

Hang Z. Yu
Blacksburg, Virginia, April 2022

Book endorsement: Additive Friction Stir Deposition

Prof. Hang Yu has created a masterpiece in this first book on additive friction stir deposition technology. Additive friction stir deposition is emerging as a very high deposit rate additive manufacturing process and has started finding real-world applications. This book provides a timely overview of this new field. As researchers and practitioners of this new technology work on various new opportunities, they will benefit from the necessary theoretical basis provided in Chapters 2–4, and 7. Prof. Yu draws from his own research and experience to provide insights. It is a very comprehensive book that covers areas ranging from process physics and materials science to tooling/applications. This is a great resource for students in this field and is a "must-read" book. Faculties teaching advanced manufacturing processes can use several chapters from the book to teach about this disruptive manufacturing process.

- Dr. Rajiv Mishra, University Distinguished Professor, University of North Texas, United States.

  This is a very timely book on this new emerging solid-state additive manufacturing technology. An excellent reference for those who are interested in this dynamic and fast-growing topic. The book covers the fundamentals of the process, technology innovations, and resulting microstructure and properties.

- Dr. Zhili Feng, Distinguished R&D Staff, and Leader Materials Joining Group, Oak Ridge National Laboratory, United States.

  Prof. Yu's research has placed him and his team at the forefront of a technology that is witnessing tremendous growth on the global additive manufacturing stage. This book expertly combines knowledge gained from years of experience into a single source that will offer the reader a keen insight into understanding the MELD process.

- Dr. Chase Cox, Vice President, MELD Manufacturing Corporation, United States.

  This is a thorough book on the emerging additive friction stir deposition technology that is poised to be very disruptive to fusion-based additive manufacturing. It sufficiently covers the benefits and challenges of using the process and provides a solid foundation for further research.

- Dr. Michael Eller, Adjunct Professor, University of New Orleans, United States.

CHAPTER 1

Introduction

The beginning is the most important part of the work.

— Plato

Shifting the manufacturing paradigm, additive manufacturing is the industrial production name for 3D printing that enables the creation of lighter, stronger, multifunctional, and multimaterial components in a layer-by-layer or voxel-by-voxel fashion. Almost all types of materials can be additively manufactured into 3D components, including polymers, ceramics, metals, composites, and even natural materials. Among these, metal additive manufacturing is of special industrial interest due to the widespread usage of metallic materials, including structural components, protective coatings, heat exchangers, and conducting wires. Like traditional metal manufacturing, which involves casting and forging, metal additive manufacturing can be implemented by fusion-based as well as solid-state approaches (Yu and Mishra, 2021). The former relies on melting and solidification, whereas the latter relies on plastic deformation or sintering. In this book, our focus is on an emerging solid-state metal additive process, *additive friction stir deposition,* which leverages high-temperature severe plastic deformation to enable location-specific metal deposition and printing.

This chapter introduces additive friction stir deposition and analyzes its place in the metal additive manufacturing spectrum. We begin with the classification of additive manufacturing technologies and the assessment of their suitability for printing metals. A distinction is drawn between the fusion-based and solid-state additive processes, wherein the benefits and limitations of each type are considered. Within the solid-state additive technologies, we introduce the principle of additive friction stir deposition and contrast its manufacturing capabilities with ultrasonic additive manufacturing and cold spray. We also compare additive friction stir deposition with another friction stir-derived technology, friction stir additive manufacturing, which is a sheet lamination process. Finally, we discuss the organization of the book and provide a brief overview of the chapters that follow—covering process fundamentals, material flow, microstructure

Additive Friction Stir Deposition
DOI: https://doi.org/10.1016/B978-0-12-824374-9.00004-6

evolution, tooling, metal matrix composites, mechanical properties, niche applications, and future perspectives.

1.1 Additive manufacturing for metals

As illustrated in Fig. 1.1, common additive manufacturing technologies can be categorized as follows (Gibson et al., 2015):

- **Powder bed fusion**, in which an energy beam is used to melt powder in a bed.
- **Directed energy deposition**, in which an energy beam and the feedstock are supplied simultaneously in the free space.
- **Binder jetting**, in which a polymeric binder is selectively deposited to glue powder together.
- **Material extrusion** or **fused deposition modeling**, in which a polymer is melted and extruded to the locations of interest.
- **Material jetting**, in which material (usually a polymer) is sprayed in droplet form.
- **Vat photopolymerization** or **stereolithography**, which leverages the interactions between UV light and photo-curable polymer.
- **Sheet lamination**, a cut-and-stack approach.

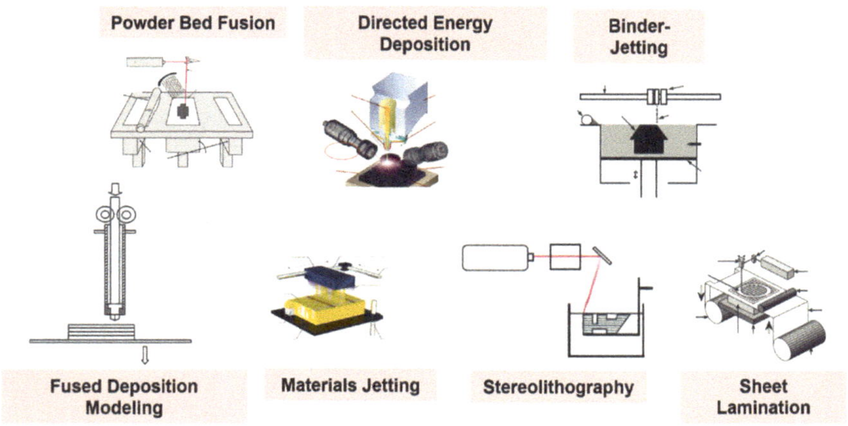

Figure 1.1 Categorization of additive technologies, including powder bed fusion, directed energy deposition, binder jetting, fused deposition modeling, material jetting, stereolithography, and sheet lamination. *Reprinted with permission from Gibson, I., Rosen, D.W., Stucker, B., 2015. Additive Manufacturing Technologies: Rapid Prototyping to Direct Digital Manufacturing. Springer US.*

For metal additive manufacturing, currently, the most popular approaches are powder bed fusion and directed energy deposition (Frazier, 2014). In powder bed fusion, high-quality metal particles are thinly laid out and subsequently selectively melted using a high-energy laser beam or electron beam. Upon cooling, a pattern of the solidified material is formed in the current layer (DebRoy et al., 2018). By repeating the powder deposition and selective melting and solidification in a layer-by-layer fashion, a specified 3D component can be built based on the G-code. In directed energy deposition, the component is built up in free space and the material (often powder or wire) and energy (laser or arc) are delivered simultaneously to the location of interest. The delivered material first melts and then solidifies upon cooling, forming metallurgical bonds with the preexisting material (Carroll et al., 2015). The key hardware is the powder feeding system, which can provide a flow of small quantities of powder with high precision.

Over the years, different terms have been used to refer to the same additive process. Selective laser melting (or sintering), direct metal laser sintering, and electron beam melting all refer to powder bed fusion. Laser metal deposition and laser engineered net shaping (i.e., LENS) are directed energy deposition processes. For laser-based powder bed fusion (Fig. 1.2), the key printer manufacturers include EOS, Renishaw, SLM Solutions, and Concept Laser (now part of GE Additive) (Gusarov et al., 2018). For electron beam-based powder bed fusion, Arcam AB (also part of GE Additive) has been dominating the manufacturing market. For directed energy deposition, Optomec is a notable industry leader and a pioneer of the LENS technology.

Although powder bed fusion and directed energy deposition differ in geometrical configuration and material feeding, they are based on the same liquid bonding mechanism, which critically relies on melting and solidification of the feedstock. These two categories are thus termed fusion-based or beam-based additive manufacturing (Yu et al., 2018). Like other fusion-based metal processes such as casting and fusion welding, control of porosity, residual stress, and hot cracking is challenging in fusion-based additive manufacturing. These problems are exacerbated by the small molten pool size, large thermal gradient, and rapid cooling rate rooted in the additive nature. Because textured, columnar grain structures naturally form along the build direction, microstructure control has also remained a persistent challenge in powder bed fusion and directed energy deposition (Basak and Das, 2016). Furthermore, melting requires higher

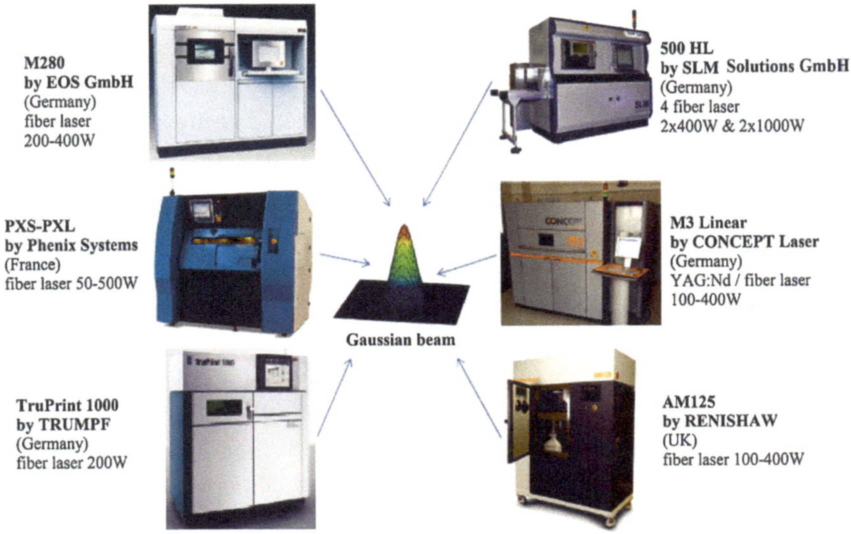

Figure 1.2 Example printers for powder bed fusion, with the manufacturers including EOS, Phenix Systems, TRUMPF, SLM Solutions, Concept Laser, and Renishaw. *Reprinted with permission from Gusarov, A.V., Grigoriev, S.N., Volosova, M.A., Melnik, Y. A., Laskin, A., Kotoban, D.V., et al., 2018. On productivity of laser additive manufacturing. Journal of Materials Processing Technology 261, 213–232.*

energy consumption and more carbon dioxide emission, making these processes unfavorable from an environmental perspective.

Then, the question is: can we print metals using other additive technologies in Fig. 1.1 while avoiding melting? Binder jetting—the technology originally termed *3D printing*—is a versatile process that selectively deposits binders to glue powders together, forming a green body that can be made of metals, ceramics, or composites (Gonzalez et al., 2016). This process involves the selective removal of binders followed by sintering. Fused deposition modeling, materials jetting, and stereolithography are widely used for printing polymers. The former is powerful for printing thermoplastics, whereas the latter two are suitable for photo-curable polymers. In principle, metal additive manufacturing can be implemented using these polymer printing processes by mixing metallic micro- or nano-particles with the polymers (Liu et al., 2020). As a result, a metal–polymer composite is first formed, followed by selective removal of the polymers (e.g., by burn-out) and sintering of the metal component. All the aforementioned processes may be termed sintering-based metal additive manufacturing, which refers to indirect approaches that initially form a metal-based composite followed by extensive postprocessing.

The last category in Fig. 1.1 is sheet lamination, which refers to hybrid processes involving both additive and subtractive procedures. In this category, there is a direct approach for metal additive manufacturing: ultrasonic additive manufacturing (Hehr and Norfolk, 2020), which leverages high-frequency sound waves for interfacial cleaning and bonding, followed by selective CNC (computer numerical control) milling of the bonded foils. Undoubtedly, ultrasonic additive manufacturing provides a low-temperature route for metal additive manufacturing; however, it is not a freeform process and cannot be used for location-specific deposition, reinforcement, and repair.

Given the limitations of fusion-based, sintering-based, as well as sheet lamination-based metal additive manufacturing, we have a strong motivation to search for or even develop new solid-state metal additive processes that go beyond the conventional list in Fig. 1.1.

1.2 Solid-state metal additive manufacturing

It is interesting to note the linkage between welding and metal additive manufacturing, both of which rely on material bonding. For example, in the category of directed energy deposition, laser engineered net shaping is an advanced version of laser welding, while wire arc additive manufacturing follows a similar principle of arc welding. The innovation in additive manufacturing lies in material feeding and print path control. Another example is ultrasonic additive manufacturing, which is simply an extension of ultrasonic welding to include CNC milling for shape control. These examples are summarized in Fig. 1.3 (Hehr and Dapino, 2017; Hu et al., 2018).

Solid-state welding or joining processes, therefore, may provide new insights into solid-state metal additive manufacturing; some existing welding processes may be advanced for additive manufacturing. Along that line, a notable option is friction stir welding (Fig. 1.4A; Gemme et al., 2010), in which a rotating nonconsumable tool is used to join two facing workpieces by leveraging the frictional heating at the tool-workpiece interface (Mishra and Ma, 2005). During welding, the material softens and undergoes severe plastic deformation due to compression and shear. The extensive material flow and rotation can lead to significant material mixing across the original boundaries, resulting in high-quality joining of the two workpieces. This welding process has shown great success in joining low-temperature materials, such as Al, Mg, and Cu (Mishra et al., 2014). The next question is: can we have a solid-state additive technology that

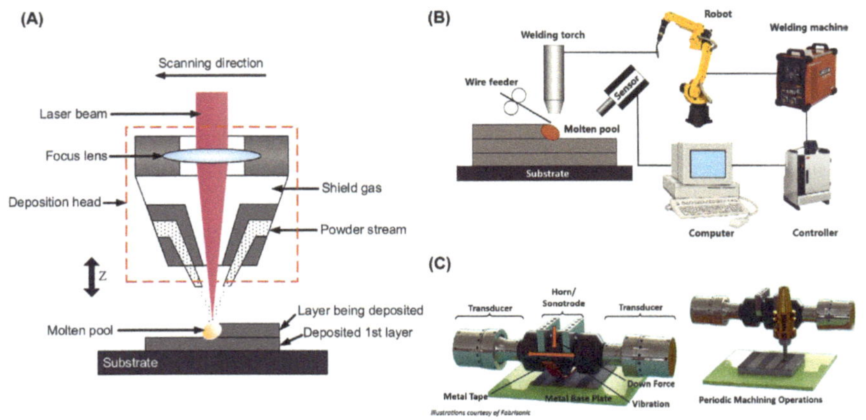

Figure 1.3 The linkage between welding and metal additive manufacturing, showing examples of (A) laser engineered net shaping, (B) wire arc additive manufacturing, and (C) ultrasonic additive manufacturing. *Reprinted with permission from Hehr, A., Dapino, M.J., 2017. Dynamics of ultrasonic additive manufacturing. Ultrasonics 73, 49–66; Hu, Y., Ning, F., Cong, W., Li, Y., Wang, X., Wang, H., 2018. Ultrasonic vibration-assisted laser engineering net shaping of ZrO2-Al2O3 bulk parts: Effects on crack suppression, microstructure, and mechanical properties. Ceramics International 44, 2752–2760.*

follows the same bonding mechanism as friction stir welding—as long as we can add an appropriate material feeding mechanism and enable automatic deposition path control?

The answer is yes. The corresponding technology is essentially the focus of this book, additive friction stir deposition, which integrates the friction stir principle with a robust material feeding mechanism to enable location-specific deposition. As shown in Fig. 1.4B, the machine's mechanical setup appears to be similar to that of fused deposition modeling: the feed material is supplied through a channel inside the print head before it is deposited onto the substrate. However, in fused deposition modeling, thermoplastic printing is implemented simply by heating and compression, whereas in additive friction stir deposition, the friction stir principle is employed. This is because the deformation of metallic feed material and formation of material–substrate bonding necessitate a much more efficient thermomechanical processing strategy. During deposition, the feed material—which is usually in the form of a metal rod—is stirred, severely deformed, and mixed with the substrate, resulting in metallurgical bond formation across the original interface. With strong interfacial bonding, the in-plane motion of the print head relative to the substrate results

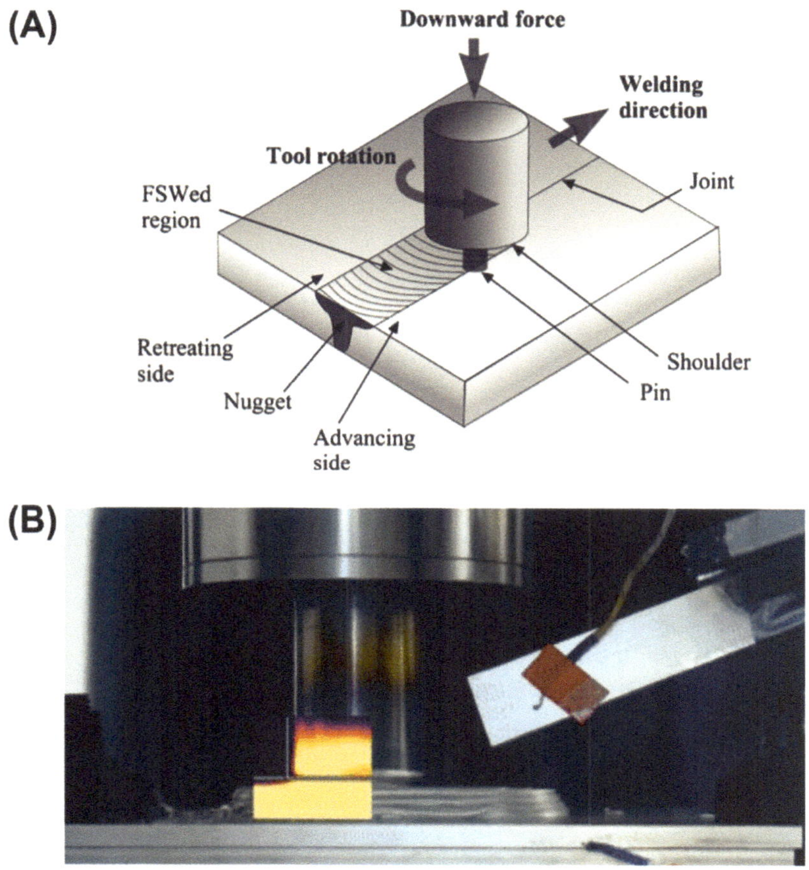

Figure 1.4 Processes based on the friction stir principle: (A) friction stir welding and (B) additive friction stir deposition. *(A) is reprinted with permission from Gemme, F., Verreman, Y., Dubourg, L., Jahazi, M., 2010. Numerical analysis of the dwell phase in friction stir welding and comparison with experimental data. Materials Science and Engineering: A 527, 4152–4160.*

in a material pattern controlled by the G-code in each layer. Through layer-by-layer deposition, 3D metal components are naturally formed. In additive friction stir deposition, the extensive flow of the feed material and its mixing with the substrate guarantee good material quality even in the as-printed state.

At the time of writing, additive friction stir deposition, ultrasonic additive manufacturing, and cold spray represent the three major solid-state metal additive manufacturing technologies; the latter was previously considered a coating process but has gradually gained more attention from the

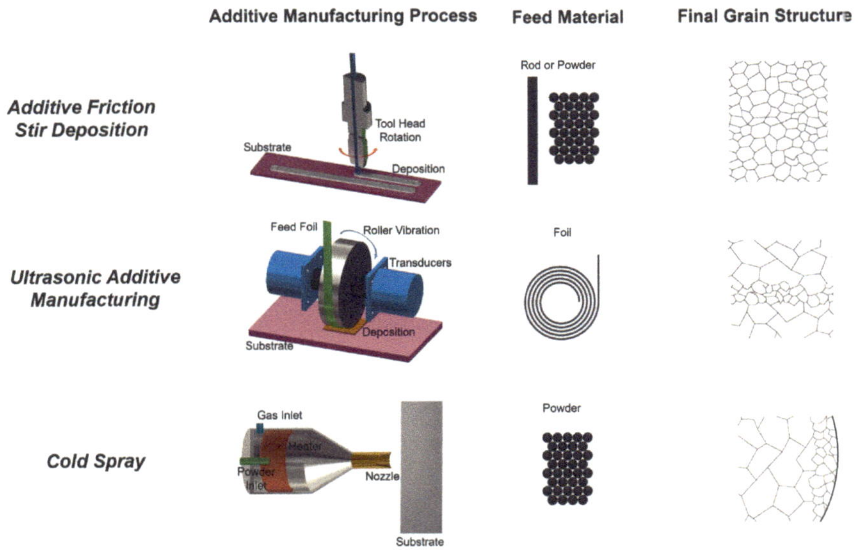

Figure 1.5 Illustration of three major solid-state metal additive processes, additive friction stir deposition, ultrasonic additive manufacturing, and cold spray. The feed material and final grain structure are included. *Reprinted with permission from Yu, H. Z., Mishra, R.S., 2021. Additive friction stir deposition: a deformation processing route to metal additive manufacturing. Materials Research Letters 9, 71–83.*

additive manufacturing community (Yin et al., 2018). All of these processes, as illustrated in Fig. 1.5 (Yu and Mishra, 2021), involve direct printing of metals without the need for extensive postprocessing, such as polymer burn-out and metal densification (e.g., by sintering or hot isostatic pressing). Both additive friction stir deposition and cold spray are freeform processes that enable location-specific deposition, reinforcement, and repair, whereas ultrasonic additive manufacturing is a sheet lamination process requiring machining. Although deformation bonding is the prevalent bonding mechanism, cold spray and ultrasonic additive manufacturing only involve local plastic deformation at the layer interface or particle contact (Tuncer and Bose, 2020). In contrast, additive friction stir deposition involves global deformation in that all the material voxels in the feed-rod undergo severe plastic deformation at high temperatures.

Thanks to the material flow and stress-state (i.e., compression and shear), additive friction stir deposition is capable of producing a fully-dense as-printed material. There is no need for postprocess annealing to eliminate the porosity, which is necessary for cold spray. In addition, the global deformation nature leads to a uniform, equiaxed, fine microstructure due to dynamic recrystallization. More importantly, the resulting mechanical properties can be comparable to wrought or forged alloys. This is a unique advantage that distinguishes additive friction stir deposition from other metal additive processes. In a certain sense, we may consider additive friction stir deposition to be a forging-based metal additive manufacturing approach.

1.3 Additive friction stir deposition

Fig. 1.6A illustrates the cross-section of the printing process by additive friction stir deposition. A metal rod rapidly rotates as it passes through the rotating print head, generating frictional heat as it contacts the substrate.

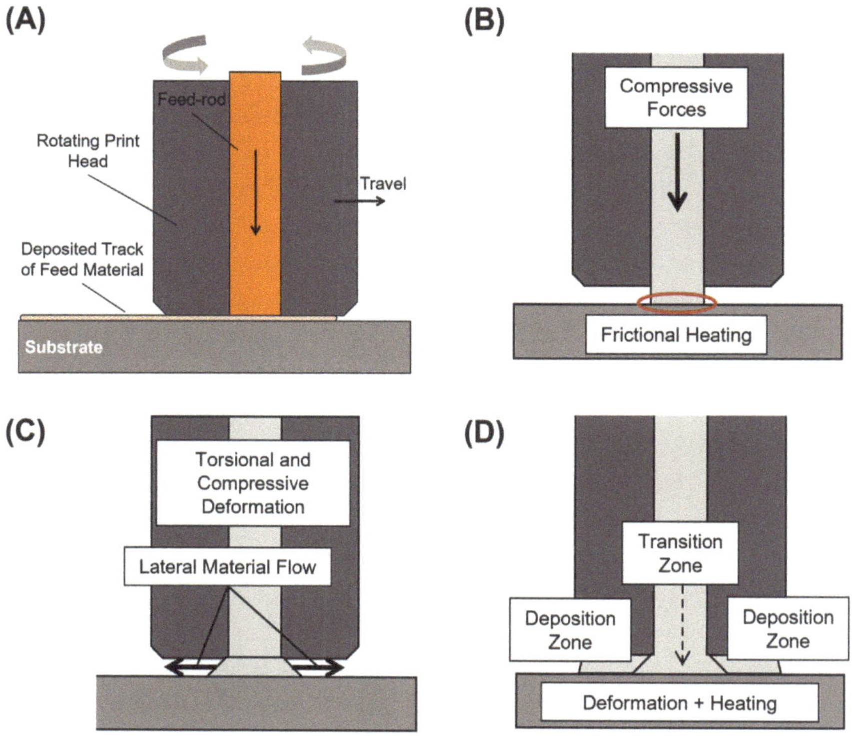

Figure 1.6 Illustration of the deformation steps in additive friction stir deposition. *Reprinted with permission from Griffiths, R.J., Garcia, D., Song, J., Vasudevan, V.K., Steiner, M.A., Cai, W., et al., 2021. Solid-state additive manufacturing of aluminum and copper using additive friction stir deposition: Process-microstructure linkages. Materialia 15, 100967.*

The feed-rod then heats up, yields, and extrudes to fill the space between the substrate and the rotating print head. The deposited material goes through several deformation steps, including initial uniaxial compression (Fig. 1.6B), compression and shear below the rotating feed-rod (Fig. 1.6C), and final shear-dominated deformation below the rotating print head (Fig. 1.6D) (Griffiths et al., 2021).

In additive friction stir deposition, the material is characterized by a macroscopic shape change from a rod to a thin disk by extrusion, followed by rotation and flow under the rotating print head. The print head regulates the layer thickness by imposing a vertical constraint. Its rotation also smears and shears the deposited material more and leads to good print quality with no porosity. Note that the surface layers of the substrate also get heated up and plastically deformed, mixing with the feed material and forming strong interfacial bonds.

Unlike friction stir welding, which produces steep strain gradients and significant microstructural differences between the stir zone, thermomechanically-affected zone, and heat-affected zone, additive friction stir deposition produces relatively uniform deformation with minor microstructural variability across the thickness or width. Additive friction stir deposition should be distinguished from another friction stir-derived technology, friction stir additive manufacturing (Palanivel et al., 2015), which is a sheet lamination process that involves stacking multiple metal layers (with a thickness of a few millimeters) and joining them by friction stir welding, followed by machining. Welding is done on the top layer of sheets using a custom-designed pin that penetrates them vertically, rotates, and traverses to create a joining line throughout the overlapping sheets. This stacking, welding, and machining process can be repeated allowing for the fabrication of large components in 3D. Although both are based on the friction stir principle for material bonding, additive friction stir deposition enables selective-area cladding, repair, and local feature buildup, which are all challenging to implement using friction stir additive manufacturing. Fig. 1.7 illustrates and compares the two processes (Yu, 2022).

No matter if the printing is based on liquid or solid bonding mechanisms, the mechanical properties of the printed material are governed by the microstructure. As a result of epitaxial solidification, the microstructure produced by fusion-based metal additive manufacturing is characterized by highly oriented, columnar grains. The ratio between the thermal gradient and solidification rate must be carefully adjusted to cause significant undercooling. With that, nucleation within the molten pool is

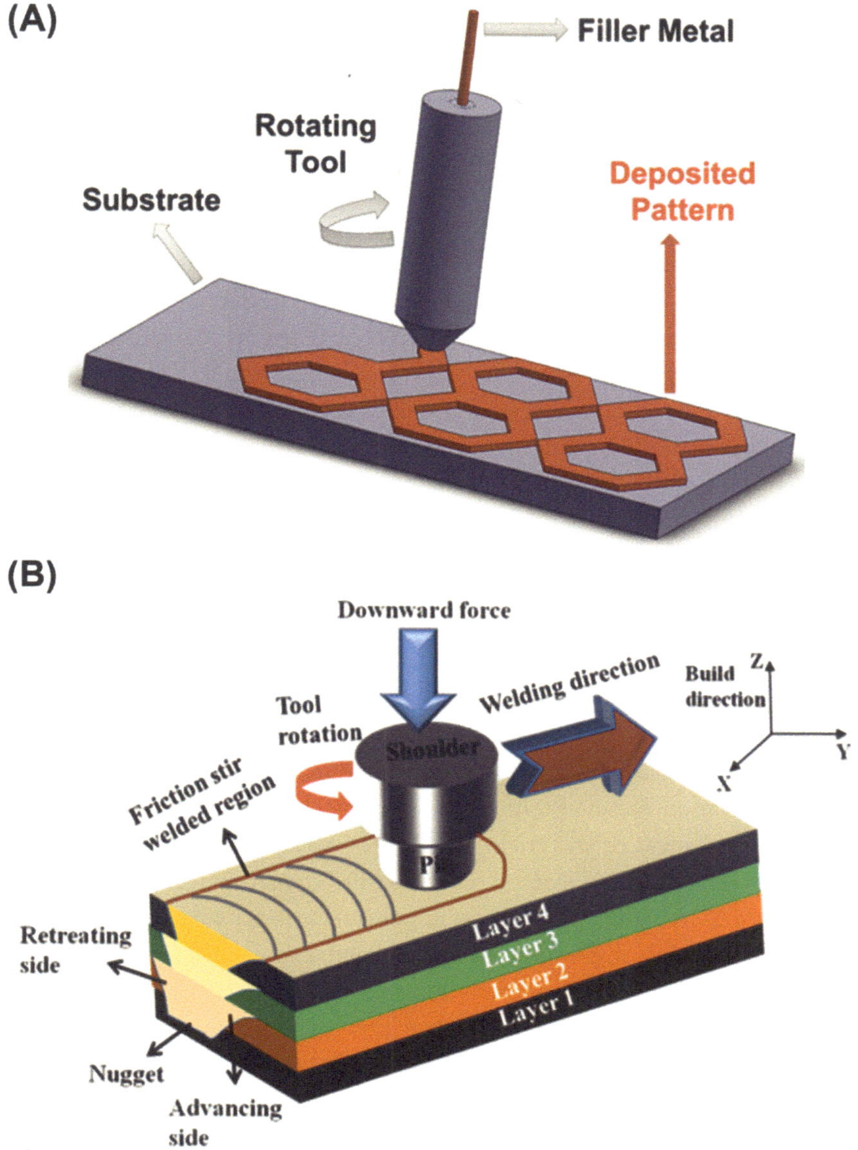

Figure 1.7 Illustration and comparison of (A) additive friction stir deposition and (B) friction stir additive manufacturing. *Reprinted with permission from Yu, H.Z., 2022. Emerging Processes—Friction Stir Based, In: Caballero, F.G. (Ed.), Encyclopedia of Materials: Metals and Alloys. Elsevier, Oxford, pp 153–161 and Palanivel, S., Nelaturu, P., Glass, B., Mishra, R.S., 2015. Friction stir additive manufacturing for high structural performance through microstructural control in an Mg based WE43 alloy. Materials & Design (1980–2015) 65, 934–952.*

enhanced to promote equiaxed microstructures. Other than adjusting the process parameters (e.g., laser power and scanning speed), nanoparticles of higher melting temperatures can be added to the feedstock and act as heterogeneous nucleation sites inside the molten pool. As shown in Fig. 1.8A, this has been demonstrated to be a viable solution for microstructure control in powder bed fusion of AA7075, an Al−Zn−Cu−Mg alloy (Martin et al., 2017).

In contrast, additive friction stir deposition offers a deformation processing route, in which the entire feed material undergoes high-temperature severe plastic deformation. The extreme thermomechanical condition during printing results in dynamic recrystallization, which can be classified into discontinuous and continuous types—depending on the processing characteristics such as peak temperature and strain rate, as well as the intrinsic material properties such as stacking fault energy. In materials with high stacking fault energy like Al alloys, additive friction stir deposition frequently results in continuous dynamic recrystallization, which is characterized by extensive dynamic recovery, subgrain formation, and a gradual increase in the misorientation angle with strain. One example is shown in Fig. 1.8B for AA2024, an Al−Cu alloy, in which the grain size has been reduced from hundreds of microns in the feed material to a few microns in the as-printed material (Perry et al., 2020). Due to significant material flow and large deformation magnitude, notably, continuous dynamic recrystallization has also been observed in additive friction stir deposition of materials with lower stacking fault energy, such as Ni alloys and stainless steel. Discontinuous dynamic recrystallization is characterized by "nucleation" (or initiation) and growth of dislocation-free grains via high angle grain boundary migration.

The result of dynamic recrystallization is the formation of fine, equiaxed microstructures and isotropic mechanical properties, which are often favored for structural applications. The mechanical properties of the printed material by additive friction stir deposition can be comparable to their wrought and forged counterparts. This has been demonstrated for fully-aged AA7075 (Yoder et al., 2021) and AA6061 (Beck et al., 2021), Inconel 625 (Rivera et al., 2017), and Ti-6Al-4V (Farabi et al., 2022). Remarkably, good mechanical properties are obtained using simple feed materials like cast or wrought rods without altering the composition of engineering alloys or adding nanoparticles, not to mention that the process requires lower energy input and causes less carbon dioxide emission than fusion-based metal additive manufacturing.

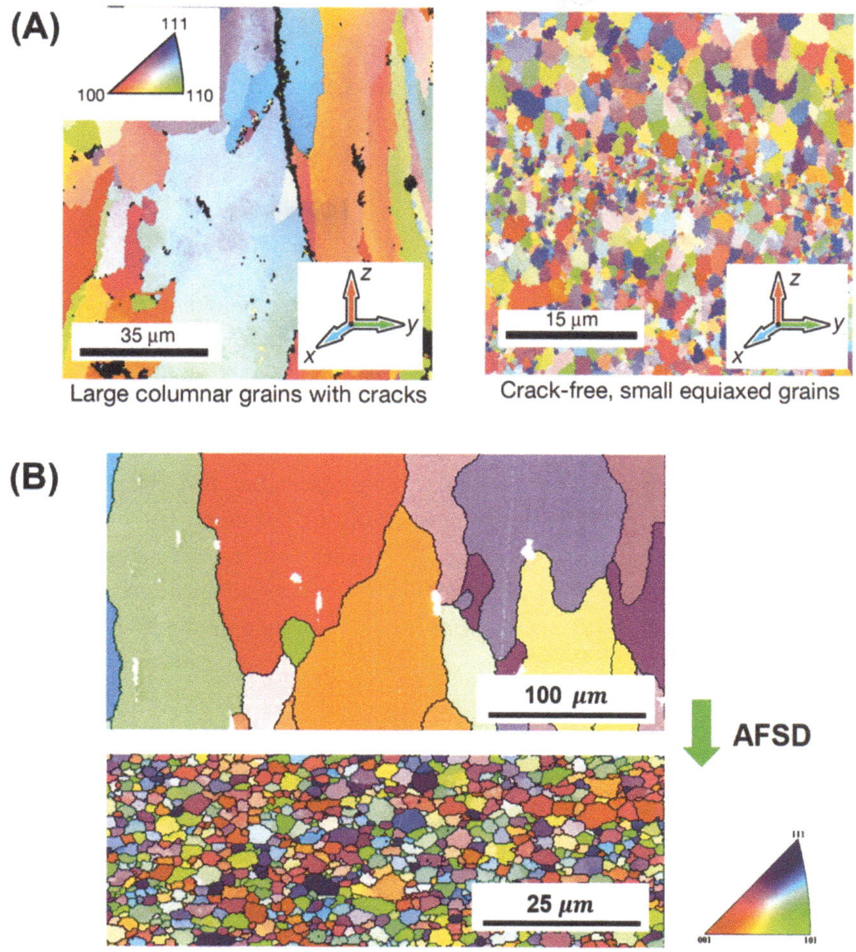

Figure 1.8 Microstructure features arising from (A) fusion-based additive manufacturing of AA7075, comparing results from (*left*) conventional feedstock and (*right*) feedstock with the addition of Zr nanoparticles, and (B) additive friction stir deposition of AA2024, comparing (*top*) the feed material and (bottom) as-printed material. *Reprinted with permission from Martin, J.H., Yahata, B.D., Hundley, J.M., Mayer, J.A., Schaedler, T.A., Pollock, T.M., 2017. 3D printing of high-strength aluminium alloys. Nature 549, 365–369. and Perry, M.E.J., Griffiths, R.J., Garcia, D., Sietins, J.M., Zhu, Y., Yu, H.Z., 2020. Morphological and microstructural investigation of the non-planar interface formed in solid-state metal additive manufacturing by additive friction stir deposition. Additive Manufacturing 35, 101293.*

The build rate in additive friction stir deposition is controlled by the size of the print head and feed-rod, and it is comparable to other large-scale additive processes like wire arc additive manufacturing. For example, with a 3/8″ diameter feed-rod, current additive friction stir deposition tooling can achieve build rates of 20 lbs/hr for Al, 10.7 lbs/hr for steels, and 5.5 lbs/hr for Ti. These build rates are significantly higher than those of fusion-based metal additive manufacturing. Additive friction stir deposition is a free-space manufacturing process, so the part size is not limited by a vacuum chamber or powder bed. This enables large-scale fabrication; components on the order of meters have been produced. Fig. 1.9A shows a large window frame, while Fig. 1.9B shows an aluminum pressure vessel printed in less than 2 hours and machined to the final surface finish. For many alloys, there is no need for atmospheric control, so rapid field repair is also possible.

Compared to alternative metal additive processes, the main advantages of additive friction stir deposition can be summarized as follows. First, thanks to the material flow as well as compression and shear, the as-printed material is typically fully dense without the need for postprocess heat treatment for porosity elimination. In addition, cast or wrought rods and billets can be used as the feedstock. Such material forms are inexpensive and convenient to prepare compared to powders or wires. Moreover, forging microstructures and mechanical properties can be achieved

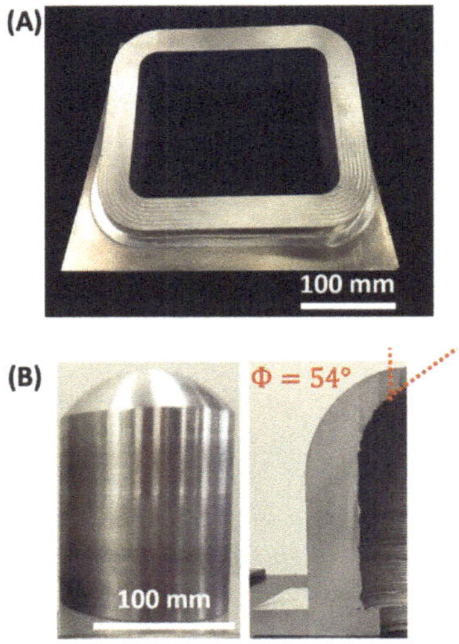

Figure 1.9 Examples of printed components by additive friction stir deposition: (A) a window frame and (B) a pressure vessel. *Reprinted with permission from Yu, H.Z., Jones, M.E., Brady, G.W., Griffiths, R.J., Garcia, D., Rauch, H.A., et al., 2018. Non-beam-based metal additive manufacturing enabled by additive friction stir deposition. Scripta Materialia 153, 122–130.*

economically without engineering the feedstock or changing the composition. The process also results in low residual stress and distortion due to the absence of melting and solidification. Using large feed materials, additive friction stir deposition enables a high build rate and meter-scale manufacturing. By offering a low-temperature processing route, additive friction stir deposition enables energy-efficient manufacturing with less impact on the environment. These advantages are summarized in Fig. 1.10.

The main limitation of additive friction stir deposition lies in its in-plane spatial resolution. The large size of feed material and flash formation along the edges of the deposition track make it difficult to print small feature sizes with precise geometric control. As a result, minor machining and surface finishing are necessary. In addition, the tooling material needs to be stiffer and stronger than the material under printing conditions. To

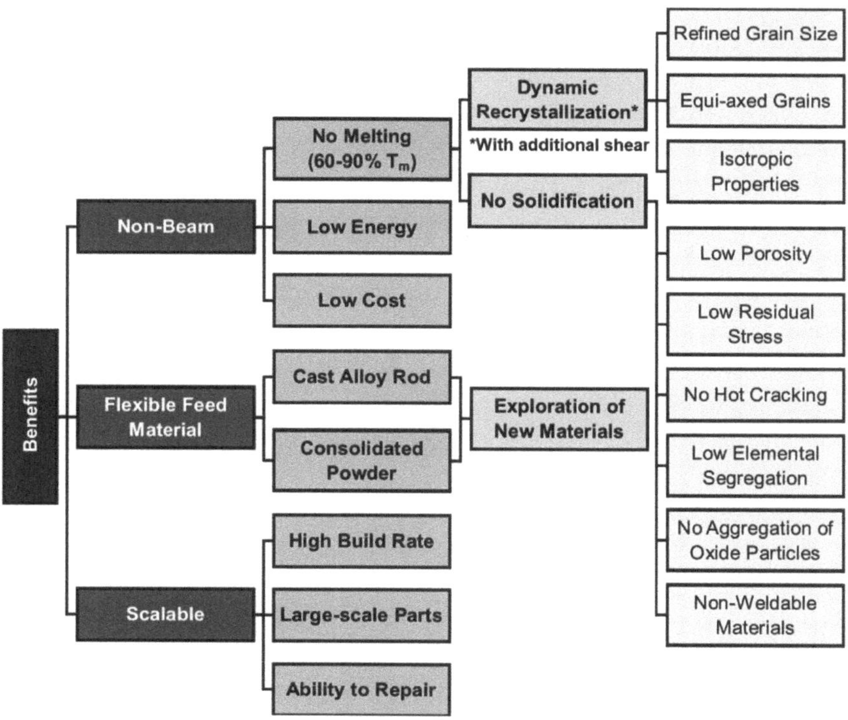

Figure 1.10 A flow chart summarizing the advantages of additive friction stir deposition. *Reprinted with permission from Yu, H.Z., Jones, M.E., Brady, G.W., Griffiths, R.J., Garcia, D., Rauch, H.A., et al., 2018. Non-beam-based metal additive manufacturing enabled by additive friction stir deposition. Scripta Materialia 153, 122–130.*

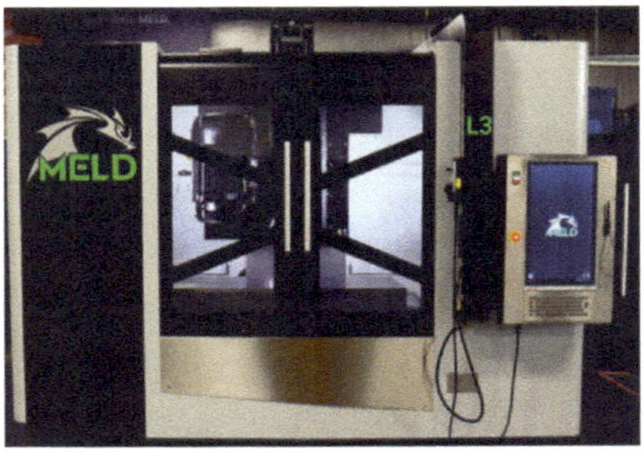

Figure 1.11 The MELD L3 printer, which enables additive friction stir deposition with a large build volume.

print relatively soft materials, such as Al, Mg, and Cu, a tool steel print head has proven to be sufficient. To print high-temperature materials like Ti–6Al–4V, Ni-based superalloys, and high strength steels, we need to employ W–Re alloys, lanthanated tungsten, or WC–Co to make the print head. Tool wear and catastrophic failure pose significant challenges for reliably printing high-temperature materials in the long term.

From a fundamental physics perspective, the phrase additive friction stir deposition may apply to a variety of processes that follow a similar basic concept. One specific process featuring a large tooling size and a high build rate has been commercialized by MELD Manufacturing Corporation, which is the sole commercial manufacturer of additive friction stir deposition machines at this moment. Therefore, additive friction stir deposition is sometimes known as the MELD technology. Fig. 1.11 shows a commercial MELD L3 printer, which enables a build volume of 4 cubic meters. Another relevant technology is termed friction surface cladding (Liu et al., 2016), which has a geometrical configuration comparable to the MELD technology. Nonetheless, the resulting edge straightness, surface finish, and excess flash all appear to be worse than that using the commercial MELD facility. This is most likely due to the difference in the tool design and the material feeding mechanism.

1.4 Organization of this book

Characterized by an interdisciplinary nature, the subject of additive friction stir deposition covers a wide variety of scientific and engineering topics, such as friction, heat and mass transfer, plasticity and flow, physical metallurgy, tool design, composite processing, mechanical properties, as well as repair, cladding, and recycling. These topics are elaborated in the following chapters. The goal is to offer a comprehensive picture of the state-of-the-art understanding of the underlying physics and materials science as well as the niche applications and potential markets.

- Chapter 2 focuses on the process fundamentals that govern the quality, microstructure, and mechanical properties of the as-printed material. Interrelated linkages are revealed among contact state, material flow, and evolution of temperature, feeding force, and torque.
- Chapter 3 focuses on the material flow phenomena in additive friction stir deposition. Tracer-based experiments disclose complex material flow patterns involving helical shape evolution and micro-ribbon formation, from which new deformation insights are gained.
- Chapter 4 focuses on dynamic microstructure evolution. Microstructure characterization along the flow path reveals specific dynamic recrystallization mechanisms. The stacking fault energy and material flow behavior are found to play an important role in defining the main evolution mechanisms and the process-microstructure linkage.
- Chapter 5 focuses on the influences of tool geometry in additive friction stir deposition. The surface features of the print head can significantly influence the material flow behavior and the degree of dynamic recrystallization.
- Chapter 6 focuses on the printing of metal matrix composites using additive friction stir deposition, which can lead to fully-dense composites in the as-printed state with good interface quality and uniform particle distribution.
- Chapter 7 focuses on the mechanical properties of the printed materials by additive friction stir deposition, showing evidence that forging-like mechanical properties can be achieved. Tension, fracture, and fatigue properties of the printed material are discussed along with the behavior of bilayer structures consisting of both the printed material and the substrate.
- Chapter 8 focuses on the niche applications of additive friction stir deposition, which are mostly related to the aerospace, automotive, and defense industries. The market lies in where mechanical performance is of top priority and postprinting machining is allowed for improving the geometry precision. Particularly promising applications include structural repair, selective-area cladding and reinforcement, recycling, large-scale additive manufacturing, and underwater printing.

- Chapter 9 focuses on the future directions of the subject, including an in-depth understanding of the process physics, materials innovation on the microscopic and mesoscopic levels, and incorporation of artificial intelligence for quality control and property optimization.

References

Basak, A., Das, S., 2016. Epitaxy and microstructure evolution in metal additive manufacturing. Annual Review of Materials Research 46, 125–149.

Beck, S.C., Rutherford, B.A., Avery, D.Z., Phillips, B.J., Rao, H., Rekha, M.Y., et al., 2021. The effect of solutionizing and artificial aging on the microstructure and mechanical properties in solid-state additive manufacturing of precipitation hardened Al–Mg–Si alloy. Materials Science and Engineering: A 819, 141351.

Carroll, B.E., Palmer, T.A., Beese, A.M., 2015. Anisotropic tensile behavior of Ti–6Al–4V components fabricated with directed energy deposition additive manufacturing. Acta Materialia 87, 309–320.

DebRoy, T., Wei, H.L., Zuback, J.S., Mukherjee, T., Elmer, J.W., Milewski, J.O., et al., 2018. Additive manufacturing of metallic components—process, structure and properties. Progress in Materials Science 92, 112–224.

Farabi, E., Babaniaris, S., Barnett, M.R., Fabijanic, D.M., 2022. Microstructure and mechanical properties of Ti6Al4V alloys fabricated by additive friction stir deposition. Additive Manufacturing Letters 100034.

Frazier, W.E., 2014. Metal additive manufacturing: a review. Journal of Materials Engineering and Performance 23, 1917–1928.

Gemme, F., Verreman, Y., Dubourg, L., Jahazi, M., 2010. Numerical analysis of the dwell phase in friction stir welding and comparison with experimental data. Materials Science and Engineering: A 527, 4152–4160.

Gibson, I., Rosen, D.W., Stucker, B., 2015. Additive Manufacturing Technologies: Rapid Prototyping to Direct Digital Manufacturing. Springer US.

Gonzalez, J.A., Mireles, J., Lin, Y., Wicker, R.B., 2016. Characterization of ceramic components fabricated using binder jetting additive manufacturing technology. Ceramics International 42, 10559–10564.

Griffiths, R.J., Garcia, D., Song, J., Vasudevan, V.K., Steiner, M.A., Cai, W., et al., 2021. Solid-state additive manufacturing of aluminum and copper using additive friction stir deposition: process-microstructure linkages. Materialia 15, 100967.

Gusarov, A.V., Grigoriev, S.N., Volosova, M.A., Melnik, Y.A., Laskin, A., Kotoban, D. V., et al., 2018. On productivity of laser additive manufacturing. Journal of Materials Processing Technology 261, 213–232.

Hehr, A., Dapino, M.J., 2017. Dynamics of ultrasonic additive manufacturing. Ultrasonics 73, 49–66.

Hehr, A., Norfolk, M., 2020. A comprehensive review of ultrasonic additive manufacturing. Rapid Prototyping Journal 26, 445–458.

Hu, Y., Ning, F., Cong, W., Li, Y., Wang, X., Wang, H., 2018. Ultrasonic vibration-assisted laser engineering net shaping of ZrO2-Al2O3 bulk parts: effects on crack suppression, microstructure, and mechanical properties. Ceramics International 44, 2752–2760.

Liu, S., Bor, T.C., Van der Stelt, A.A., Geijselaers, H.J.M., Kwakernaak, C., Kooijman, A.M., et al., 2016. Friction surface cladding: an exploratory study of a new solid state cladding process. Journal of Materials Processing Technology 229, 769–784.

Liu, B., Wang, Y., Lin, Z., Zhang, T., 2020. Creating metal parts by fused deposition modeling and sintering. Materials Letters 263, 127252.
Martin, J.H., Yahata, B.D., Hundley, J.M., Mayer, J.A., Schaedler, T.A., Pollock, T.M., 2017. 3D printing of high-strength aluminium alloys. Nature 549, 365–369.
Mishra, R.S., Ma, Z.Y., 2005. Friction stir welding and processing. Materials Science and Engineering: R: Reports 50, 1–78.
Mishra, R.S., De, P.S., Kumar, N., 2014. Friction Stir Welding and Processing: Science and Engineering. Springer International Publishing.
Palanivel, S., Nelaturu, P., Glass, B., Mishra, R.S., 2015. Friction stir additive manufacturing for high structural performance through microstructural control in an Mg based WE43 alloy. Materials & Design (1980–2015) 65, 934–952.
Perry, M.E.J., Griffiths, R.J., Garcia, D., Sietins, J.M., Zhu, Y., Yu, H.Z., 2020. Morphological and microstructural investigation of the non-planar interface formed in solid-state metal additive manufacturing by additive friction stir deposition. Additive Manufacturing 35, 101293.
Rivera, O.G., Allison, P.G., Jordon, J.B., Rodriguez, O.L., Brewer, L.N., McClelland, Z., et al., 2017. Microstructures and mechanical behavior of Inconel 625 fabricated by solid-state additive manufacturing. Materials Science and Engineering: A 694, 1–9.
Tuncer, N., Bose, A., 2020. Solid-state metal additive manufacturing: a review. JOM 72, 3090–3111.
Yin, S., Cavaliere, P., Aldwell, B., Jenkins, R., Liao, H., Li, W., et al., 2018. Cold spray additive manufacturing and repair: fundamentals and applications. Additive Manufacturing 21, 628–650.
Yoder, J.K., Griffiths, R.J., Yu, H.Z., 2021. Deformation-based additive manufacturing of 7075 aluminum with wrought-like mechanical properties. Materials & Design 198, 109288.
Yu, H.Z., 2022. Emerging processes—friction stir based. In: Caballero, F.G. (Ed.), Encyclopedia of Materials: Metals and Alloys. Elsevier, Oxford, pp. 153–161.
Yu, H.Z., Mishra, R.S., 2021. Additive friction stir deposition: a deformation processing route to metal additive manufacturing. Materials Research Letters 9, 71–83.
Yu, H.Z., Jones, M.E., Brady, G.W., Griffiths, R.J., Garcia, D., Rauch, H.A., et al., 2018. Non-beam-based metal additive manufacturing enabled by additive friction stir deposition. Scripta Materialia 153, 122–130.

CHAPTER 2

Process fundamentals

Most of the fundamental ideas of science are essentially simple, and may, as a rule, be expressed in a language comprehensible to everyone.

—Albert Einstein

Additive friction stir deposition is implemented by integrating the friction stir principle with a material feeding mechanism. At each time step, a small volume of the feed material undergoes severe plastic deformation at elevated temperatures. The final component is a result of the addition of these small volumes. The resulting porosity, microstructure, mechanical properties, and the level of residual stresses are all dependent on the thermal and deformation history of the deposited material, which is governed by the multi-physics coupling of heat and mass transfer. As illustrated in Fig. 2.1, the corresponding fundamental physical processes are the cornerstone to understanding and controlling additive friction stir deposition. These processes include, but are not limited to, heat generation via interfacial friction and plastic energy dissipation, the material deformation flow beneath the print head, and the contact phenomena at the print head–material interface. Convolution of these intertwined processes leads to complex physics and mechanics, which are the focus of this chapter.

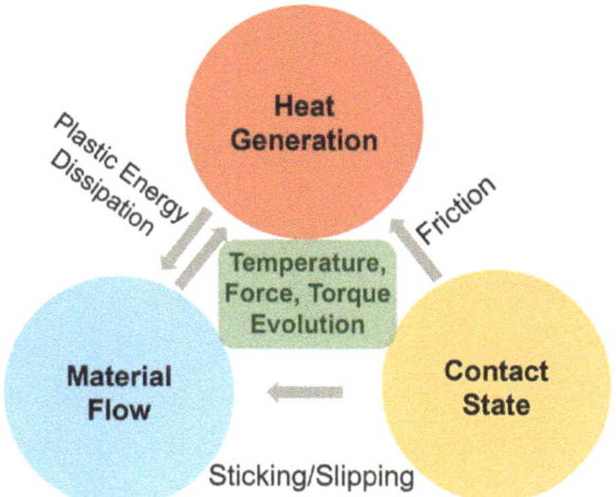

Figure 2.1 A summary of the key physical processes in additive friction stir deposition.

Additive Friction Stir Deposition
DOI: https://doi.org/10.1016/B978-0-12-824374-9.00002-2

In this chapter, we start by introducing the elements of friction theory and the basics of heat and mass transfer. We then describe the printing principle of additive friction stir deposition, explaining how the metallic feed material is heated up and rapidly deformed to enable location-specific deposition. We emphasize the importance of the underlying physical processes, such as heat generation and material flow, and demonstrate how these processes may be better understood through an integrated in situ monitoring platform. The in situ monitoring results are presented for the deposition of two materials, an Al-Mg-Si alloy (AA6061) and Cu, under various processing conditions. Through semi-quantitative analysis, we discuss the generic physical trends and the drastic differences between Al-Mg-Si and Cu in thermal and deformation aspects during friction stirring. By employing such a comparative approach, this chapter intends to present a clear picture of the interrelated linkages of contact state, material flow, and evolution of temperature, feeding force, and torque. Such linkages arise from the extreme thermomechanical processing nature of additive friction stir deposition, which is characterized by rapid severe plastic deformation with the full coupling of heat and mass transfer.

Before delving into the principle of additive friction stir deposition and in situ monitoring investigation, we review some background physics relevant to the process fundamentals, including the *elements of friction theory* (Section 2.1) and the *fundamentals of heat and mass transfer* (Section 2.2).

2.1 Elements of friction theory

As the name suggests, friction, especially friction between rotating pieces, is key to implementing additive friction stir deposition. While ubiquitous in daily life and engineering design, the phenomenon of friction is far more complex than that described by simple empirical rules like Coulomb friction. A variety of different friction mechanisms have been discovered, which are sensitively dependent on surface roughness (down to the nanoscale), surface contamination, surface topography, plastic deformation of asperities, fracture of asperities or wear, lubrication, and stick-slip conditions (Tabor, 1981). Owing to its complex nature, a general macroscopic theory for predicting the frictional forces between any given components does not exist. Indeed, our understanding of fundamental physics is rather limited, and we continue to rely heavily on empirical friction knowledge. The goal of this section is to provide a state-of-the-art summary of the friction theory, with the hope of shedding light on the appropriate modeling of the friction forces in additive friction stir deposition—which remains an open question.

Friction on the macroscopic scale can take place either between dry contacts or lubricated surfaces. Dry friction is also called Coulomb friction and has been described by the following laws (Gao et al., 2004):

1. The force of friction is directly proportional to the applied load (Amontons' 1st law, 1699).
2. The force of friction is independent of the apparent area of contact (Amontons' 2nd law, 1699).
3. Kinetic friction is independent of the sliding velocity (Coulomb's law, 1790).

These laws can be summarized by one equation:

$$\mu = \frac{F}{N}. \tag{2.1}$$

Here, μ is the coefficient of friction (COF), which is independent of the apparent contact area and sliding velocity. F and N refer to the friction force and normal force, respectively. The above Amontons & Coulomb's laws are shown to be effective for the vast majority of dry surfaces, no matter if the surfaces are ductile or brittle, or rough or smooth. However, it should be emphasized that these laws are purely empirical rather than derived from fundamental principles. Eq. (2.1) is thus not universal. For example, it is invalid if the loads or the sliding velocities cover a large range; it completely breaks down for atomically smooth surfaces with strong contact adhesion.

There have been many attempts to explain the empirical Amontons & Coulomb's laws. Early efforts modeled the sliding process as a climbing process of the asperities on one surface over another, so the lateral force F would equal the normal load multiplied by the maximum mean slope of the asperities. This physics picture is illustrated in Fig. 2.2A and B. Such a simple explanation was challenged by various theoretical arguments and experimental evidence. In particular, in the 1940s Bowden and Tabor pointed out that the true area of contact between two solids is only a small fraction of the apparent contact area due to surface roughness. Such roughness exists even in polished metals with mirror surface flatness, in which the roughness is significantly below the wavelength of light but may still be notable on the nanometer scale (Fig. 2.2C). Bowden and Tabor's theory also predicts strong friction forces between atomically clean metal surfaces, as illustrated in Fig. 2.2D.

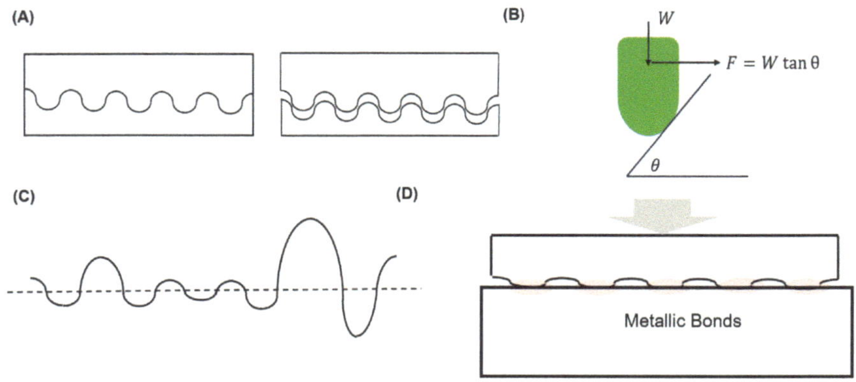

Figure 2.2 Amontons & Coulomb's laws versus the theory by Bowden and Tabor. (A) Coulomb's view of interlocking surface asperities. (B) For a slope characterized by an angle θ, COF is predicted to be equal to tanθ based on the asperity climbing model. (C) The real contact area depends on asperity deformation with the surface topography described in the Greenwood–Williamson model. (D) Illustration of atomic and near-atomic contact of clean metals.

The conclusion by Bowden and Tabor was supported by a classical experiment involving the measurement of the electrical conduction across two metallic surfaces in contact. The conduction was found to be proportional to the load applied to press the two surfaces both under stationary and sliding conditions. From Bowden and Tabor's theory, the two surfaces are in contact only on the microscopic scale, at some asperities or junctions. The friction force is equal to the force necessary to shear these junctions and is therefore given by

$$F = \tau_c A_{real}. \tag{2.2}$$

Here τ_c is the yield stress during shear, which is the maximum lateral stress the contact can bear. If, as suggested by the conduction measurement, the real contact area A_{real} scales with the applied normal force N, then Eq. (2.2) can be rewritten in the same form as Eq. (2.1). Accordingly, the empirical laws by Amontons and Coulomb are well explained.

Despite the observation from the electrical conduction experiment, the relationship between A_{real} and N deserves a more detailed mechanics analysis. In most practical circumstances, the real area of contact may be accurately approximated by assuming that plastic deformation has happened at each junction and that all junctions are in a state of incipient plastic flow. For a single junction, the real contact area is

$\Delta A_{real} = \Delta N/\sigma_c$, where ΔN is the force exerted on that junction and σ_c is the largest compressive stress that the material can bear without plastic yielding (i.e., the critical compressive stress). The total real contact area is a sum of the real contact area of each junction, $A_{real} = \sum_i \Delta A_{real}|_i$. The macroscopically applied normal force is a sum of the force exerted on each junction $N = \sum_i \Delta N|_i$. This leads to a linear relationship between the two terms: $A_{real} \propto N$, making Bowden and Tabor's theory consistent with the laws by Amontons and Coulomb.

However, if friction mostly involves elastic deformation of contacting asperities, the Hertzian contact theory (Johnson, 1987) gives $A_{real} \propto N^{2/3}$. This would make the results from Bowden and Tabor's theory deviate from the Amontons and Coulomb's laws. Based on the observation that the height profile distribution of engineering surfaces often resembles a Gaussian distribution, Greenwood and Williamson proposed an explanation for this discrepancy in 1966 (Greenwood et al., 1966). They approximated the surface asperities using spherical caps with the same radius of curvature but following a Gaussian height distribution, wherein an exponential distribution can be used to approximate the number of asperities that are higher than the average surface level. With this geometrical assumption, the Greenwood–Williamson model gives the result that A_{real} is, in fact, proportional to N for the elastic deformation case based on the Hertzian contact theory.

Using an atomic force microscope and surface forces apparatus (SFA), a series of innovative single-junction friction experiments were conducted around the early 2000s (Gao et al., 2004). For example, Israelachvili and coworkers designed experiments to compare load-controlled and adhesion-controlled scenarios. For nonadhering surfaces (load-controlled), the friction force was found to obey the classical, empirical Amontons & Coulomb's laws (Fig. 2.3). Since only a single junction was involved, the Greenwood–Williamson model was irrelevant. The Hertzian contact theory would predict that A_{real} forms a sublinear (rather than linear) function with the normal load N in elastic deformation of *a single junction*, which was indeed confirmed in the SFA experiment (Berman et al., 1998). This made Bowden and Tabor's view questionable, which would predict that the real contact area forms a linear relationship with the normal load in a single junction.

For adhering surfaces, the friction force was found to scale with the real contact area but not the normal force, so Amontons & Coulomb's laws were seen to break down. This is consistent with Bowden and

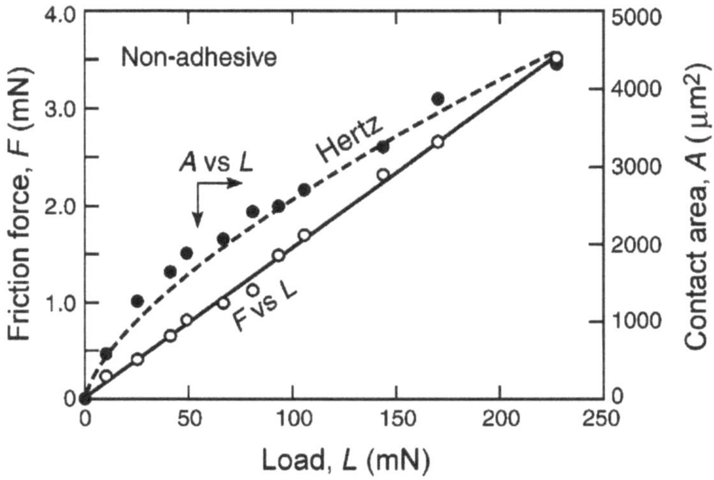

Figure 2.3 Measured friction force F and real contact area A against externally applied load L for two molecularly smooth mica surfaces sliding in 0.5 M KCl solution at T = 22°C. *Reprinted with permission from Berman, A., Drummond, C., Israelachvili, J. 1998. Amontons' law at the molecular level. Tribology Letters 4, 95–101.*

Tabor's view—because in the presence of adhesion, the real contact area of a single junction still forms a nonlinear relationship with the normal force. This conclusion is achieved using various contact mechanics models, such as the DMT (Derjaguin-Muller-Toporov) model (Derjaguin et al., 1975), JKR (Johnson-Kendall-Roberts) model (Johnson et al., 1971), or Maugis-Dugdale model (Maugis, 1992). These models are based on different assumptions. For example, in the DMT model, the contact area coincides with that obtained from the nonadhesive problem; the adhesive tractions are thus confined in the separated zone. In contrast, the JKR solution allows both compressive and adhesive tractions to be exchanged within the contact area (Ciavarella et al., 2019). This is illustrated in Fig. 2.4.

The observations in single-junction friction experiments suggested that Bowden and Tabor's theory works well when there is strong adhesion between the contact, but it cannot explain the nonadhering friction of a single junction, for which the Amontons & Coulomb's laws were found to generally apply well.

Using large-scale molecular dynamics simulations, Mo et al. considered the friction laws at the nanoscale (Mo et al., 2009). As shown in Fig. 2.5, one of their main findings was that the real contact area A_{real} in

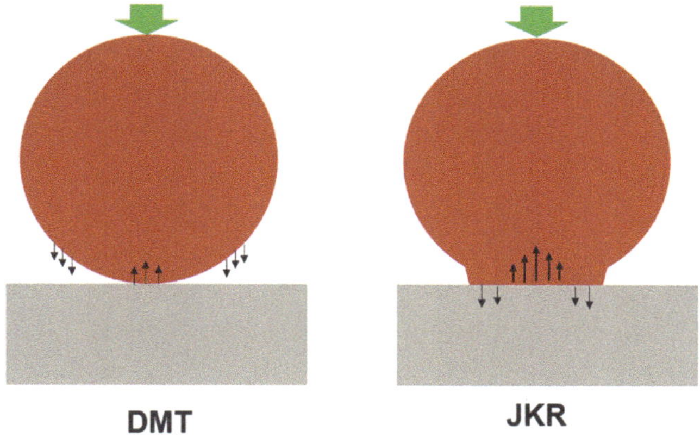

Figure 2.4 Sketch of traction distributions in DMT and JKR types of solution.

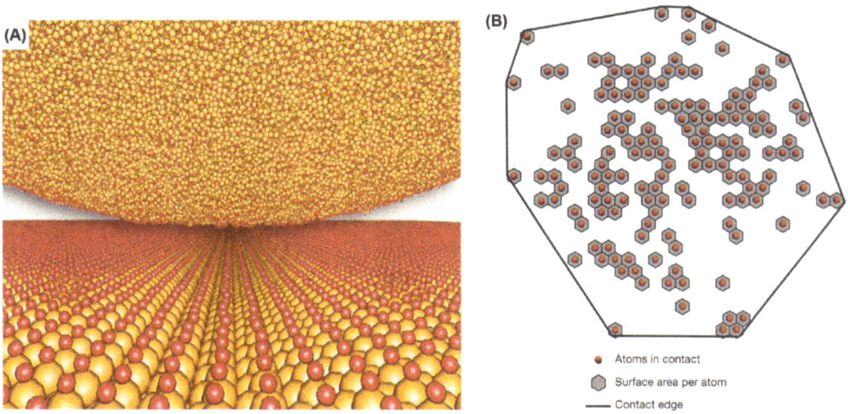

Figure 2.5 Large-scale molecular dynamics simulations of the contact between an amorphous carbon tip and a diamond sample. (A) Far view showing the contact geometry. (B) Contact area definitions, wherein red circles represent the sample atoms within the range of chemical interactions from the tip atoms; the contact area per atom is represented by grey hexagons. *Reprinted with permission from Mo, Y., Turner, K.T., Szlufarska, I. 2009. Friction laws at the nanoscale. Nature 457, 1116–1119.*

single-junction friction is not the same as the junction (or asperity) contact area A_{asp}. Instead, the real contact area should be defined as $A_{real} = A_{at}N_{at}$, where A_{at} is the average surface area per atom and N_{at} is the number of atoms having chemical interactions with the tip atoms. The simulation showed that A_{real} is considerably smaller than A_{asp}. For nonadhering surfaces, the friction force was found to scale with the applied normal force, so the Amontons & Coulomb's laws are well applied. More importantly, while A_{asp} was found to form a sublinear relationship with the normal force—consistent with the Hertzian contact theory, the real contact area does form a linear relationship with the normal force. This finding supports the view by Bowden and Tabor, essentially establishing a bridge between Bowden and Tabor's theory and the empirical Amontons & Coulomb's laws. For adhering surfaces, the friction force was found to scale with the real contact area, but forms a nonlinear relationship with the normal force. This again supports the view by Bowden and Tabor.

Overall, with the explanation from molecular dynamics simulations by Mo and coworkers, Bowden and Tabor's theory well explains both non-adhering and adhering surfaces. Amontons & Coulomb's laws work well for dry contacts between nonadhering surfaces, in which the friction force is governed by the applied normal force. For systems with strong adhesion, the frictional force no longer scales with the applied normal force but is proportional to the real contact area, which is usually considerably smaller than the apparent contact area. A special case in friction is between soft materials and rough surfaces. In the extreme case that the soft material can deform so easily to fill the valleys between asperities (i.e., conforming to surface roughness), the true contact area could become even larger than the apparent one, resulting in unusually high friction.

Given the complex nature of friction, the COF values from the literature should always be used with caution, because these values may strongly depend on the exact surface condition, including microscale conditions that are not well characterized. The COF of pure metals (without adsorbates, e.g., under ultrahigh vacuum) is 3–7. The value drops to 0.6–1.0 after oxidation. The COF value might also vary depending on the load. Under low loads, the friction is governed by the oxide coating; under high loads, microcontacts may penetrate the oxide layer. Usually, the friction of dry and clean surfaces does not change much with roughness. Strong adhesive interaction will occur only for very flat and clean surfaces with atomic contact over significant areas, leading to high friction and wear.

In addition to dry sliding, lubricant friction is of great importance in many industries. Depending on the thickness of the lubricating layer, we may distinguish between two different lubrication regimes, hydrodynamic and boundary lubrication. In hydrodynamic lubrication, the lubrication layer is thicker than the maximum height of the surface asperities; this leads to a complete separation of the friction partners. In this regime, the lubricating layer, rather than the surface junctions, carries the load. The friction force is thus fully determined by the viscous friction of the lubricant. In boundary lubrication, the lubrication layer is typically only a few molecular layers thick and therefore thinner than the surface roughness. In many practical applications, the scenario is between the two extremes, which is referred to as mixed lubrication.

In additive friction stir deposition, friction initially occurs between the rotating feed-rod and stationary substrate; in the steady-state deposition, friction mostly occurs between the rotating print head and deposited material. The state-of-the-art friction theory can help us model the friction force and frictional heat generation at these interfaces, although the exact physics of the friction stir state are not well understood and the friction model may depend on specific material systems.

2.2 Fundamentals of heat and mass transfer

2.2.1 Heat transfer

Additive friction stir deposition is implemented by severe plastic deformation at elevated temperatures in a concentrated region. Heat and mass transfer are central to determining the thermal history (i.e., the plot of T as a function of t) and deformation history (i.e., the plot of $\dot{\varepsilon}$ as a function of t) of the deposited material. In this section, we review the fundamentals of heat and mass transfer, paving the road towards better understanding the evolution of the temperature field and material velocity field in additive friction stir deposition.

Heat transfer is defined as thermal energy in transit because of the temperature differences. The heat transfer may proceed by three modes, including conduction, convection, and radiation (see Fig. 2.6). Conduction occurs across a medium of a solid or a stationary fluid (i.e.,

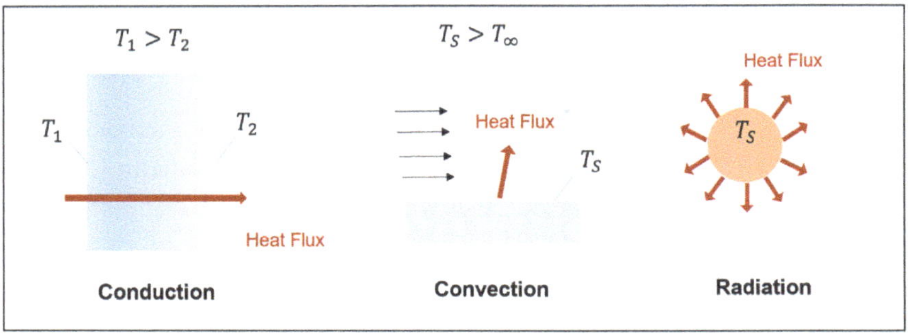

Figure 2.6 Illustration of the three heat transfer mechanisms.

liquid or gas) via microscopic collisions of particles (like molecules, atoms, and electrons) and movement of electrons within a body. The transfer of disorganized kinetic and potential energy via microscopic collision leads to diffusion of energy. For heat transfer problems with a temperature distribution $T = T(x, y, z)$, Fourier's law gives

$$\Phi_{conduction} = -\kappa \cdot \nabla T, \tag{2.3}$$

where $\Phi_{conduction}$ is the heat flux or the heat transfer rate per unit area perpendicular to the transfer direction; κ is a transport property known as the thermal conductivity, which is an intrinsic material property with a unit of $W/(m \cdot K)$.

Convection occurs between a surface and a moving fluid when they are at different temperatures. Convection is comprised of two mechanisms. The first mechanism is the diffusion of energy via random molecular motion. The second mechanism is known as advection, and it involves the transfer of energy via the bulk motion of fluid in the presence of temperature differences. Heat is effectively transferred in space by the collective motion of molecules. Newton's law of cooling gives

$$\Phi_{convection} = h \cdot (T_s - T_\infty), \tag{2.4}$$

where $\Phi_{convection}$ is the convective heat flux, T_s and T_∞ are the temperatures of the surface and the fluid, respectively, and h is the convection heat transfer coefficient with a unit of $W/(m^2 \cdot K)$.

Thermal radiation occurs because any surface with finite temperature emits energy via electromagnetic waves. While the occurrence of conduction and convection necessitates a material medium, radiation readily occurs in vacuum; in fact, radiation in vacuum is more efficient than radiation in the presence of any material medium. The rate of energy released per unit area is called the surface emissive power. For a blackbody, the emissive power is given by the Stefan-Boltzmann law:

$$\Phi_{blackbody} = \sigma T_s^4, \tag{2.5}$$

where T_s is the absolute temperature of the surface, and $\sigma = 5.67 \times 10^{-8} W/(m^2 \cdot K^4)$, which is the Stefan-Boltzmann constant. For the real surface, the heat flux is lower by a factor and Eq. (2.5) should be rewritten as $\Phi_{radiation} = \varepsilon \sigma T_s^4$, where ε is the emissivity $(0 \leq \varepsilon \leq 1)$. The emissivity depends on the temperature of the surface, the wavelength and angle of the emission, and the surface roughness. Considering the radiation from the surroundings at a temperature of T_{amb}, the net rate of heat transfer is

$$\Phi_{radiation} = \varepsilon\sigma\left(T_s^4 - T_{amb}{}^4\right). \tag{2.6}$$

2.2.2 Mass transfer

During additive friction stir deposition, the feed material undergoes continuous plastic deformation with intricate flow patterns, in which complex heat and mass transfer phenomena arise. No matter how complex they are, these phenomena follow the universal laws of conservation. The first law of conservation is related to mass, giving rise to the continuity equation. If ρ is the material density and $\mathbf{v}$ is the velocity, we have

$$\frac{\partial \rho}{\partial t} + \nabla \cdot (\rho \mathbf{v}) = 0. \tag{2.7}$$

The symbols in bold are vectors or tensors. If the material density during deposition remains constant and the material is incompressible, Eq. (2.7) reduces to

$$\nabla \cdot \mathbf{v} = 0. \tag{2.8}$$

The second law of conservation is related to momentum and is based on Newton's second law. It can be expressed using the Cauchy momentum equation:

$$\frac{\partial}{\partial t}(\rho \mathbf{v}) = -\nabla \cdot (\rho \mathbf{v} \otimes \mathbf{v}) - \nabla \mathrm{p} + \nabla \cdot \boldsymbol{\tau} + \rho \mathbf{g}. \tag{2.9}$$

Here, p is the pressure; $\mathbf{g}$ is the body accelerations, such as gravity, inertial accelerations, electrostatic accelerations, *etc.*; $\mathbf{v} \otimes \mathbf{v} = \mathbf{v}\mathbf{v}^T$ is the dyadic tensor of the velocity and $\boldsymbol{\tau}$ is the deviatoric stress tensor. For constant density ρ and viscosity η, by writing $\boldsymbol{\tau}$ in terms of Newton's law: $\boldsymbol{\tau} = \eta(\nabla \mathbf{v} + \nabla \mathbf{v}^T)$, Eq. (2.9) is simplified to the classical Navier–Stokes equation. Under friction stir conditions, however, the material flow behavior rarely follows the Newtonian fluid behavior. Rather than directly using the Navier–Stokes equation, therefore, Eq. (2.9) should be used in combination with the constitutive model best describing the specific material flow behavior.

The third law of conservation is related to energy. The heat transfer mechanisms inside the deposited material include thermal conduction and thermal convection via material flow. The governing equation of heat transfer can be written as:

$$\frac{\partial}{\partial t}(\rho C_P T) + \nabla \cdot \rho \mathbf{v}(C_P T) + \nabla \cdot \kappa \nabla T = \dot{q}. \tag{2.10}$$

Here C_P is the specific heat capacity, T is the temperature, κ is the thermal conductivity, and $\dot{q}$ is the volumetric heat generation or dissipation rate.

In addition to bulk motion, which is termed advection, mass transfer can take place on the microscopic level through diffusion. Diffusion is time-dependent, originating from the random motion of individual particles. For a given period t, the diffusion distance is

$$L = \sqrt{Dt}, \tag{2.11}$$

where D is the diffusion coefficient or diffusivity. In the presence of a concentration gradient, long range diffusion can happen following Fick's law:

$$\boldsymbol{J}_{Diff} = -D\nabla c, \tag{2.12}$$

where c is the concentration. The advection flux can be written as $\boldsymbol{J}_{Adv} = c\mathbf{v}$. Considering both advection and diffusion and ignoring any source or sink, the continuity equation for such a scenario can be written as

$$\frac{\partial c}{\partial t} + \nabla \cdot \boldsymbol{J}_{Total} = 0. \tag{2.13}$$

Here, the total flux $\boldsymbol{J}_{Total} = \boldsymbol{J}_{Diff} + \boldsymbol{J}_{Adv}$. Then, Eq. (2.13) can be written as the following, which is the general advection–diffusion equation:

$$\frac{\partial c}{\partial t} = \nabla \cdot (D\nabla c) - \nabla \cdot (\mathbf{v}c). \tag{2.14}$$

The ratio of transport by advection and by diffusion can be characterized by the Peclet number, Pe, which is defined as

$$\mathrm{Pe} = \mathrm{v}L/D. \tag{2.15}$$

Pe can also be interpreted as the ratio of timescales between diffusion and advection. For a given distance L, the time scale for diffusion is L^2/D, and the time scale for advection is L/v. Their ratio equals $\mathrm{v}L/D$, which is the Peclet number. If we consider chemical reaction or phase

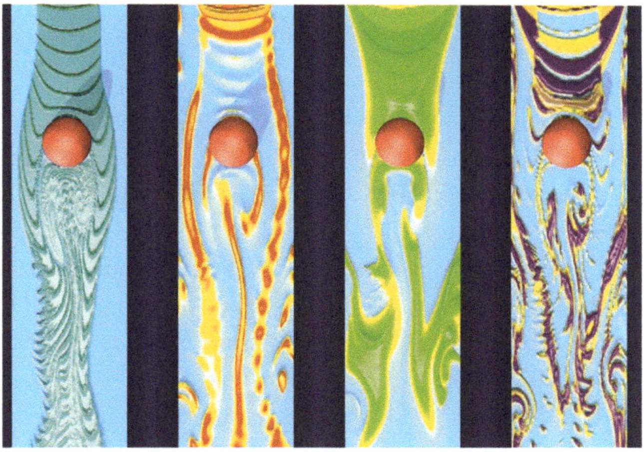

Figure 2.7 Simulation of advection-reaction-diffusion in the presence of a hard sphere barrier. From the left to the right: advection only; Gray–Scott reaction diffusion with advection; FitzHugh–Nagumo reaction diffusion with advection; Barkley reaction diffusion with advection. *Reprinted with permission from Kim, T., Lin, M. 2007. Stable advection-reaction-diffusion with arbitrary anisotropy. Computer Animation and Virtual Worlds 18, 329–338.*

transformation, we can add a source or sink term in Eq. (2.14) to make it an advection-reaction-diffusion equation. This type of equation can be used to simulate complex mixing phenomena and pattern formation in engineering or biological systems. One example of the related simulation results is shown in Fig. 2.7 (Kim and Lin, 2007).

In additive friction stir deposition, advection, diffusion, and chemical reaction (or phase transformation) can all occur simultaneously. Similar to friction stir welding, additive friction stir deposition often results in significant material flow, which is driven by the compression and shear imposed by the feeding apparatus and print head. This flow is advection, following the momentum conservation law. For a heterogeneous feed material, diffusion occurs during printing following Fick's law. In terms of phase transformation, dissolution, precipitation, and mechanically-induced new phase formation all need to be considered in additive friction stir deposition.

2.3 Basic principle of additive friction stir deposition

Additive friction stir deposition allows for printing of metals and metal matrix composites in a solid-state additive manner. Its configuration is similar to fused deposition modeling in that a feed-rod is delivered through a channel inside the print head as shown in Fig. 2.8. What makes it special is that the feed-rod and print head rapidly rotate at a high rate of $\Omega \sim 10^2 - 10^3$ revolutions per minute (RPM). When the rotating feed-rod comes into contact with the stationary substrate, the dynamic interaction creates frictional heat, which leads to thermal softening of the feed-rod. Owing to compressive force from the feeding apparatus and shear force due to the rotation rate gradient along the vertical direction, the feed-rod yields and extrudes into the space between the substrate and the print head. The rotating print head then imposes compression and shear forces onto the deposited material, which plasticizes and mixes with the substrate surface.

Note that features like protrusions and scrolls can be introduced to the bottom of the print head. The influence of tool geometry is elaborated in Chapter 5. These protrusions typically insert into the substrate during printing to further enhance the material flow and layer mixing. Once good mixing and strong interfacial bonding are achieved, in-plane motion of the substrate relative to the print head will lead to deposition of a single track of material. The thickness of the deposition track is set by the distance between the bottom of the print head and the substrate surface, and can range from several hundreds of microns to a few millimeters. Arbitrary material patterns can be created by programming the print path based on the G-code. With this process repeated layer by layer, a three-dimensional component can be built up. The printing process in additive friction stir deposition is illustrated in Fig. 2.9 (Perry et al., 2021). The key processing parameters include the print head rotation rate Ω, travel velocity V, material feed rate F, and layer height h.

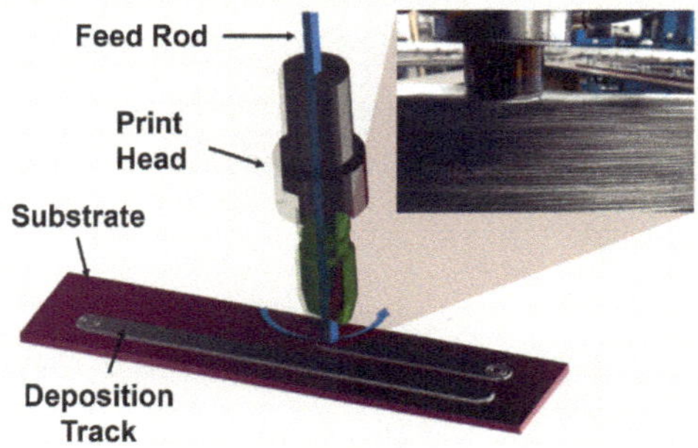

Figure 2.8 An illustration of the additive friction stir deposition process.

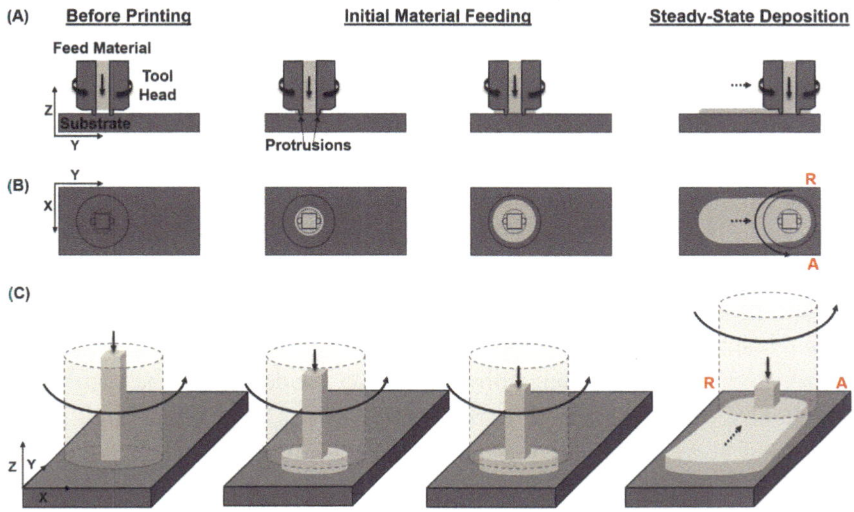

Figure 2.9 Schematics of the main processing steps in additive friction stir deposition (from left to right): before printing, during initial material feeding, and during steady-state deposition. This includes the (A) side view, (B) top view, and (C) isometric view. R and A refer to the retreating and advancing side, respectively. *Reprinted with permission from Perry, M.E.J., Rauch, H.A., Griffiths, R.J., Garcia, D., Sietins, J.M., Zhu, Y., et al. 2021. Tracing plastic deformation path and concurrent grain refinement during additive friction stir deposition. Materialia 18, 101159.*

Initially, frictional heat is generated at the interface between the substrate and feed-rod through dynamic contact. As the material extrudes beneath the print head, frictional heat will be generated at the interface between the print head and material, with a much larger contact area. Insertion of the rotating protrusions into the substrate also generates frictional heat. Beyond the *interfacial* heat generated via friction, rapid plastic deformation of the deposited material results in plastic energy dissipation, which is a *volumetric* heat source. Similar to other metal additive manufacturing processes, temperature evolution is critical in additive friction stir deposition; but more than that, the distinctive consideration of *material flow* is necessary. On one hand, the material flow determines the volumetric energy produced from plastic deformation and contributes to heat transfer via convective heat flow. On the other hand, the material flow is influenced by the temperature distribution inside the deposited material, because the constitutive material properties—such as dynamic viscosity, strain hardening component, and flow stress—are strongly temperature dependent. Therefore, the thermal and deformation aspects are fully coupled in additive friction stir deposition.

The heat transfer mechanisms inside the deposited material include thermal conduction and thermal convection via material flow. The governing equation of heat transfer follows Eq. (2.10). To model the material flow, one may treat the deposited material as a viscous fluid based on computational fluid dynamics using an Eulerian mesh, which is inspired by the friction stir welding literature (Nandan et al., 2007). Eqs. (2.8) and (2.9) are the key equations for modeling material flow. Beyond Eqs. (2.8)–(2.10), we need to know the constitutive relationship of the specific material that links the effective strain rate to the effective stress. This relationship is typically modeled by existing relationships, such as the Sellars and Tegart model, Johnson-Cook model, and the Buffa law (Kuykendall et al., 2013). For printing Al alloys, the Sellars and Tegart model has been widely used for modeling the friction stir condition:

$$\sigma = \frac{1}{\alpha}\sinh^{-1}\left[\left(\frac{Z}{A}\right)^{\frac{1}{n}}\right] = \frac{1}{\alpha}\ln\left\{\left(\frac{Z}{A}\right)^{\frac{1}{n}} + \left[1+\left(\frac{Z}{A}\right)^{\frac{2}{n}}\right]^{\frac{1}{2}}\right\}. \tag{2.16}$$

Here the Zener-Hollomon parameter $Z = \dot{\varepsilon}\exp\left(\frac{Q}{RT}\right)$. A, α, Q, n can be determined from hot compression and torsion tests. After that, the temperature field and velocity field can be established for given boundary conditions.

As an alternative to computational fluid dynamics, a solid mechanics-based formulation can be used to model additive friction stir deposition too. In order to reduce the computational costs while still capturing the large plastic deformation and material mixing, we can employ an arbitrary Lagrangian–Eulerian formulation (Al-Badour et al., 2013) or a mesh-free solid-mechanics formulation, such as smoothed particle hydrodynamics (Stubblefield et al., 2021). When coupled with crystal plasticity, one can predict the grain-level deformation and texture formation. To account for the material addition (i.e., deposition) in transient thermal analysis, techniques like quiet or inactive elements can be integrated with the finite element modeling. This kind of strategy has been widely employed in modeling other additive manufacturing processes (Li et al., 2018).

In addition to the governing equations for heat and mass transfer, the heat flow and material flow boundary conditions are critical in modeling. It is instructive to compare the boundary conditions between friction stir welding or processing and additive friction stir deposition. In the former, the stirred material is surrounded and constrained by the material in the thermomechanically affected zone, so conduction is a dominant mechanism of heat loss at the

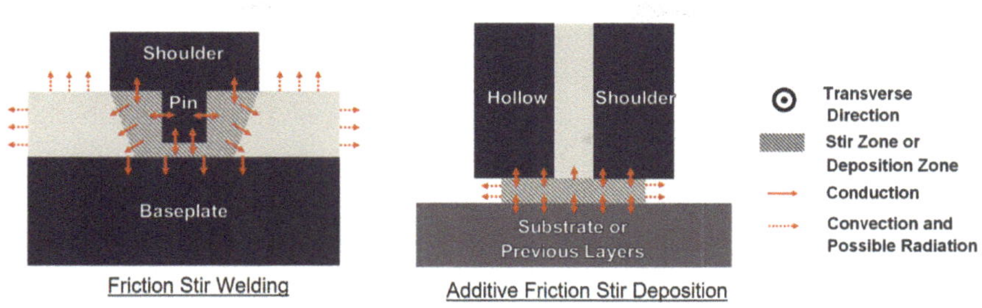

Figure 2.10 A comparison of the boundary conditions for heat flow between friction stir welding and additive friction stir deposition. *Reprinted with permission from Yu, H. Z., Jones, M.E., Brady, G.W., Griffiths, R.J., Garcia, D., Rauch, H.A., et al. 2018. Non-beam-based metal additive manufacturing enabled by additive friction stir deposition. Scripta Materialia 153, 122–130.*

boundaries of the stir zone. In the latter, the deposited material is in direct contact with air, so convection and radiation should be considered. The boundary heat input in the former is at the shoulder-material and pin-material interfaces, whereas in the latter the heat is generated at the interfaces between print head and material, the substrate and material, as well as the tool protrusions and material. Fig. 2.10 compares the different boundary conditions between friction stir welding and additive friction stir deposition (Yu et al., 2018).

Despite the availability of various modeling tools, it is important to note that our understanding of the physical processes in additive friction stir deposition is rather inadequate; neither the computational fluid dynamics formulation nor the solid mechanics formulation fully and accurately capture the underlying physics. For example, the velocity field at the material-print head interface is unclear because a thorough understanding of the contact state is still lacking. The velocity field and heat generation rate at the material-substrate interface are ill defined because little is known about the details of co-plastic deformation and mixing. Addressing these issues calls for an integrated in situ monitoring system to enable direct experimental characterization of the *temperature evolution* and *material flow* during additive friction stir deposition. This is the focus of the next section.

2.4 Establishment of an integrated in situ monitoring system: real-time measurement of temperature, force, torque, and material flow

To understand the process fundamentals and gain new physical insights, we introduce an integrated in situ monitoring system that has been incorporated into the additive friction stir deposition facility in the author's laboratory at Virginia Tech. This system provides direct characterization of the temperature evolution, force and torque evolution, as well as the material flow and interface contact. Key in situ monitoring capabilities of this system include:

1. *Thermal field evolution*. Similar to friction stir welding or processing, direct measurement of temperature evolution in the deposited material is challenging in additive friction stir deposition. Thermocouples, for example, are difficult to place in the deposition zone due to the damage caused by the mechanical forces during deposition. In friction stir welding or processing, the tool shoulder and pin obscure the stir zone; in additive friction stir deposition, the deposition zone is covered by the print head, making it difficult to track the evolution of temperature from the top view. However, because the deposited material lies on top of the substrate, additive friction stir deposition allows for direct measurement of the deposited material's thermal field evolution from the side view, using noncontact methods such as infrared cameras or pyrometers. As a result, one infrared camera has been mounted onto the bed of the printing system, monitoring the temperature evolution of the deposited material from the side view. Another infrared camera is mounted at a higher level to capture the thermal evolution from an isometric angle. Along with thermographic measurement, a thermocouple array (K-type) is used to continuously monitor the local temperatures in the substrate directly beneath the deposited material.
2. *Force and torque evolution*. The standard force and torque in-line measurement is integrated into the additive friction stir deposition facility. The compressive force exerted by the linear feed apparatus on the feed-rod is determined by a strain-based load cell. An ABB ACS-880 spindle drive implicitly measures the torque at the print head. The current applied to the motor is measured and then converted to a torque value based on the motor characteristics and a gear ratio of 1.775 between the motor and the print head. The torque for driving the print head rotation is also measured without material deposition; this value is then subtracted from the measured values to account for the net torque used in additive friction stir deposition.
3. *Material flow*. Like thermal imaging, the material flow can be monitored from the side through in situ surface imaging based on optical video recording. The cameras may be placed on the leading edge, advancing side, and retreating side of the deposition track. During

printing, a high-temperature black primer is sprayed on the surface of the deposit to measure the motion of the deposited material relative to the print head. The black speckles provide contrast and are monitored individually to obtain the velocity field of the exterior surface of the deposited material. This gives rise to new insights into the stick-slip phenomena between the deposited material and print head.

Fig. 2.11 schematically illustrates how the additive friction stir deposition facility and the in situ monitoring system are integrated, including infrared thermal imaging, thermocouple measurement, force and torque measurement, and optical video recording. This system provides the basis for data-driven closed loop control in future work, wherein the process parameters can be automatically adjusted if a process anomaly is in situ detected. This may involve the employment of multi-channel data fusion and development of learning-based defect mitigation strategies (Jafari-Marandi et al., 2019).

For the rest of this chapter, the in situ monitoring results are presented based on the printing of two specific material systems, Al-Mg-Si and Cu, which are not easy to print using beam-based or fusion-based additive processes. The challenges are mostly from their high thermal conductivity and high reflectivity; for high strength Al alloys, susceptibility to hot cracking is an additional concern. These materials, however, are readily printed using additive friction stir deposition with high quality and good mechanical properties. One important reason for selecting Al-Mg-Si and Cu is their distinct thermomechanical response and interaction with the print head, which is made of tool steel in this case. Through a comparative study of the thermal and deformation behavior of these two materials, we can further explore the role of intrinsic material properties in additive friction stir deposition and unravel the correlation between the material flow behavior and the temperature, force, and torque evolution.

Printing of AA6061 (an Al-Mg-Si alloy) and commercially pure Cu is performed using a MELD R2 system based on a 38-mm-wide steel print head. Two protrusions with rounded surfaces and a height of 1.5 mm are located directly around the feed-rod exit hole (see Fig. 2.12). In each experiment, a stack of two deposited layers is created with a single-layer thickness of approximately 0.8 mm and a nominal length of 70.2 mm. The substrate and the deposited material are of the same composition. Substrate plates are of 6.35 mm thick, while the feed-rod has a square section of 9.525 $\times$ 9.525 mm. Additive friction stir deposition is performed systematically at various processing conditions with the print head rotation

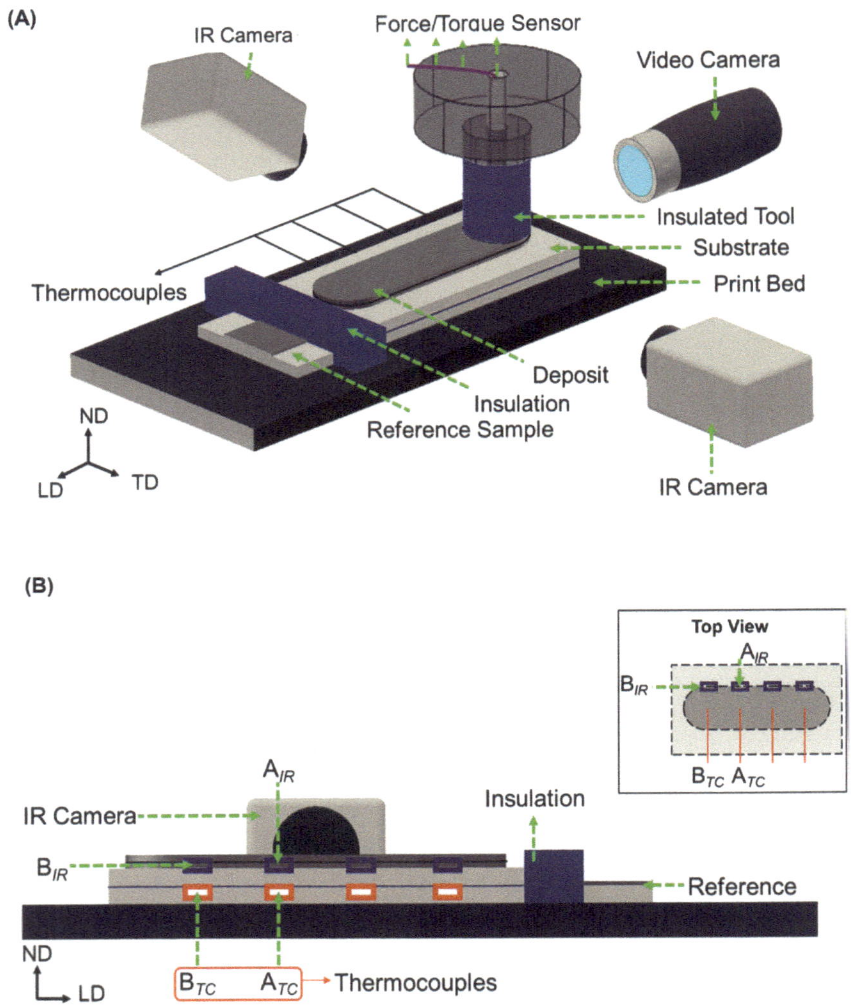

Figure 2.11 A schematic of the integrated in situ monitoring system, including temperature measurement of the deposited material and substrate, force and torque measurement, and surface imaging of the material flow. Top: Isometric view of the system. Bottom: Side view illustrating the thermocouple locations. A_{TC} and B_{TC} are two locations in the substrate, where the temperature is measured by the thermocouples. A_{IR} and B_{IR} are two locations in the first layer of the deposited material, where the temperature is measured by the IR camera. A_{IR} and B_{IR} are directly above A_{TC} and B_{TC}.

rate of 300 RPM, 600 RPM and 900 RPM and travel velocity of 1.00, 2.00, and 3.00 mm/s. To control the volumetric deposition rate, the ratio between the material feed rate and the in-plane velocity is held constant at a value of 1:3 during all experiments.

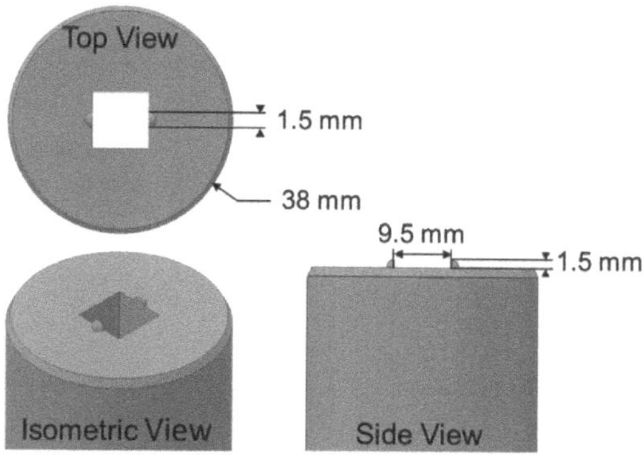

Figure 2.12 Geometrical details of protrusions on the bottom surface of the print head. *Reprinted with permission from Garcia, D., Hartley, W.D., Rauch, H.A., Griffiths, R. J., Wang, R., Kong, Z.J., et al. 2020. In situ investigation into temperature evolution and heat generation during additive friction stir deposition: a comparative study of Cu and Al-Mg-Si. Additive Manufacturing 34, 101386.*

2.5 Temperature evolution in the deposited material and substrate

For accurate thermographic measurement, calibration of the infrared camera temperature is necessary. Here, the Al-Mg-Si and Cu samples previously printed by additive friction stir deposition are heated using a Fisher Science hot plate. The emittance of a given substance (Al-Mg-Si or Cu) is adjusted to match the infrared camera outputs with the readings from a pre-calibrated thermocouple. The process is repeated at 50 K intervals. During additive friction stir deposition, the rotating print head is coated with fiberglass insulation to eliminate undesirable radiative disturbance (e.g., reflections). A piece of Al-Mg-Si or Cu held at known elevated temperatures is positioned in the field of view of the camera to act as a temperature guide.

2.5.1 Thermal history of the deposited materials

An overview of the measured thermal field evolution during deposition is shown in Fig. 2.13, which consists of several representative snapshots from a thermal video for a two-layer Cu deposition. The presented thermographic measurements are performed using a Micro-Epsilon TIM 400 infrared camera, which is positioned perpendicular to the side of the deposit at a working distance of 11 cm. The pixel size provides sufficient resolution to distinctly measure the temperature of each layer while

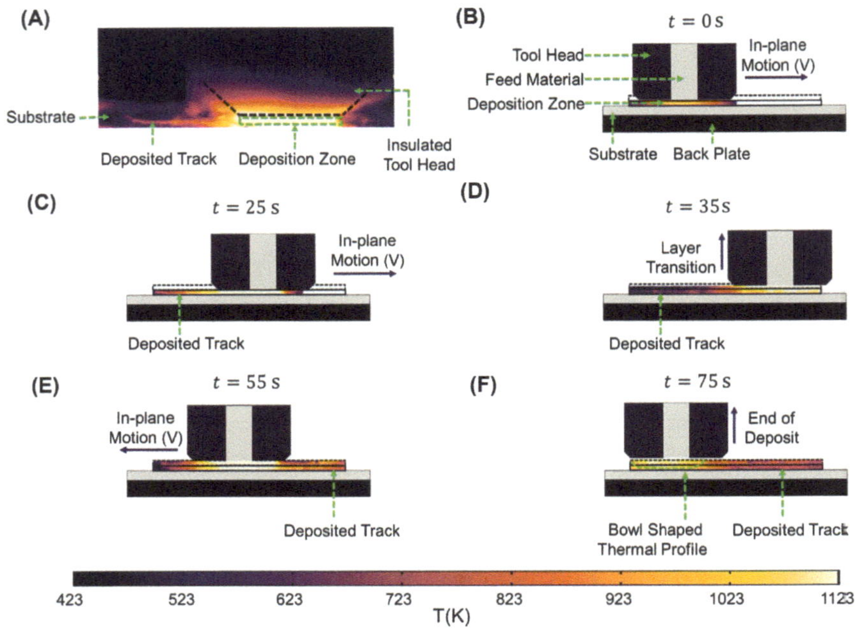

Figure 2.13 An overview of the thermal profile during additive friction stir deposition of Cu. The print head rotation rate is 600 RPM and the travel velocity is 1 mm/s. (A) A representative thermal image of the entire field of view. (B)–(F) The measured temperature field at time step $t = 0$, 25, 35, 55, and 75 s. *Reprinted with permission from Garcia, D., Hartley, W.D., Rauch, H.A., Griffiths, R.J., Wang, R., Kong, Z.J., et al. 2020. In situ investigation into temperature evolution and heat generation during additive friction stir deposition: a comparative study of Cu and Al-Mg-Si. Additive Manufacturing 34, 101386.*

providing a large field of view to capture the entire length of the deposit. Time $t = 0$ s corresponds to the thermal profile of the plunge process and the print head starts to move in-plane. At $t = 25$ s, the print bed has translated relative to the rotating print head. The trail behind shows a lower temperature than the material directly below the print head. The layer transition starts at $t = 35$ s. At $t = 55$ s, some material has been deposited in the second layer; the relative movement between the print bed and the print head is in the opposite direction to the first layer deposition. At $t = 75$ s, the desired two-layer geometry is complete, and the print bed is lowered to detach the feed-rod from the deposited material. In the final thermal picture, a bowl-shaped high temperature region is seen at the end, which is consistent with the predicted temperature distribution resulting from a moving heat source.

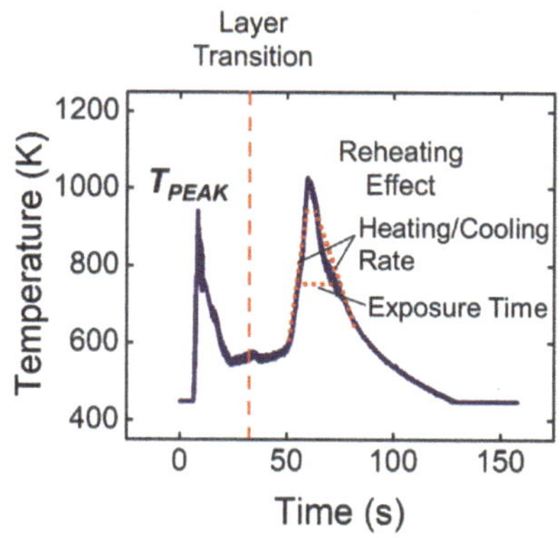

Figure 2.14 A typical T-t plot measured for a single spot of the deposit, showing two peaks. The first peak is caused by deposition; the second peak corresponds to a substantial reheating effect. *Reprinted with permission from Garcia, D., Hartley, W.D., Rauch, H.A., Griffiths, R.J., Wang, R., Kong, Z.J., et al. 2020. In situ investigation into temperature evolution and heat generation during additive friction stir deposition: a comparative study of Cu and Al-Mg-Si. Additive Manufacturing 34, 101386.*

Fig. 2.14 presents the typical *T-t* plot or the temperature evolution history of the deposited material. This is based on the temperature measurement at a given spot. There are two distinct peaks. The first peak refers to the deposition of fresh material at Spot A_{IR} (see its position in Fig. 2.11). Because it is an intrinsic thermal feature that is governed by the processing parameters and print head-material interactions, this peak is denoted T_{Peak}. Since no substance is initially present in the field of view, the heating component of the curve in the first peak reflects the infrared camera reaction time rather than the temperature rise rate in the material. The second peak is caused by the second layer of deposition immediately above Spot A_{IR}, clearly indicating a significant reheating effect between layers. The value of the second peak temperature is not an intrinsic thermal feature in additive friction stir deposition, because it depends on the geometry of the print, the print path, and the transition time between the deposition of the first and second layer.

2.5.2 Dependence of thermal features on the processing conditions in additive friction stir deposition

In situ temperature measurements are obtained on the exterior surface of the deposited Al-Mg-Si and Cu throughout the processing space, which is defined by the print head rotation rate Ω and in-plane velocity V. For all successful Al-Mg-Si depositions with good print quality, T_{Peak} lies in the range of 76% to 92% of T_M, which is consistent with the peak temperature range of Al alloys in friction stir welding. For all successful Cu depositions, the peak temperature T_{Peak} lies in the range of 49% to 79% of the melting temperature T_M. The yield strength of Cu (Miller et al., 1991) drops more rapidly than Al alloys (Muraca and Whittick, 1972) as the temperature increases, so Cu can be processed at a lower homologous temperature in friction stirring.

Fig. 2.15 plots the peak temperature T_{Peak} as a function of Ω and V. Al-Mg-Si and Cu show the same trend: T_{Peak} decreases when the in-plane velocity V increases and rotation rate Ω decreases. Similar to the observation in the friction stir welding literature, this trend is expected. As the rotation rate Ω increases, the interfacial frictional heat generation rate likely increases, resulting in a positive correlation between T_{Peak} and Ω. With a higher in-plane velocity V, the material feed rate is higher; for a given amount of time, more material is deposited. As a result, the heat input is lower per unit of material volume, resulting in a negative correlation between T_{Peak} and V.

The T-t plot also offers the information of exposure time, reheating rate, and cooling rate. The exposure time is the length of time that the deposited material is at elevated temperatures, which is experimentally determined by finding the full width at half maximum of the peak temperature. Since the heating component of the first peak only represents

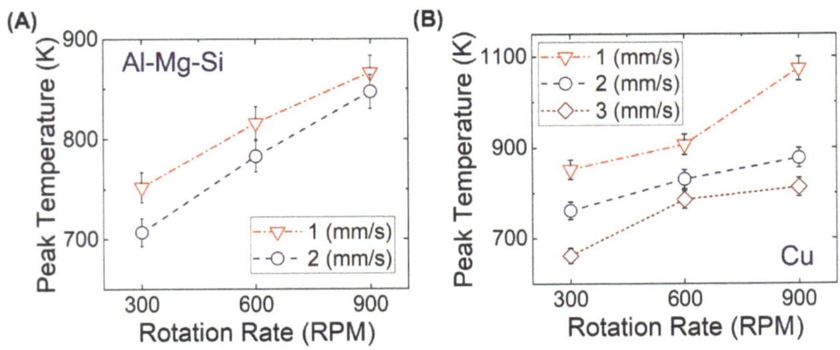

Figure 2.15 Dependence of the peak temperature on the print head rotation rate and travel velocity. (A) Al-Mg-Si and (B) Cu show consistent trends that lower travel velocity and higher rotation rate produce higher peak temperatures. *Reprinted with permission from Garcia, D., Hartley, W.D., Rauch, H.A., Griffiths, R.J., Wang, R., Kong, Z.J., et al. 2020. In situ investigation into temperature evolution and heat generation during additive friction stir deposition: a comparative study of Cu and Al-Mg-Si. Additive Manufacturing 34, 101386.*

the reaction time of the infrared camera, the exposure time is estimated based on the second peak. The instantaneous reheating and cooling rates are time dependent. From an engineering point of view, therefore, the average reheating and cooling rates are described based on the linear regions in the second peak of the *T-t* plot. This is illustrated using the pink dashed lines in Fig. 2.14.

Fig. 2.16 plots the exposure time as a function of the processing parameters during printing of Al-Mg-Si and Cu. Apparently, the exposure time decreases with an increase of V. This trend can be explained by considering the time of material interaction with the print head, which depends on the ratio between the deposition radius and the travel velocity V. With an increase of V, the print head-material interaction time and the exposure time should both decrease. The exposure time is also seen to increase with the rotation rate Ω, which likely originates from the positive correlation between Ω and T_{Peak}. It generally takes more time to cool down from a higher peak temperature; when Ω is larger, T_{Peak} is higher, and the exposure time is longer.

Fig. 2.17 plots the reheating rate and cooling rate under various processing conditions, which are generally in the range of 20–80 K/s and 5–25 K/s, respectively. For both materials, a positive correlation is observed between the reheating rate and the velocity V. This trend can be explained as follows. At high in-plane velocity V, the heat source

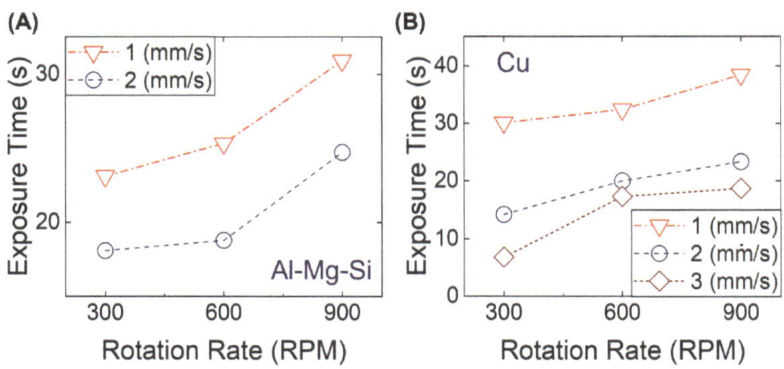

Figure 2.16 The influences of print head rotation rate and travel velocity on the exposure time. The data is measured during additive friction stir deposition of (A) Al-Mg-Si and (B) Cu. *Reprinted with permission from Garcia, D., Hartley, W.D., Rauch, H.A., Griffiths, R.J., Wang, R., Kong, Z.J., et al. 2020. In situ investigation into temperature evolution and heat generation during additive friction stir deposition: a comparative study of Cu and Al-Mg-Si. Additive Manufacturing 34, 101386.*

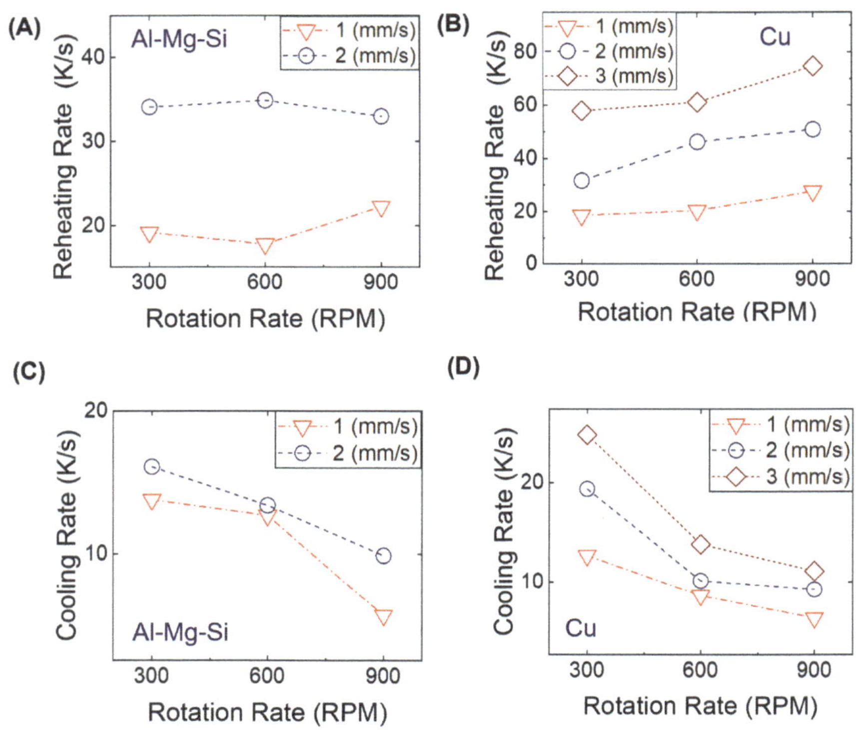

Figure 2.17 The influences of print head rotation rate and travel velocity on thermal features in additive friction stir deposition. This includes the reheating rate of (A) Al-Mg-Si and (B) Cu, and the cooling rate of (C) Al-Mg-Si and (D) Cu. *Reprinted with permission from Garcia, D., Hartley, W.D., Rauch, H.A., Griffiths, R.J., Wang, R., Kong, Z.J., et al. 2020. In situ investigation into temperature evolution and heat generation during additive friction stir deposition: a comparative study of Cu and Al-Mg-Si. Additive Manufacturing 34, 101386.*

quickly approaches a given point and results in a rapid transition from the idle temperature to the peak temperature. The cooling rate is found to be lower at higher Ω or lower V. These conditions lead to a larger heat input, which likely results in more preheating in the substrate material ahead of the traveling print head. As a result, when the new material is being deposited under the print head, the substrate temperature directly below that is already elevated. This will lead to a lower temperature gradient ∇T across the newly deposited material and the substrate. If the thermal conduction into the substrate is a major mechanism for heat loss in the deposition, a lower cooling rate is then expected.

2.5.3 Power law relationships of peak temperature and processing parameters

For all the thermal features plotted in Figs. 2.15–2.17, additive friction stir deposition of Al-Mg-Si and Cu exhibits the same physical trends with respect to the key processing parameters, Ω and V. However, this does not mean that the two materials have the exact same thermal behavior. One major difference lies in the quantitative relationship between the peak temperature T_{Peak} and processing parameters. For friction stir welding of Al alloys, Arbegast (2003) shows that the peak temperature is well described by an empirical power law equation: $T_{Peak}/T_M = K\left(\Omega^2/V\right)^a$, where K and a are fitting constants and Ω^2/V is a pseudo-heat index described by the key processing parameters. An analysis of the in situ measured data shows that the peak temperature in additive friction stir deposition of Al-Mg-Si follows a similar relationship (see Fig. 2.18A):

$$\frac{T_{Peak}}{T_M} = K_{Al}\left(\frac{\Omega^2}{V}\right)^{0.07}, \text{for Al-Mg-Si.} \tag{2.17A}$$

In Arbegast's empirical fitting, the power a is found to be in the range of 0.04–0.06, whereas Eq. (2.17A) shows a comparable power value of 0.07. This implies that the heat generation mechanisms are similar between friction stir welding and additive friction stir deposition of Al alloys.

However, the peak temperature in the additive friction stir deposition of Cu cannot be described by the pseudo-heat index Ω^2/V. As shown in Fig. 2.18B, linear fitting of $\ln\left(\frac{T_{Peak}}{T_M}\right)$ and $\ln(\Omega^2/V)$ shows an R-squared value of only 0.81 for Cu; as a comparison, this value is 0.98 for Al-Mg-Si. Instead, the peak temperature of Cu forms a power-law relationship with another pseudo-heat index of Ω/V(see Fig. 2.18C). In other words, $T_{Peak}/T_M = K\left(\Omega/V\right)^a$. An analysis of the measured data gives rise to:

$$\frac{T_{Peak}}{T_M} = K_{Cu}\left(\frac{\Omega}{V}\right)^{0.2}, \text{for Cu.} \tag{2.17B}$$

Linear fitting of $\ln\left(\frac{T_{Peak}}{T_M}\right)$ and $\ln(\Omega/V)$ shows a high R-squared value of 0.95 for Cu; but for Al-Mg-Si, this value drops to 0.87 (see Fig. 2.18D).

To conclude, the peak temperature in Cu forms a power law relationship with Ω/V, whereas the peak temperature in Al-Mg-Si forms a power law relationship with Ω^2/V. This observation reveals differences in heat generation during additive friction stir deposition of Al-Mg-Si and

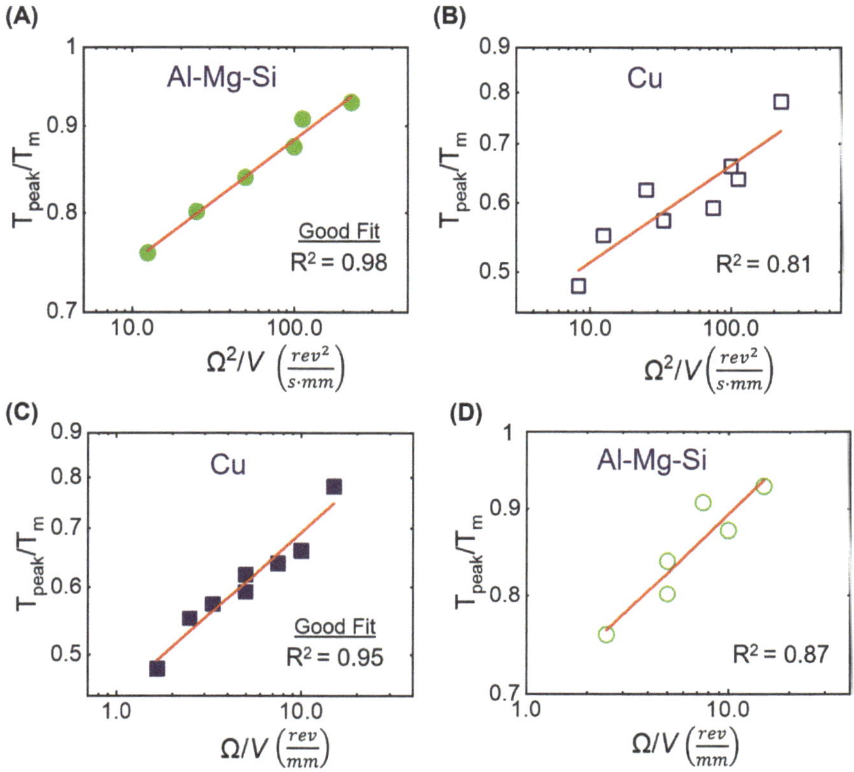

Figure 2.18 Plots of peak temperature as a function of the pseudo-heat index, including the fitting results of (A) Al-Mg-Si and (B) Cu data using pseudo-heat index Ω^2/V as well as the fitting results of (C) Cu and (D) Al-Mg-Si data using Ω/V. Plots in (A) and (C) show good fitting and a high R-squared value. Revolution per second (rev/s) is used as the unit for Ω and millimeter per second (mm/s) is used as the unit for V. *Reprinted with permission from Garcia, D., Hartley, W.D., Rauch, H.A., Griffiths, R. J., Wang, R., Kong, Z.J., et al. 2020. In situ investigation into temperature evolution and heat generation during additive friction stir deposition: a comparative study of Cu and Al-Mg-Si. Additive Manufacturing 34, 101386.*

Cu. The origin of this difference will be discussed in Section 2.8 in the context of interfacial contact and the material flow behavior.

2.5.4 Temperature evolution of the substrate

The embedded thermocouples offer an accurate and reliable measurement of the substrate temperature directly beneath the deposited material. Fig. 2.19 shows the typical temperature evolution in Al-Mg-Si and Cu

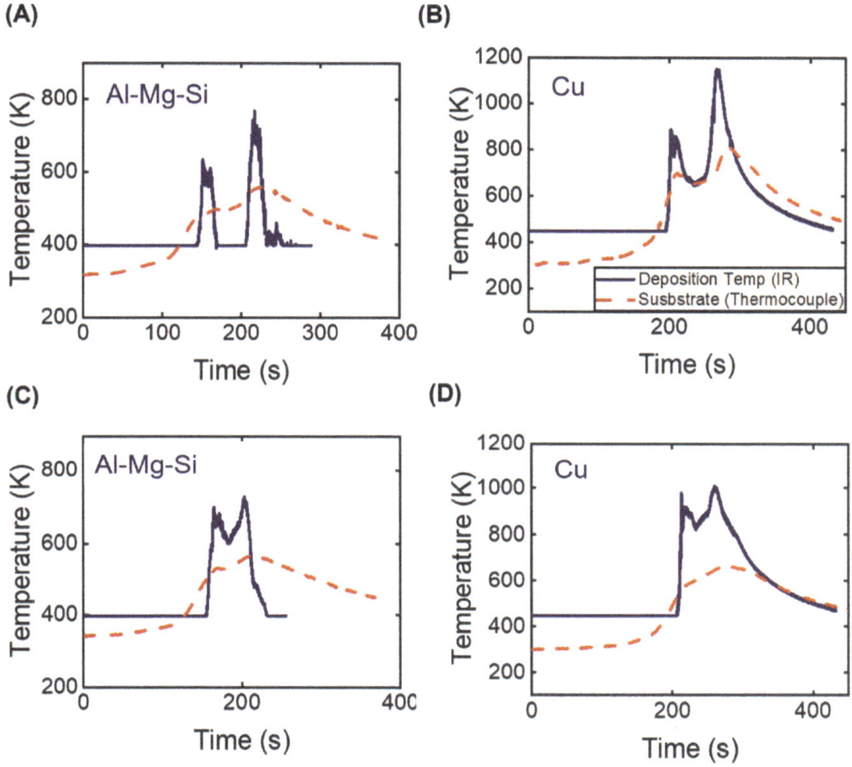

Figure 2.19 Temperature evolution of the substrate measured by the thermocouples: spot A_{TC} for (A) Al-Mg-Si at 300 RPM and 2 mm/s and (B) Cu at 300 RPM and 1 mm/s; spot B_{TC} for (C) Al-Mg-Si at 300 RPM and 2 mm/s and (D) Cu at 300 RPM and 1 mm/s. The curves are overlaid with the results measured at A_{IR} and B_{IR} using a thermal camera. *Reprinted with permission from Garcia, D., Hartley, W.D., Rauch, H.A., Griffiths, R.J., Wang, R., Kong, Z.J., et al. 2020. In situ investigation into temperature evolution and heat generation during additive friction stir deposition: a comparative study of Cu and Al-Mg-Si. Additive Manufacturing 34, 101386.*

measured by thermocouples at positions A_{TC} and B_{TC}. The corresponding T-t plots are overlaid with the results of the deposited material selected at spots A_{IR} and B_{IR}, which are measured using the infrared camera. As illustrated in Fig. 2.11, A_{IR} and B_{IR} are directly above A_{TC} and B_{TC}.

The thermocouple data for A_{TC} and B_{TC} results from the heat conduction from the deposited material into the substrate, whereas the infrared imaging data for A_{IR} and B_{IR} is related to the heat generation from print head-material interactions as well as rapid plastic deformation of the material. Even before the arrival of the print head at Point A_{IR}, Point A_{TC} in the substrate experiences a gradual pre-heating effect due to the material deposited before Point A_{IR}. The preheating and gradual cooling due to conduction result in a broader temperature profile than Point A_{IR}. Compared to the substrate, therefore, the temperature evolution in the deposited material is characterized by a sharper peak with a higher magnitude, a smaller exposure time, a higher heating rate, and a higher cooling rate. The temperature profiles are also highly site-specific. Locations far from the layer transition, such as Spot A_{TC} and A_{IR}, exhibit two different peaks. Spot B_{TC} and B_{IR} are closer to the layer transition, so the two peaks are seen to overlap with each other to some extent, approximately converging to a single peak.

With a comparison of the temperature field in the deposited layers and substrate, the cooling mechanisms in additive friction stir deposition can be analyzed as follows (recall Section 2.2):

1. First, the heat flux by conduction can be estimated using Fourier's conduction law $\Phi_{conduction} = -\kappa \cdot \Delta T/\Delta \text{x}$, where κ is the thermal conductivity. ΔT is the difference in temperature between the deposited material (A_{IR}) and the substrate below it (A_{TC}), and Δx is the distance between the two measured locations. Using $\kappa = 170\ W/(m \cdot K)$ for Al-Mg-Si and $401\ W/(m \cdot K)$ for Cu (Garcia et al., 2020), with $\Delta \text{x} = 6.8 \times 10^{-3}$ m (dimension measured between A_{TC} and A_{IR}) $\Phi_{conduction}$ is calculated as $\sim 10^7\ W/m^2$ or above.
2. Second, the heat flux of convection can be estimated using Newton's law of cooling $\Phi_{convection} = -h \cdot (T - T_{amb})$. Here, h is the heat transfer coefficient of air assumed to be in the range of $10-100\ \text{W}/(\text{m}^2 \cdot \text{K})$, T is the peak temperature measured by the IR camera, and T_{amb} is the ambient temperature set to be 298K. $\Phi_{convection}$ is estimated to be $\sim 10^3-10^4\ \text{W}/\text{m}^2$.
3. Third, the heat flux by radiation is $\Phi_{radiation} = \varepsilon\sigma(T^4 - T^4_{amb})$, where the Stefan-Boltzmann constant $\sigma = 5.67 \times 10^{-8}\ \text{W}/(\text{m}^2 \cdot \text{K}^4)$ and ε is the emissivity. $\Phi_{radiation}$ is estimated to be $\sim\ 10^4\ \text{W}/\text{m}^2$.

Overall, the heat flux by conduction is much more significant than that by air convection and radiation. The total heat flow rate by each mechanism is a product of the heat flux and the cooling area. Factoring in the cooling area of each mechanism, the corresponding analysis concludes that thermal conduction is the dominant cooling mechanism in additive friction stir deposition for both Al-Mg-Si and Cu.

2.6 Force and torque evolution

For processes based on the friction stir principle, the force and torque have been shown to be vital for the resulting material quality (Shahi and Barmouz, 2014). The force and torque evolution processes have been studied in friction stir welding, in which the key measured quantities include the normal force F_z applied by the tool to the surface and the torque M_z imparted into the base material. The torque is dependent on the tool rotation rate, the sticking coefficient, and the normal force. F_x and F_y are reactionary forces that develop from material drag and the mechanical constraints as the tool moves across the build plate. This is illustrated in Fig. 2.20A.

In additive friction stir deposition, the in-plane forces are generated by the same mechanisms as in friction stir welding, but the mechanical constraint conditions are conspicuously different. As shown in Fig. 2.20B, the print head travels in the x-direction. F_x has no mechanical constraint along the leading edge but is constrained by the deposited material on the trailing edge. F_y has no constraint in either the leading or trailing edge. F_z is the normal force directly applied to the feed-rod that causes material extrusion into the deposition zone. The normal force applied by the print head is a reactionary force that depends on the material mixing and flow as well as the spacing between the print head and the substrate. It is also important to point out the dimension differences between additive friction stir deposition and friction stir welding. Typically, the in-plane dimension of the feed-rod in additive friction stir deposition is slightly

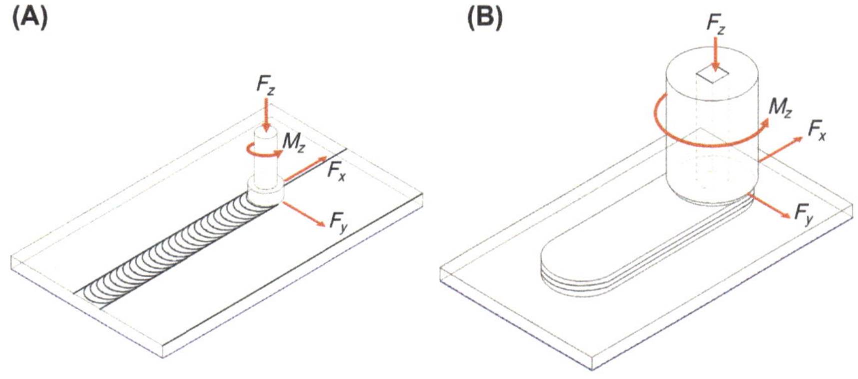

Figure 2.20 A schematic of (A) friction stir welding and (B) additive friction stir deposition showing the key differences in the mechanical loading conditions of the processed material.

larger than the pin in friction stir welding (e.g., 9 mm vs. 6 mm). However, the diameter of the print head in additive friction stir deposition can be substantially larger than the tool shoulder in friction stir welding (e.g., 40 mm vs. 15 mm). There is also a difference in layer thickness by additive friction stir deposition compared to the plate thickness in friction stir welding. Fig. 2.20 shows a schematic comparing the two processes, wherein the key differences in geometrical configuration and mechanical constraint are highlighted.

2.6.1 Multiple phases of force and torque evolution

In additive friction stir deposition, the force and torque boundary conditions determine the effective strain and strain rate applied to the material and thus have a great influence on the print quality and final microstructure. We employ in situ force and torque measurement to unravel important insights into the underlying physics. Similar to the temperature evolution, several distinct phases of additive friction stir deposition can be described based on the operating conditions and the associated force and torque evolution.

Fig. 2.21 shows the typical plots of compression force F_z and torque M_z versus time during deposition of Al-Mg-Si and Cu (Garcia et al., 2022). Prior to deposition, the feed-rod and print head rotate together at a rapid rate without an applied normal force. Once they are up to speed, the print head moves vertically to set the appropriate layer height such that the surface protrusions are plunged into the substrate surface (Phase I in Fig. 2.21). At this moment, no force is applied to the feed-rod; however, there is a baseline torque due to the interaction of the tool protrusions with the substrate surface. With the tool protrusions inserting into the substrate, one may name this phase the *tool protrusion plunge phase*. Note that the tool protrusions ($\sim$ 1 mm) have a much smaller size than the pin ($\sim$ 6 mm diameter) in friction stir welding. The next stage (Phase II) is the *feed-rod plunge phase*, which is described by deformation of the feed-rod under uniaxial compression. In the elastic deformation regime, the input stress can be approximated based on the change in the beam length ΔL: $F_z = E\frac{\Delta L}{L}A$. Here, E is the Young's modulus, A is the cross-sectional area, and L is the original length of the feed-rod. Heat is generated by a slipping-dominated contact state at the interface between the feed-rod and substrate, with the heat generation rate strongly depending on F_z.

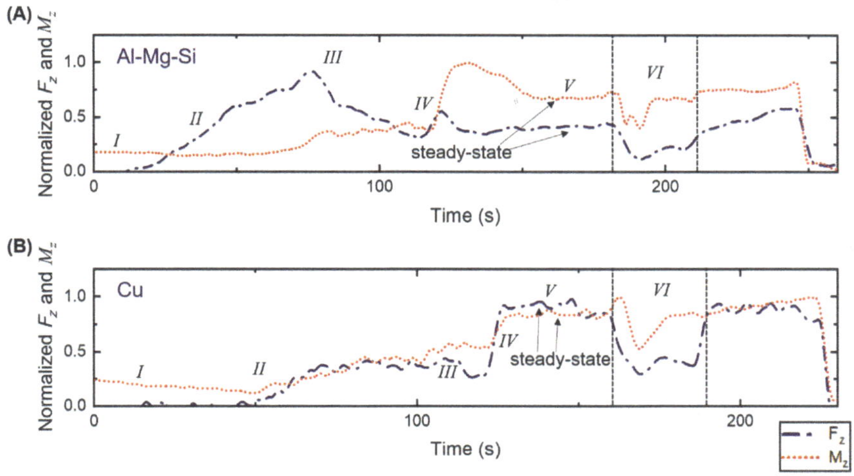

Figure 2.21 Typical plots of the normal force and torque for (A) Al-Mg-Si and (B) Cu. Roman numerals correspond to different phases of operation for additive friction stir deposition including the initial tool protrusion plunge, the feed-rod plunge, the initial yielding, onset of in-plane motion, steady-state deposition, and layer transition phases. *Adapted from Garcia, D., Griffiths, R., Joey, Y., Hang Z., 2022. In situ monitoring of force and torque evolution during additive friction stir deposition: a comparative study of Al-Mg-Si and Cu. Under review.*

Frictional heating softens the feed-rod and substrate at the contact interface; the continuous buildup of F_z causes both materials to yield (Phase III). One may name this phase the *initial yielding phase*. As soon as the feed material begins to yield and enter the deposition zone underneath the print head, a combination of yielding at elevated temperatures and shearing at the print head-material interface leads to a drop-off in F_z. The torque gradually rises as the material fills the space beneath the print head and the interaction volume increases. At this point, the *in-plane motion phase* is initiated (Phase IV). The start of the in-plane motion phase requires an increase in the material feed rate to match the in-plane velocity. For the results presented in Fig. 2.21, the ratio between the feed rate and in-plane velocity is set at 1:3, which has been found to consistently produce depositions with good quality. Shortly after that, F_z reaches a relatively stable value denoted as the steady-state force with only minor oscillations in magnitude.

This is thus termed the *steady-state deposition phase* (Phase V). Once a layer is finished, the *layer transition phase* (Phase VI) creates a large drop in F_z and M_z as the tool assembly moves away from the substrate along the vertical direction by the desired layer height. A gradual return to

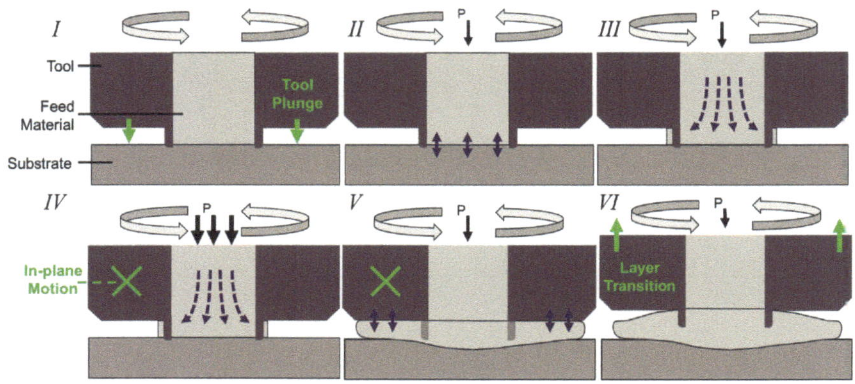

Figure 2.22 Illustration of various processing phases of force and torque evolution in additive friction stir deposition. *Adapted from Garcia, D., Griffiths, R., Joey, Y., Hang Z., 2022. In situ monitoring of force and torque evolution during additive friction stir deposition: a comparative study of Al-Mg-Si and Cu. Under review.*

steady-state conditions occurs as a second layer of material is deposited. At the end of deposition, the print head raises, and the feed material breaks from the deposit, resulting in a final drop of the measured compressive force and torque. These six processing phases are schematically illustrated in Fig. 2.22.

Closer scrutiny between the plots in Fig. 2.21A and B reveals that the torque evolution curves are fairly similar between Al-Mg-Si and Cu, but the force evolution trends are differentiated. The peak force observed during the initial yielding phase (Phase III) for Al-Mg-Si is higher than the steady-state force (Phase V) required during deposition. In Cu, the opposite trend is observed—the steady-state force is higher than the peak plunge force. What is the key difference between Phase III and Phase V with regards to the operating conditions? The latter requires an increase in the material feed rate compared to the former, and therefore a higher strain rate. One may thus explain this phenomenon by considering the relative low yield strength and strong strain hardening effect in Cu as compared to Al-Mg-Si. The heat generation and material flow behavior should also be considered to fully reveal the origin of this difference, which will be discussed in Section 2.8.

2.6.2 Dependence of steady-state force and torque on processing conditions

Fig. 2.23 plots the measured steady-state normal force at various processing conditions. For steady-state deposition, the normal stress applied to

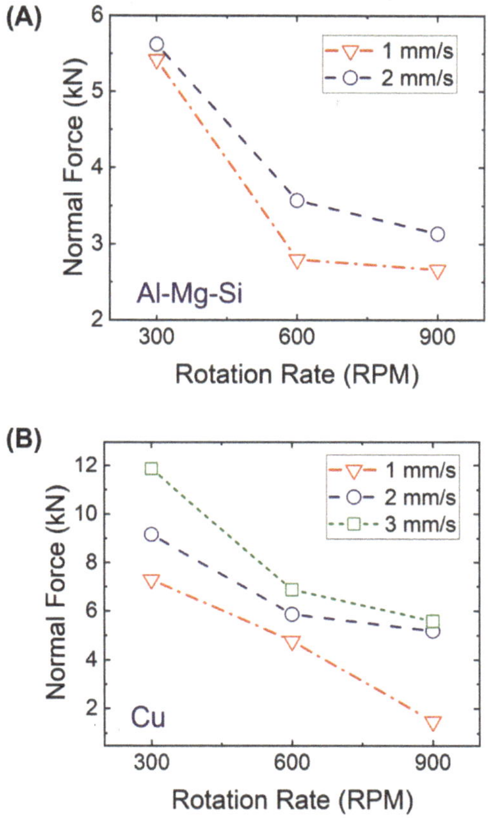

Figure 2.23 Influences of the print head rotation rate and travel velocity on the steady-state normal force. The data is collected during additive friction stir deposition of (A) Al-Mg-Si and (B) Cu. *Adapted from Garcia, D., Griffiths, R., Joey, Y., Hang Z., 2022. In situ monitoring of force and torque evolution during additive friction stir deposition: a comparative study of Al-Mg-Si and Cu. Under review.*

the feed-rod is a measure of the yield strength at that specific thermomechanical condition. For both Cu and Al-Mg-Si, the steady-state compressive stress decreases with an increase of Ω or a decrease of V. From the previous section, we know that a high Ω or a low V corresponds to a higher peak temperature. The trends shown in Fig. 2.23 can thus be explained by noting that the yield strength of both material systems decreases with increasing the temperature.

Fig. 2.24 shows the measured steady-state torque as a function of Ω and V. For Al-Mg-Si, the steady-state torque decreases with an increase of Ω or a decrease of V, which corresponds to a higher peak temperature.

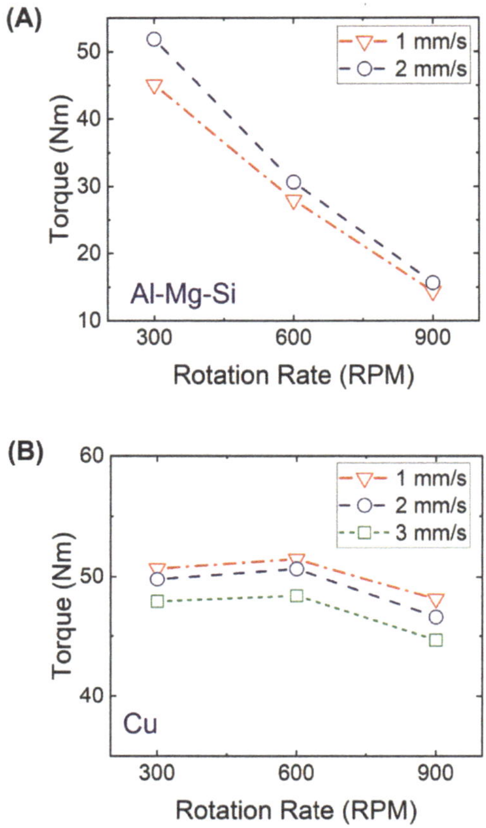

Figure 2.24 Influences of the print head rotation rate and travel velocity on the steady-state torque. The data is collected during additive friction stir deposition of (A) Al-Mg-Si and (B) Cu. *Adapted from Garcia, D., Griffiths, R., Joey, Y., Hang Z., 2022. In situ monitoring of force and torque evolution during additive friction stir deposition: a comparative study of Al-Mg-Si and Cu. Under review.*

If the material is under a sticking condition with the rotating print head, the torque is expected to scale with the shear flow stress. Then, this is again consistent with the expected temperature trend because the flow stress decreases at higher temperatures. In Cu, however, there is little influence of the equipment processing variables on the torque during deposition. This difference reminds us about the different power-law relationships found in Fig. 2.18. Both differences could originate from distinct thermomechanical properties and tool-material interactions between Al-Mg-Si and Cu. The origin of this phenomenon will be discussed in Section 2.8.

Quantitatively, trends for the force and torque variation with respect to the processing conditions can be analyzed by an empirical fitting of $Y = A(X_1)^p(X_2)^q$, where Y denotes the output process parameter (normal force, torque, etc.), X_1 and X_2 are the input process parameters, and A, p, and q are fitting constants (Arora et al., 2010). Eq. (2.18) shows the fitting results for F_z and M_z for Al-Mg-Si and Cu. R-squared values are 0.948, 0.938, 0.965, and 0.601, respectively (Garcia, 2020).

$$F_z(Al\text{-}Mg\text{-}Si) \propto (\Omega^{-0.647})(V^{0.1484}) \tag{2.18A}$$

$$F_z(Cu) \propto (\Omega^{-0.716})(V^{0.4987}) \tag{2.18B}$$

$$M_z(Al\text{-}Mg\text{-}Si) \propto (\Omega^{-0.916})(V^{0.181}) \tag{2.18C}$$

$$M_z(Cu) \propto (\Omega^{-0.0465})(V^{-0.0548}) \tag{2.18D}$$

For both material systems, F_z positively correlates with the in-plane velocity V, and negatively correlates with the rotation rate Ω. An increase in V leads to reduced total heat input and tool interaction time. In turn, this reduces the expected peak temperature and subsequent thermal softening. Therefore, the normal force exerted on the feed-rod will decrease. For M_z similar trends, are found for Al-Mg-Si; however, Cu shows a poor fitting in Eq. (2.18D) with a low R-squared value of 0.601. Furthermore, the value of the coefficients is an order of magnitude smaller than any of the other fitting coefficients. This suggests that neither processing parameter has a significant influence on the measured torque at the print head-material interface in Cu.

2.7 In situ visualization of material rotation and flow

Friction stir welding relies on the intense deformation and significant material flow to ensure good joining quality. However, visualization of material flow is challenging: from the top view, the tool shoulder obscures the stir zone; from the side view, the stir zone is laterally constrained by the workpieces. In additive friction stir deposition, because the material is deposited on top of the substrate without lateral constraint, the exterior surface of the deposited material can be directly monitored from the side view. Below we present the material flow results during additive friction stir deposition of Al-Mg-Si and Cu based on in situ surface visualization, in which a Dino-Lite Edge AM4115ZT Microscope is positioned perpendicular to the leading edge of the deposit at a distance of 8 cm. The camera has a spatial resolution of 1280 × 1024 pixels with a pixel size of 0.03 × 0.03 mm at the given field of view.

2.7.1 Footprint and material rotation

During additive friction stir deposition, Al-Mg-Si and Cu exhibit drastically different deposition footprints, as illustrated in Fig. 2.25. In Al-Mg-Si, the deposited material extends significantly toward the leading direction with a small asymmetry between the leading and trailing side. Overall, the in-plane shape is close to a circle. In situ monitoring can only capture the material movement on the exterior surface of the deposit. In Cu, the leading edge of the deposition track is only slightly ahead of the leading edge of the feed-rod. Without significant forward extension, the footprint on the leading and trailing side is highly asymmetric. If we place an optical camera on the leading side, in situ surface monitoring can

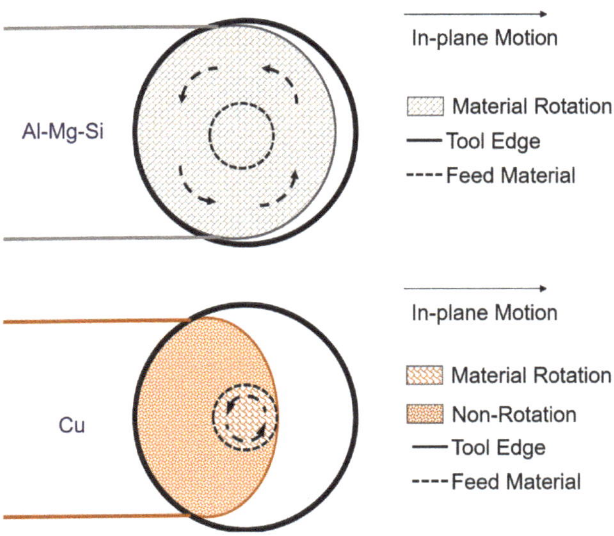

Figure 2.25 Illustration of the deposition footprint based on in situ observation, with a comparison of Al-Mg-Si and Cu. *Reprinted with permission from Garcia, D., Hartley, W.D., Rauch, H.A., Griffiths, R.J., Wang, R., Kong, Z.J., et al. 2020. In situ investigation into temperature evolution and heat generation during additive friction stir deposition: a comparative study of Cu and Al-Mg-Si. Additive Manufacturing 34, 101386.*

provide the material flow and rotation information deep inside, almost reaching the region directly beneath the rotating feed-rod.

The material rotation behavior is also distinctive between Al-Mg-Si and Cu. The exterior surface of Al-Mg-Si is seen to rotate with the print head, but the rotation rate is much lower. In other words, the rapidly rotating print head is able to cause some material rotation in the whole deposition track but not at the same speed. In Cu, however, the material does not stick to the rotating print head at all; it is found to remain effectively stationary after being extruded beneath the print head. As examples, Figs. 2.26 and 2.27 present snapshots of the optical videos captured on the leading edge, wherein the flow paths of two distinct spots on the Al-Mg-Si or Cu surface are investigated. In the snapshots of Al-Mg-Si, Speckle P and Q are both on the exterior surface of the deposition and are seen to move from left to right without observable changes in the relative distance. Over the 1.51 s, the linear velocity of Spot P and Spot Q is measured as ~1% of the print head velocity at the same radius.

In Cu, Spot M is below the print head and is seen to be stationary. Spot N is inside the deposition track almost reaching the feed-rod; it is seen to rapidly move from the left to the far right in 0.13 s. Given the radius for the outer edge of motion and the time between the snapshot

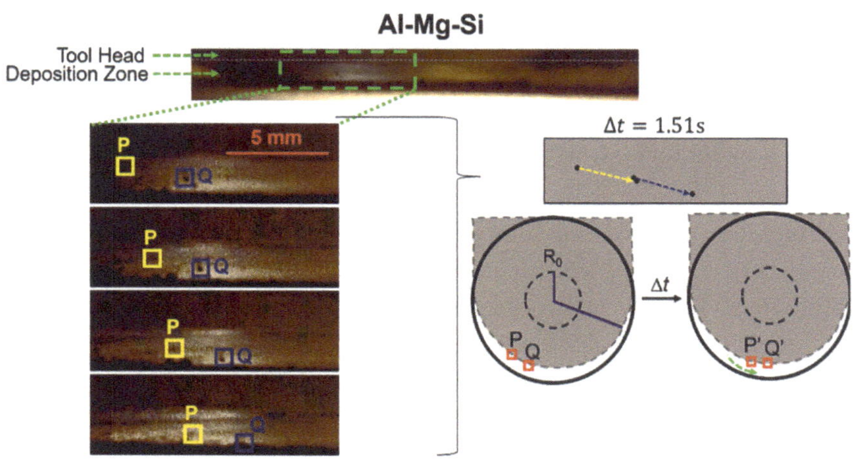

Figure 2.26 The observed material flow features of Al-Mg-Si at 300 RPM and 2 mm/s in-plane velocity. The Al-Mg-Si snapshots show the rotation of the entire deposition zone. *Reprinted with permission from Garcia, D., Hartley, W.D., Rauch, H.A., Griffiths, R. J., Wang, R., Kong, Z.J., et al. 2020. In situ investigation into temperature evolution and heat generation during additive friction stir deposition: a comparative study of Cu and Al-Mg-Si. Additive Manufacturing 34, 101386.*

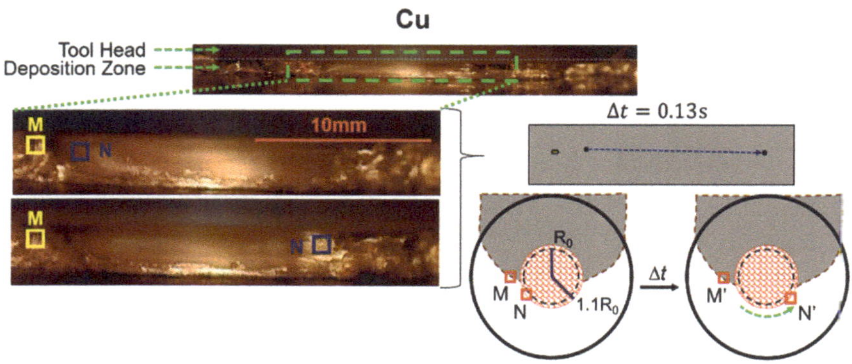

Figure 2.27 The observed material flow features of Cu at 300 RPM and 2 mm/s in-plane velocity. The Cu snapshots show two distinct regions where Point M remains stationary and Point N rotates with the print head. *Reprinted with permission from Garcia, D., Hartley, W.D., Rauch, H.A., Griffiths, R.J., Wang, R., Kong, Z.J., et al. 2020. In situ investigation into temperature evolution and heat generation during additive friction stir deposition: a comparative study of Cu and Al-Mg-Si. Additive Manufacturing 34, 101386.*

images, the linear velocity of Spot N is measured to be 90% of the print head velocity at the same radius. This is not surprising because Spot N is so close to the feed-rod. In Al-Mg-Si deposition, for any spot close to the feed-rod, we also expect a high rotation rate; however, due to the large footprint of Al-Mg-Si, such information cannot be directly captured via surface imaging.

2.7.2 Contact state and sticking coefficient

Assume that we start with a cylindrical feed-rod with a radius of R_0. With compression, the material first moves downwards to the region directly below the feed-rod (without lateral constraint from the print head), and then moves laterally to fill the print head-substrate gap. This involves a macroscopic material shape change from a cylindrical shape to a disk-like shape via plastic deformation. This also involves a transition from a compression-dominated state to a shear-dominated state. After material extrusion, the resulting disk has a radius of R and a thickness h, which is pre-defined by the gap between the print head and the substrate. The steady-state configuration is illustrated in Fig. 2.28. The region directly below the rotating feed-rod is defined as the *transition zone* (radius of R_0) since it is an intermediate step between the feed-rod inside the print head and deposition track beneath the print head. The region below the

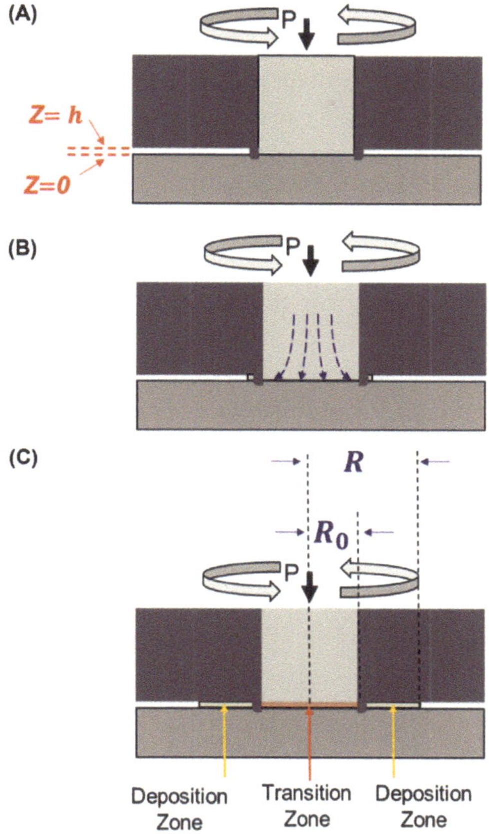

Figure 2.28 Description of geometry and zones in additive friction stir deposition. *Reprinted with permission from Garcia, D., Hartley, W.D., Rauch, H.A., Griffiths, R.J., Wang, R., Kong, Z.J., et al. 2020. In situ investigation into temperature evolution and heat generation during additive friction stir deposition: a comparative study of Cu and Al-Mg-Si. Additive Manufacturing 34, 101386.*

rotating print head is defined as the *deposition zone* (radius ranging from R_0 to R).

We next introduce the concept of contact state of an interface, which can be characterized by the ratio of linear velocity of the two materials across the interface. This ratio, which is often referred to as the sticking coefficient, ranges from 0 to 1. A value of 0 corresponds to a full slipping condition, a value of 1 corresponds to a full sticking condition, and a value between 0 and 1 corresponds to a partial slipping/sticking condition. The sticking coefficient depends on the dynamic friction coefficient between the print head material and the deposited material; a larger dynamic friction coefficient likely results in a larger sticking coefficient. Another relevant definition is the fractional slip, which focuses on the degree of slipping in friction. The sum of the fractional slip and sticking coefficient is equal to 1.

We may use δ to denote the fractional slip, then $(1 - \delta)$ is the sticking coefficient. For the case of additive friction stir deposition, the value of sticking coefficient or fractional slip is also location dependent: $\delta = \delta(r)$. Close to the rotation axis of the print head, the linear velocity of the print head is small, and it is easier for the deposited material to keep the same rotation rate with the print head, resulting in a higher sticking coefficient. Moving along the radial direction towards the edge of the print head, the linear velocity of the print head increases. This likely results in a decrease of the sticking coefficient (Nandan et al., 2007). Fig. 2.29 schematically

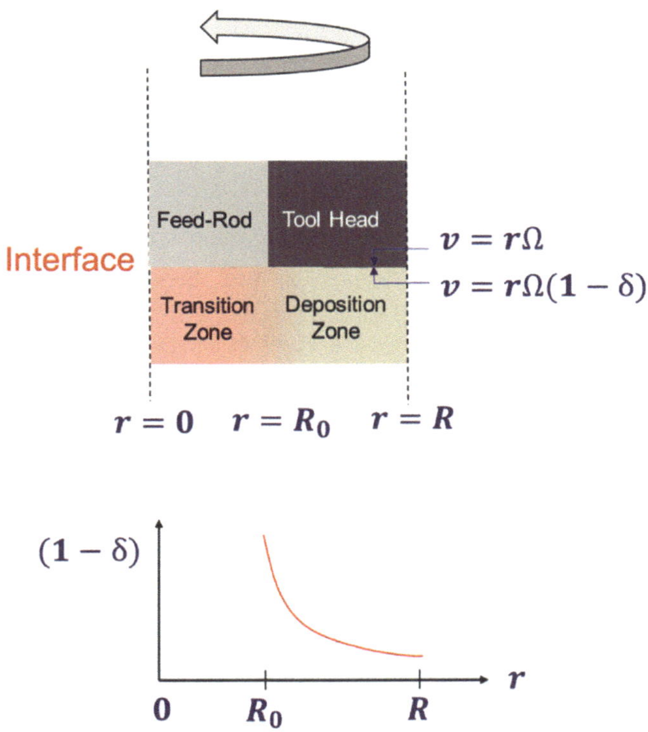

Figure 2.29 Contact state and sticking coefficient at the interface between the print head and deposited material. *Reprinted with permission from Garcia, D., Hartley, W.D., Rauch, H.A., Griffiths, R.J., Wang, R., Kong, Z.J., et al. 2020. In situ investigation into temperature evolution and heat generation during additive friction stir deposition: a comparative study of Cu and Al-Mg-Si. Additive Manufacturing 34, 101386.*

shows how the sticking coefficient may evolve as a function of the radial location.

Based on the definitions and concepts introduced above, we can reexamine the material flow snapshots in Figs. 2.26 and 2.27. For Al-Mg-Si, the sticking coefficient $(1-\delta)$ is 0.01 at the exterior of the deposition zone. Since the sticking coefficient is close to 1 in the transition zone, we will expect a gradual decrease of the sticking coefficient from R_0 to R. This means that the material rotates and flows to some extent throughout the entire deposition zone ($R_0 < r < R$), and the contact state is characterized by a partial sticking/slipping condition. In Cu, the material rotation and intensive flow are almost limited to the transition zone. In the deposition zone, there is no material rotation, so the print head-material interface is characterized by a full slipping condition. At the boundary between the transition zone and deposition zone, the sticking coefficient rapidly drops from 0.9 to 0. Based on the radial location of Points P, Q, M, and N, Fig. 2.30 plots the radial dependence of the sticking coefficient for the additive friction stir deposition of Al-Mg-Si and Cu.

Fig. 2.31 compares the top-view and side-view appearance of typical Al-Mg-Si and Cu deposition tracks. The top-view image of Al-Mg-Si shows straight and clear edges. The side-view image shows that the edge is round on the side with a smooth curvature, while the flash is indistinguishable from the rest of the deposit. In contrast, with substantial material

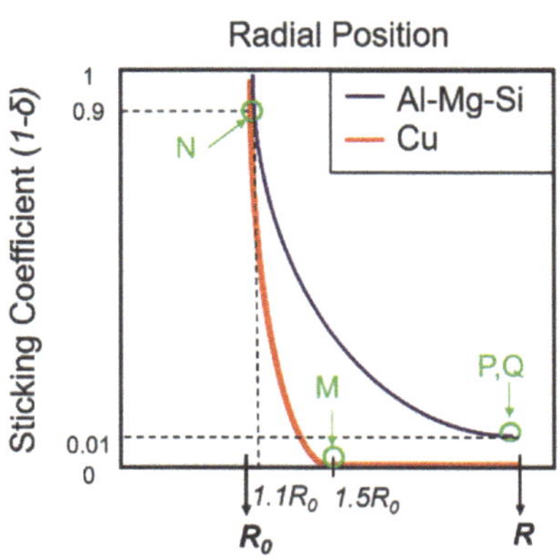

Figure 2.30 A schematic showing the differences between Al-Mg-Si and Cu in the observed sticking coefficient as a function of the radial position. *Reprinted with permission from Garcia, D., Hartley, W.D., Rauch, H.A., Griffiths, R.J., Wang, R., Kong, Z.J., et al. 2020. In situ investigation into temperature evolution and heat generation during additive friction stir deposition: a comparative study of Cu and Al-Mg-Si. Additive Manufacturing 34, 101386.*

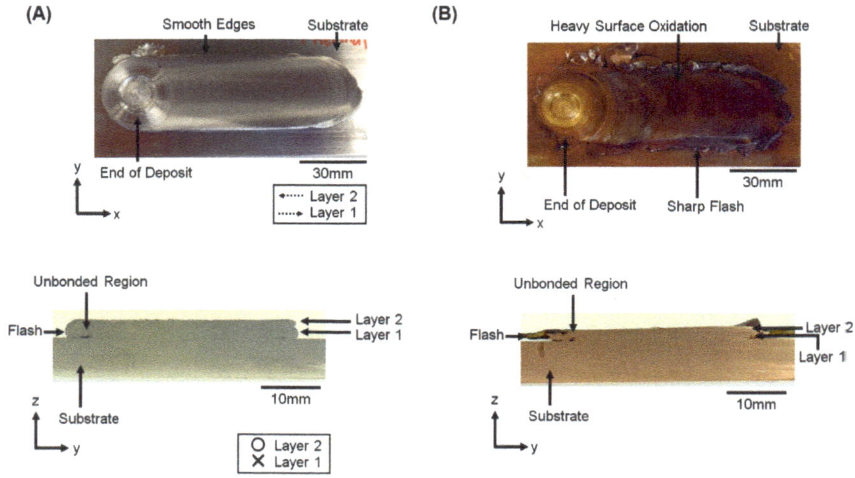

Figure 2.31 Pictures of the as-deposited tracks for (A) Al-Mg-Si and (B) Cu, from the top-down view and side view. A key difference between the two material systems lies in the shape and continuity of the flash. *Reprinted with permission from Garcia, D., Hartley, W.D., Rauch, H.A., Griffiths, R.J., Wang, R., Kong, Z.J., et al. 2020. In situ investigation into temperature evolution and heat generation during additive friction stir deposition: a comparative study of Cu and Al-Mg-Si. Additive Manufacturing 34, 101386.*

flash, Cu is seen to develop discrete, sharp edges from the side-view. These observations are consistent with the in situ monitoring results of material rotation and flow in the deposition zone. In Al-Mg-Si, the constant rotation in the entire deposition zone ($R_0 \leq r \leq R$) facilitates continuous and smooth flash flow without geometric discontinuities. In Cu, the material is simply extruded into the deposition zone and the flash begins to form as discontinuous sheets. In the absence of subsequent shear-induced rotation and flow, the previously deposited material is pushed outward upon the arrival of new material, resulting in rough edges.

2.8 Correlation of the material flow behavior to temperature, force, and torque evolution

From the in situ monitoring results of temperature evolution, force and torque evolution, as well as material rotation and flow, a complete picture of the physical processes can be established for additive friction stir deposition of Al-Mg-Si and Cu. The two materials display the same physical trends in several aspects:

1. Qualitative dependence of thermal features, which include peak temperature, exposure time, reheating rate, and cooling rate on the processing parameters, which include the print head rotation rate and travel velocity.
2. Multiple phases in force and torque evolution, including (I) tool protrusion plunge, (II) feed-rod plunge, (III) initial yielding, (IV) in-plane motion, (V) steady-state deposition, and (VI) layer transition.
3. Qualitative dependence of steady-state compression force on the processing parameters, which include the print head rotation rate and travel velocity.

In advanced manufacturing, any behavior or trend should originate from convolution of the process and material. There are salient differences between Al-Mg-Si and Cu, for example, in forgeability, stacking fault energy, and interactions with the print head, but they still show some similar trends. This means that the process factor dominates over the material factor, and these physical trends reflect the nature of additive friction stir deposition rather than the thermomechanical responses of the specific material. For additive friction stir deposition of materials other than Al-Mg-Si and Cu, therefore, we will likely observe similar trends in the aforementioned aspects.

When the process factor is less dominant, the material properties can play a vital role in the physical processes during additive friction stir deposition. Thus, several conspicuous differences are observed between Al-Mg-Si and Cu:

1. During additive friction stir deposition, the peak temperature T_{Peak} is correlated with Ω^2/V in Al-Mg-Si but with Ω/V in Cu, showing quantitative differences in the power-law relationships.
2. In Al-Mg-Si, the steady-state force during deposition is lower than the force to cause initial yielding, but in Cu an opposite trend is observed.
3. During additive friction stir deposition of Al-Mg-Si, the steady-state torque decreases as the print head rotation rate increases or the travel velocity decreases. However, no clear trend is observed for Cu.
4. After the feed material extrudes beneath the print head, Al-Mg-Si shows material rotation throughout the entire deposition zone, with the print head-material interface characterized by a partial sticking/slipping condition. However, Cu effectively remains stationary in the

deposition zone, with the print head-material interface characterized by a full slipping condition.

The differences in peak temperature, force, and torque between Al-Mg-Si and Cu can all be explained by their distinct material flow behavior, which are elaborated as follows.

2.8.1 Influences of the contact state and material flow on heat generation

In the steady state of additive friction stir deposition, heat can be generated by the interfacial friction caused by slipping of the print head on the deposited material. The interfacial heat generation rate $\dot{Q}_\delta$ is a function of the interfacial contact stress $\tau_{contact}$ and the velocity difference between the print head and the top surface of the deposit. The velocity difference can be written as $\Delta v(r) = r\Omega - (1-\delta)r\Omega = \delta r\Omega$, where r is the radial position, that is, the distance to the central rotating axis. The total heat generation rate at the print head-material interface can be written as

$$\dot{Q}_\delta = \int \tau_{contact}(v_{tool} - v_{material})d\tilde{A} = \int \tau_{contact} r\Omega\delta d\tilde{A}. \tag{2.19}$$

Here $d\tilde{A}$ denotes an infinitesimal interface area, and the integration covers the entire deposition zone: $R_0 \leq r \leq R$.

In addition to the slipping-dominated interface condition, it is possible to have the sticking condition dominate especially at small r values. This corresponds to a scenario where the material in the deposition zone rotates together with the print head, rendering a volume of material with intensive flow. The material velocity gradient in this volume results in shear deformation and plastic energy dissipation. The corresponding volumetric heat generation rate $\dot{Q}_V$ depends on the shear strain rate and the shear stress at yielding:

$$\dot{Q}_V = \int \zeta \dot{\varepsilon}_{ij}^{p} \sigma_{ij} d\tilde{V}. \tag{2.20}$$

Here $\dot{\varepsilon}_{ij}^{p}$ and σ_{ij} are the components of the plastic strain rate tensor and the Cauchy stress tensor, respectively. The product of them defines the density of plastic energy dissipation (Simar et al., 2012). ζ denotes the ratio of the heat converted from plastic deformation; it typically ranges from 0.80 to 0.99 (Rosakis et al., 2000). $d\tilde{V}$ is an infinitesimal volume, and the integration covers the volume of material with intensive flow.

During additive friction stir deposition of Al-Mg-Si, even the exterior surface is found to rotate, meaning that the print head effectively drives the material to flow in the deposition zone. This suggests significant shear deformation and considerable volumetric heat generation in addition to the interfacial heat source due to friction. During additive friction stir deposition of Cu, the material rotation and intensive material flow are mostly observed within the transition zone ($r < R_0$), whereas the majority of the material in the deposition zone ($R_0 < r < R$) stays stationary. As a result, the volumetric heat generation is insignificant in the deposition zone, and interfacial heat generation is the dominant mechanism. The differences in the heat generation mechanisms between Al-Mg-Si and Cu are illustrated in Fig. 2.32, in which the interfacial heat generation is

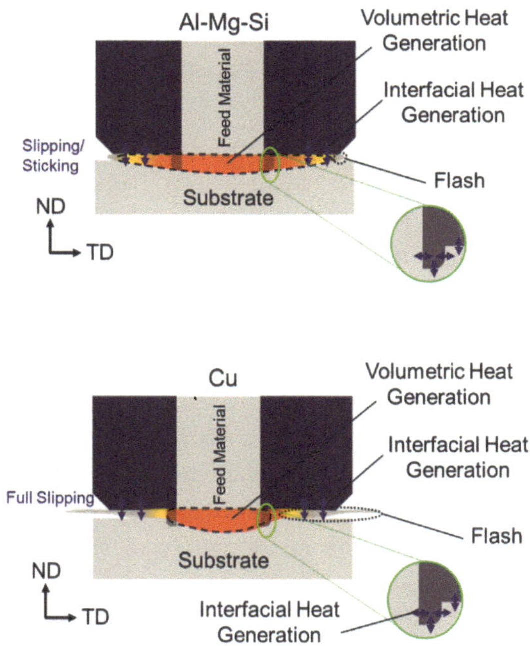

Figure 2.32 A diagram showing the heat generation mechanisms for Al-Mg-Si and Cu. The volumetric heat generation regions are shown in red and the interfacial friction is highlighted with blue arrows. The Cu system has the most significant volumetric heat generation directly beneath the feed-rod and the flash is large. In Al-Mg-Si, the volumetric heat generation is prevalent due to substantial material flow. *Reprinted with permission from Garcia, D., Hartley, W.D., Rauch, H.A., Griffiths, R.J., Wang, R., Kong, Z.J., et al. 2020. In situ investigation into temperature evolution and heat generation during additive friction stir deposition: A comparative study of Cu and Al-Mg-Si. Additive Manufacturing 34, 101386.*

marked by blue arrows and the regions with significant volumetric heat generation is highlighted in red. The color gradient of the figure corresponds to the differences in material flow and rotation, rather than the actual temperature distribution.

Now let us consider how the interface contact state and material flow behavior influence the heat generation rates described by Eqs. (2.19) and (2.20). For the interfacial heat generation rate by friction, Al-Mg-Si is characterized by a partial sticking/slipping condition, so $\dot{Q}_\delta^{Al-Mg-Si} = \int \tau_{contact} r\Omega\delta d\tilde{A}$, $0 < \delta < 1$. It should be emphasized that δ is not a constant in this case, but depends on the linear velocity of the print head: $\delta = \delta(r\Omega)$. For Cu, the interface is characterized by a full slipping condition, so δ is a constant at $\delta = 1$. Thus Eq. (2.19) reduces to $\dot{Q}_\delta^{Cu} = \int \tau_{contact} r\Omega d\tilde{A}$ for Cu. From these equations, the interfacial heat generation is expected to increase with the print head rotation rate Ω in both Al-Mg-Si and Cu due to the term $r\Omega$. However, there is an additional term $\delta = \delta(r\Omega)$ in $\dot{Q}_\delta^{Al-Mg-Si}$, which also increases with the rotation rate Ω. Considering the appearance of Ω in both terms of $r\Omega$ and $\delta = \delta(r\Omega)$, the rotation rate should have a more significant influence on the interfacial heat generation rate in Al-Mg-Si than in Cu.

For the volumetric heat generation rate by plastic deformation, Eq. (2.20) applies to both Al-Mg-Si and Cu, but the total volume of intensive flow $\tilde{V}^{Flow}$ is saliently different. For Al-Mg-Si, $\tilde{V}^{Flow}$ is larger than the volume of the transition zone, and likely increases with the tool rotation rate Ω because a higher material velocity at the top surface can potentially drive a larger volume with material flow. This trend has been suggested by previous studies. In friction stir welding of Al alloy 7020-T6, Lorrain et al. experimentally observed an increase of the stir zone size with the tool rotation rate increasing from 300 RPM to 600 RPM (Lorrain et al., 2010). Using the methodology of scaling, Mendez et al. analyzed the heat transfer and plastic deformation during friction stir welding and predicted a positive correlation between the thickness of the shear layer, which is directly related to the volume with intensive flow, and the tool rotation rate (Mendez et al., 2010). In Cu, $\tilde{V}^{Flow}$ is almost equal to the volume of the transition zone without a strong dependence on the processing parameters. Therefore, $\tilde{V}^{Flow}$ stays almost constant in Cu but increases at a higher rotation rate in Al-Mg-Si. The print head rotation rate should thus have a more significant influence on the volumetric heat generation rate in Al-Mg-Si than in Cu.

Al-Mg-Si and Cu show different power laws of peak temperature: T_{Peak} is controlled by Ω^2/V in Al-Mg-Si but is controlled by Ω/V in Cu. In other words, the relative dominance of tool rotation rate Ω over travel velocity V is stronger in Al-Mg-Si than Cu. This is consistent with the above heat generation analyses, which suggest that the print head rotation rate Ω should have more influence on the interfacial and volumetric heat generation rates in Al-Mg-Si than that in Cu. Therefore, the distinct contact states and material flow behavior between Al-Mg-Si and Cu render drastic differences in the heat generation mechanisms and heat generation rates (especially their dependence on the processing parameters), eventually leading to the quantitative difference in the power laws of peak temperature.

2.8.2 Influences of the contact state and material flow on force and torque

In additive friction stir deposition, the feed rate and therefore the effective strain rate have a sudden increase from Phase III, which corresponds to initial yielding without in-plane motion, to Phase V, which corresponds to steady-state deposition. In Al-Mg-Si, the steady-state force during deposition is lower than the force to cause initial yielding. This may be attributed to two factors. First, Al-Mg-Si has a high stacking fault energy ($\sim$160 mJ/m^2 or more) and excellent recovery capability during deformation (Muzyk et al., 2018). Although an increase of the strain rate would lead to a higher dislocation generation rate, the dynamic recovery rate is high so that dislocation annihilation and reorganization are sufficient to maintain the dislocation density to a low level. Second, there is significant material flow in the deposition zone, which gives rise to heat generation via plastic deformation. At higher strain rates, the volumetric heat generation rate increases, and the temperature of the feed material increases. The force required to yield the feed-rod thus decreases.

During additive friction stir deposition of Cu, the deformation during all phases is primarily driven by the normal force applied to the feed material as opposed to shearing at the print head-material interface. Since Cu has a much lower stacking fault energy [$\sim$80 mJ/m^2 or less (Heidarzadeh et al., 2019)] than Al alloys, dynamic recovery is less sufficient; the increase in material feed-rate may lead to considerable strain hardening. As a result, the Cu feed-material requires a higher force for steady-state deposition. Since the material flow of Cu is negligible in the deposition zone, an increase of the strain rate is unlikely to cause a significant temperature increase via plastic energy dissipation. Therefore, we see an increase of the compression force in the steady-state deposition of Cu than initial yielding due to an increase of the strain rate.

In Section 2.6, at the steady-state deposition of both Al-Mg-Si and Cu, we show that the compressive stress decreases as the print head rotation rate increases or the travel velocity decreases. For Al-Mg-Si, similar correlation is found between the steady-state torque and the processing parameters. However, for the steady-state deposition of Cu, no such correlation is found. Based on the peak temperature measured at various processing conditions in Section 2.5, Fig. 2.33 plots the steady-state normal force and torque as a function of the peak temperature for Al-Mg-Si and Cu. The normal force trend for both materials can be well explained by

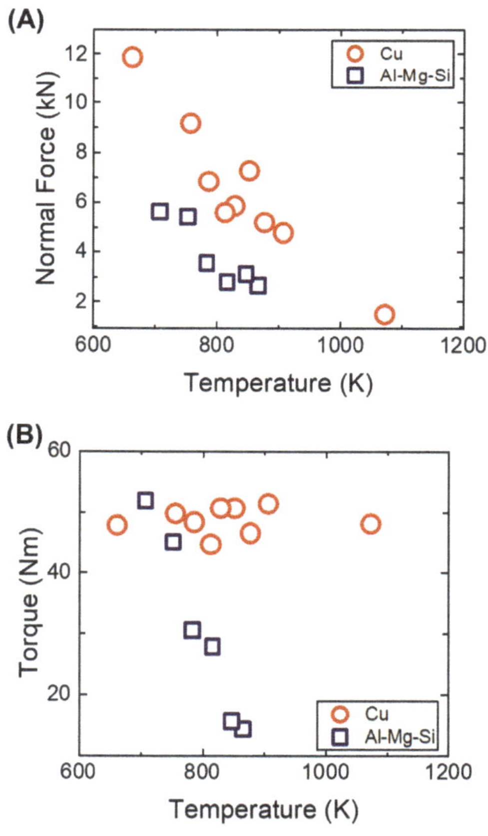

Figure 2.33 Steady-state normal force and torque during Al-Mg-Si and Cu as a function of peak temperature.

thermal softening, because the yield strength required to plasticize the material decreases at elevated temperatures. For the steady-state torque, while Al-Mg-Si shows a negative correlation with the peak temperature, Cu implies no influence from the peak temperature.

The measured torque at the tool interface should take the form of

$$M_z = \int \vec{r_A} \times \left(\vec{\tau}_{contact} d\widetilde{A}\right), \tag{2.21}$$

where r_A is the position vector and $d\widetilde{A}$ denotes an infinitesimal interfacial area. The contact stress at the interface $\tau_{contact}$ may be given by Nandan et al. (2007)

$$\tau_{contact} = \left[\delta\mu_f P + (1-\delta)\tau_Y\right], \tag{2.22}$$

where τ_y is the yield stress in shear at a given position, μ_f is the dynamic friction coefficient between the tool and working material, and P is the normal stress imposed on the material (Arora et al., 2009). For additive friction stir deposition, P is determined by the reactionary force at the print head-material interface rather than the applied force.

For the case of Cu, $\delta = 1$ in the bulk of the deposition zone, so the contact stress at the interface may be simplified to $\tau_{Cu} = \mu_f P$ according to Eq. (2.22). There is only one in-plane mechanical constraint on the trailing edge, so the material should flow preferentially outward towards the longitudinal or transverse directions. The value of P is probably weakly dependent on the processing conditions, so the steady-state torque in Cu shows a rather independent behavior with respect to the peak temperature. For Al-Mg-Si, τ_y noticeably decreases as temperature increases, so the torque drops as the peak temperature increases.

2.8.3 Factors governing the contact state and material flow behavior

So far, we show that the key difference between additive friction stir deposition of Al-Mg-Si and Cu is the distinct contact states observed in the deposition zone, which lead to differences in temperature, force, and torque evolution. The contact state can be well described by the fractional slip δ or the sticking coefficient $(1-\delta)$. For Cu, there is a very high sticking fraction (i.e., $\delta \sim 0$) in a small volume that surrounds the original feed-rod diameter. The majority of the deposition zone is in a full slipping condition, that is, $\delta = 1$. For Al-Mg-Si, the bulk of the deposition zone is observed to be in a partial-sticking condition $0 < \delta < 1$. To tailor the process for any given feed material, we must understand what factors lead to such drastic differences.

For solid materials, the sticking period between two surfaces is dependent on the kinetic friction coefficient. Here, the print head is made of tool steel. The kinetic friction coefficient between Cu and Fe (Nuruzzaman and Chowdhury, 2012) is known to be lower as compared to Al and Fe (Rumble, 2019). As a result, there is a lack of sticking in the Cu to the rotating steel as compared to Al-Mg-Si. To improve the sticking in Cu or reduce the sticking in Al-Mg-Si, the bottom of the print head may be coated with another material. This can alter the contact state, the steady-state force and torque, as well as the heat generation mechanisms.

The sticking coefficient is also dependent on the processing condition. For example, its value is expected to change with temperature, the normal force imposed on the material, and the print head rotation rate. The slipping/sticking phenomenon is key to understanding the physics in friction stirring. However, this phenomenon is very challenging to measure or quantify through any direct observation in friction stir welding (Wang et al., 2021). Thanks to the absence of workpieces on the side, additive friction stir deposition provides a unique opportunity to directly study the contact state in friction stirring, for example, by measuring the sticking coefficient at the outer edge of the deposition zone,

For a given contact state, the degree of material flow depends on the intrinsic material properties. From a fluid dynamics perspective, we may consider viscosity. The material velocity is expected to decrease along the z-direction since the substrate does not rotate. Far away from the rotation axis, the velocity likely decreases along the radial direction due to the drop of sticking coefficient. As a result, the material volume with intense deformation may have a bowl shape. A higher viscosity likely results in a lower velocity gradient under a given shear stress. Then, the material flow volume is larger. From a metal working perspective, we may use the term forgeability to predict the material flow behavior. A better inherent forgeability means more plastic flow and easier development of flow lines under the same thermomechanical condition. From Al alloys, to Cu alloys, stainless steel, Ni alloys, and Ti alloys, the forgeability ranks in descending order. Here, we have already investigated Al and Cu thoroughly; for future work, it will be interesting to examine the latter three, especially the contact state and material flow behavior and their correlation with forgeability.

2.9 Summary

This chapter focuses on the process fundamentals of additive friction stir deposition, especially the temperature evolution, heat generation, force and torque evolution, contact state, and material flow behavior. We first introduce the basics of friction theory as well as heat and mass transfer, which are key to understanding the underlying physics. After presenting the basic printing principle and the coupled analysis for heat and material flow, we introduce an integrated in situ monitoring system that has been incorporated into the additive friction stir deposition facility. Through process mapping and comparisons of the in situ monitoring results from two materials, Al-Mg-Si and Cu, important physical insights have been revealed.

There are similarities between the printing of Al-Mg-Si and Cu, such as (1) the trends of peak temperature, exposure time, reheating rate, and cooling rate with respect to the processing parameters; (2) the trend of steady-state normal force with respect to the processing parameters; (3) the existence of multiple phases in force and torque evolution. These aspects are process-dominated and likely remain for printing other materials.

By examining the differences between Al-Mg-Si and Cu, we conclude that the contact state at the print head-material interface is a vital factor in determining the physical processes in additive friction stir deposition. A full slipping condition will lead to no material rotation and flow in the deposition zone, so interfacial friction is the main heat generation mechanism, like the case in Cu. Under this condition, the steady-state torque does not link to the yield strength of the material, showing weak or no dependence on the peak temperature and processing condition. If the contact state is characterized by a sticking or partial sticking/slipping condition, we will expect substantial material rotation and flow under the print head, like the case in Al-Mg-Si. Both volumetric heat source due to plastic energy dissipation and interfacial heat source due to friction should be considered. Under this condition, the steady-state torque depends on the yield stress and will decrease when the peak temperature is higher.

Because of their distinct contact states, the overall heat generation rate in Al-Mg-Si is more controlled by the print head rotation rate as compared to Cu. This leads to different pseudo heat indices in printing Al-Mg-Si and Cu, so the two materials exhibit distinct power law relationships between the peak temperature and the processing parameters.

References

Al-Badour, F., Merah, N., Shuaib, A., Bazoune, A., 2013. Coupled Eulerian Lagrangian finite element modeling of friction stir welding processes. Journal of Materials Processing Technology 213, 1433–1439.

Arbegast, W.J. 2003. Modeling friction stir joining as a metal working process, TMS Annual Meeting, San Diego, CA, pp. 313–327.

Arora, A., Nandan, R., Reynolds, A.P., DebRoy, T., 2009. Torque, power requirement and stir zone geometry in friction stir welding through modeling and experiments. Scripta Materialia 60, 13–16.

Arora, K.S., Pandey, S., Schaper, M., Kumar, R., 2010. Effect of process parameters on friction stir welding of aluminum alloy 2219-T87. The International Journal of Advanced Manufacturing Technology 50, 941–952.

Berman, A., Drummond, C., Israelachvili, J., 1998. Amontons' law at the molecular level. Tribology Letters 4, 95–101.

Ciavarella, M., Joe, J., Papangelo, A., Barber, J.R., 2019. The role of adhesion in contact mechanics. Journal of The Royal Society Interface 16, 20180738.

Derjaguin, B.V., Muller, V.M., Toporov, Y.P., 1975. Effect of contact deformations on the adhesion of particles. Journal of Colloid and Interface Science 53, 314–326.

Gao, J., Luedtke, W.D., Gourdon, D., Ruths, M., Israelachvili, J.N., Landman, U., 2004. Frictional forces and Amontons' Law: from the molecular to the macroscopic scale. The Journal of Physical Chemistry. B 108, 3410–3425.

Garcia, D. 2020. An Investigation of the Processing History During Additive Friction Stir Deposition Using In-process Monitoring Techniques. (Ph.D. thesis) Virginia Tech.

Garcia, D., Hartley, W.D., Rauch, H.A., Griffiths, R.J., Wang, R., Kong, Z.J., et al., 2020. In situ investigation into temperature evolution and heat generation during additive friction stir deposition: a comparative study of Cu and Al-Mg-Si. Additive Manufacturing 34, 101386.

Garcia, D., Griffiths, R., Joey, Yu, Hang Z., 2022. In situ Monitoring of Force and Torque Evolution During Additive Friction Stir Deposition: A Comparative Study of Al-Mg-Si and Cu. Under Review.

Greenwood, J.A., Williamson, J.B.P., Bowden, F.P., 1966. Contact of nominally flat surfaces. Proceedings of the Royal Society of London. Series A: Mathematical and Physical Sciences 295, 300–319.

Heidarzadeh, A., Saeid, T., Klemm, V., Chabok, A., Pei, Y., 2019. Effect of stacking fault energy on the restoration mechanisms and mechanical properties of friction stir welded copper alloys. Materials & Design 162, 185–197.

Jafari-Marandi, R., Khanzadeh, M., Tian, W., Smith, B., Bian, L., 2019. From in-situ monitoring toward high-throughput process control: cost-driven decision-making framework for laser-based additive manufacturing. Journal of Manufacturing Systems 51, 29–41.

Johnson, K.L., 1987. Contact Mechanics. Cambridge University Press.

Johnson, K.L., Kendall, K., Roberts, A.D., Tabor, D., 1971. Surface energy and the contact of elastic solids. Proceedings of the Royal Society of London. A. Mathematical and Physical Sciences 324, 301–313.

Kim, T., Lin, M., 2007. Stable advection-reaction-diffusion with arbitrary anisotropy. Computer Animation and Virtual Worlds 18, 329–338.

Kuykendall, K., Nelson, T., Sorensen, C., 2013. On the selection of constitutive laws used in modeling friction stir welding. International Journal of Machine Tools and Manufacture 74, 74–85.

Li, J., Jin, R., Yu, H.Z., 2018. Integration of physically-based and data-driven approaches for thermal field prediction in additive manufacturing. Materials & Design 139, 473–485.

Lorrain, O., Favier, V., Zahrouni, H., Lawrjaniec, D., 2010. Understanding the material flow path of friction stir welding process using unthreaded tools. Journal of Materials Processing Technology 210, 603–609.
Maugis, D., 1992. Adhesion of spheres: the JKR-DMT transition using a dugdale model. Journal of Colloid and Interface Science 150, 243–269.
Mendez, P.F., Tello, K.E., Lienert, T.J., 2010. Scaling of coupled heat transfer and plastic deformation around the pin in friction stir welding. Acta Materialia 58, 6012–6026.
Miller, T.J., Zinkle, S.J., Chin, B.A., 1991. Strength and fatigue of dispersion-strengthened copper. Journal of Nuclear Materials 179–181, 263–266.
Mo, Y., Turner, K.T., Szlufarska, I., 2009. Friction laws at the nanoscale. Nature 457, 1116–1119.
Muraca, R.F., Whittick, J.S. 1972. Materials Data Handbook: Aluminum Alloy 6061.
Muzyk, M., Pakieła, Z., Kurzydłowski, K.J., 2018. Generalized stacking fault energies of aluminum alloys–density functional theory calculations. Metals 8.
Nandan, R., Roy, G.G., Lienert, T.J., Debroy, T., 2007. Three-dimensional heat and material flow during friction stir welding of mild steel. Acta Materialia 55, 883–895.
Nuruzzaman, D.M., Chowdhury, M.A., 2012. Friction coefficient and wear rate of copper and aluminum sliding against mild steel. International Transaction Journal of Engineering, Management, & Applied Sciences & Technologies 4, 29–40.
Perry, M.E.J., Rauch, H.A., Griffiths, R.J., Garcia, D., Sietins, J.M., Zhu, Y., et al., 2021. Tracing plastic deformation path and concurrent grain refinement during additive friction stir deposition. Materialia 18, 101159.
Rosakis, P., Rosakis, A.J., Ravichandran, G., Hodowany, J., 2000. A thermodynamic internal variable model for the partition of plastic work into heat and stored energy in metals. Journal of the Mechanics and Physics of Solids 48, 581–607.
Rumble, J.R., 2019. CRC Handbook of Chemistry and Physics, 100 ed. CRC Press/ Taylor & Francis, Boca Raton, FL.
Shahi, P., Barmouz, M. 2014. Force and Torque in Friction Stir Welding, pp. 459–497.
Simar, A., Bréchet, Y., de Meester, B., Denquin, A., Gallais, C., Pardoen, T., 2012. Integrated modeling of friction stir welding of 6xxx series Al alloys: process, microstructure and properties. Progress in Materials Science 57, 95–183.
Stubblefield, G.G., Fraser, K., Phillips, B.J., Jordon, J.B., Allison, P.G., 2021. A meshfree computational framework for the numerical simulation of the solid-state additive manufacturing process, additive friction stir-deposition (AFS-D). Materials & Design 202, 109514.
Tabor, D., 1981. Friction—the present state of our understanding. Journal of Lubrication Technology 103, 169–179.
Wang, X.L., Gao, Y., Liu, X., McDonnell, M., Feng, Z., 2021. Tool-workpiece stick-slip conditions and their effects on torque and heat generation rate in the friction stir welding. Acta Materialia 213, 116969.
Yu, H.Z., Jones, M.E., Brady, G.W., Griffiths, R.J., Garcia, D., Rauch, H.A., et al., 2018. Non-beam-based metal additive manufacturing enabled by additive friction stir deposition. Scripta Materialia 153, 122–130.

CHAPTER 3

Material flow phenomena

Be still like a mountain and flow like a river.

— ***Laozi***

While powder bed fusion and directed energy deposition rely on melting and solidification, additive friction stir deposition offers a deformation processing route to metal additive manufacturing. The plastic deformation in additive friction stir deposition is severe, rapid, and global. The *global* character of deformation differs from most other solid-state metal additive manufacturing technologies, which only involve *local* deformation. For example, in ultrasonic additive manufacturing, the plastic deformation is mostly in the interfacial regions between layers; in cold spray additive manufacturing, more extensive deformation occurs close to the particle contact regions as compared to the particle center. In additive friction stir deposition, the entire feed material undergoes severe plastic deformation at elevated temperatures.

The material deformation and flow are the most salient features that distinguish additive friction stir deposition from alternative solid-state metal additive processes. Fig. 3.1 illustrates the potential theoretical and

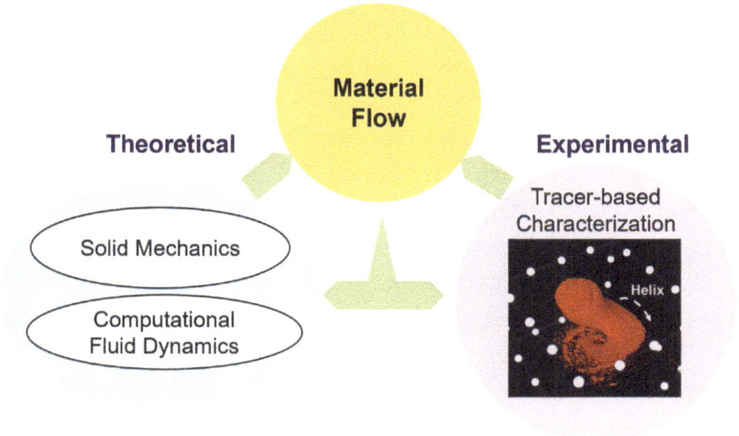

Figure 3.1 Potential experimental and theoretical approaches for material flow studies in additive friction stir deposition.

Additive Friction Stir Deposition
DOI: https://doi.org/10.1016/B978-0-12-824374-9.00001-0

experimental approaches for material flow studies in additive friction stir deposition. Theoretically, both solid mechanics and computational fluid dynamics are useful and have been employed to investigate the material flow phenomena in friction stir welding (Assidi and Fourment, 2009; Su et al., 2015). Experimentally, tracer-based characterization is an effective approach for tracking the material flow path (Schmidt et al., 2006; Seidel and Reynolds, 2001). The focus of this chapter is to develop a quantitative understanding of the material flow phenomena in additive friction stir deposition, which will provide insight into the control of the microstructure, microstructure heterogeneity, and mechanical characteristics of the as-printed materials.

In this chapter, we first provide a high-level summary of the plasticity and fluid mechanics theories. This is followed by a review of previous experimental studies on material flow in friction stir welding. We then present state-of-the-art experimental studies on the material flow phenomena in additive friction stir deposition, involving the design of tracer-based feed material and 3D characterization using X-ray computed tomography. By placing the tracer material in the center or at the edge of the feed-rod, the material flow path during initial material feeding and steady-state deposition can be elucidated. From that, mesoscopic shape evolution is revealed, and the strain levels can be estimated accordingly. Following the same philosophy, dissimilar material printing is used to explore the interfacial material flow phenomena and interface morphology formation.

3.1 Plasticity and finite deformation theory

Additive friction stir deposition is characterized by plastic deformation with a large strain magnitude. Since no melting is involved, finite deformation-based solid mechanics can be employed to study the material flow phenomena. In this section, we provide an overview of the plasticity and finite deformation theory (Lubliner, 2013), from which experimental findings can be fully interpreted and a solid mechanics modeling framework may be established for additive friction stir deposition in the future.

Plasticity describes materials that change shape easily when sufficient external forces are applied and largely retain their new shape when the external forces are removed. Plastic deformation is primarily a distortion rather than dilation or contraction as the volume of the body during plastic deformation only shows very slight, if any, changes. The plastic deformation is thus governed by the deviatoric components of the stress tensor. The well-known von Mises yielding criterion states that the material will yield when the second invariant of the stress deviator J_2 exceeds a critical value. This is equivalent to stating that yielding occurs when the distortion energy density U_D reaches a critical value, because $U_D = \frac{1}{2G}|J_2|$, where G is the shear modulus.

Using the stress terms, the criterion can be rephrased as: yielding occurs when the von Mises stress reaches the yield strength. The von Mises stress is defined as

$$\sigma_{vM} = \frac{1}{\sqrt{2}}\sqrt{[(\sigma_1-\sigma_2)^2 + (\sigma_2-\sigma_3)^2 + (\sigma_3-\sigma_1)^2]}$$
$$= \frac{1}{\sqrt{2}}\sqrt{\left[(\sigma_x-\sigma_y)^2 + (\sigma_y-\sigma_z)^2 + (\sigma_z-\sigma_x)^2\right] + 6(\tau_{xy}^2 + \tau_{yz}^2 + \tau_{zx}^2)} \tag{3.1}$$

σ_x, σ_y, and σ_z are the normal stress components; $\tau_{xy}\tau_{yz}$, and τ_{zx} are the shear stress components. The subscript 1,2,3 denotes the three principal directions. The von Mises stress is also used to describe the effective stress σ_e in plastic deformation.

Regarding the strain in plastic deformation, it is important to recognize its history dependence or path dependence. Rather than simply comparing the initial and final geometry, the intermediate deformation steps should be considered. For example, we may first bend a very ductile bar to create an arch shape, and then bend it in the opposite direction to approximately restore the original geometry. If we just compare the initial and final state, we will conclude that little deformation occurs as the shape barely changes. However, the bar indeed undergoes two-step significant plastic deformation with possibly salient microstructural changes. To reflect the history dependence, instead of talking about strain, the incremental strain (or its derivative with time, the strain rate) is a more appropriate term in plasticity (Dieter, 1976). Under complex strain states, the effective incremental plastic strain can be written as

$$d\varepsilon_e^p = \frac{\sqrt{2}}{3}\sqrt{(d\varepsilon_1^p - d\varepsilon_2^p)^2 + (d\varepsilon_2^p - d\varepsilon_3^p)^2 + (d\varepsilon_3^p - d\varepsilon_1^p)^2}$$
$$= \frac{\sqrt{2}}{3}\sqrt{(d\varepsilon_x^p - d\varepsilon_y^p)^2 + (d\varepsilon_y^p - d\varepsilon_z^p)^2 + (d\varepsilon_z^p - d\varepsilon_x^p)^2 + \frac{3}{2}\left[(d\gamma_{xy}^p)^2 + (d\gamma_{yz}^p)^2 + (d\gamma_{zx}^p)^2\right]}. \tag{3.2}$$

Here, $\varepsilon_x, \varepsilon_y$, and ε_z are the normal strain components; γ_{xy}, γ_{yz}, and γ_{zx} are the engineering shear strain components. The superscript p denotes plastic strain, and the subscript 1,2,3 denotes the three principal directions.

With volume staying constant during plastic deformation, we have $\mathrm{d}\varepsilon_1^p + \mathrm{d}\varepsilon_2^p + \mathrm{d}\varepsilon_3^p = 0$. This simplifies Eq. (3.2) to be

$$d\varepsilon_e^p = \sqrt{\frac{2}{3}\left[\left(\mathrm{d}\varepsilon_1^p\right)^2 + \left(\mathrm{d}\varepsilon_1^p\right)^2 + \left(\mathrm{d}\varepsilon_1^p\right)^2\right]}. \tag{3.3}$$

Once the incremental plastic strain or strain rate is known, the total strain can be obtained by integration throughout the loading path: $\varepsilon_e^p = \int d\varepsilon_e^p$ or $\varepsilon_e^p = \int \dot{\varepsilon}_e^p dt$.

For ideal plastic solids with large plastic deformation and negligible elastic strain, the stress and incremental strain are linked by the Levy-Mises equations:

$$\begin{aligned}
\mathrm{d}\varepsilon_1 &= \frac{d\varepsilon_e}{\sigma_e}\left[\sigma_1 - \frac{1}{2}(\sigma_2 + \sigma_3)\right] \\
\mathrm{d}\varepsilon_2 &= \frac{d\varepsilon_e}{\sigma_e}\left[\sigma_2 - \frac{1}{2}(\sigma_3 + \sigma_1)\right] \\
\mathrm{d}\varepsilon_3 &= \frac{d\varepsilon_e}{\sigma_e}\left[\sigma_3 - \frac{1}{2}(\sigma_1 + \sigma_2)\right]
\end{aligned}. \tag{3.4}$$

Here the superscript p is removed because the total strain can be approximated as the total plastic strain for ideal plastic solids. For elastic-plastic solids, the elastic strain needs to be considered. The total incremental strain is a sum of the (1) incremental elastic strain, which can be expressed using stress components, the Young's modulus, and the Poisson's ratio; and (2) the incremental plastic strain, which is formulated in the Levy-Mises equations. This will lead to a linkage between the stress and the total incremental strain, which is expressed by the Prandtl-Reuss equations (Bruhns, 2014).

The next question is how to theoretically treat the material deformation. First, let us consider the simplest case when the deformation is small. In linear elasticity, the strain is defined based on the *small displacement kinematics* (Fung, 1977; Sadd, 2019), with the strain components written as

$$\begin{aligned}
&\varepsilon_x = \frac{\partial u}{\partial x}, \varepsilon_y = \frac{\partial v}{\partial y}, \varepsilon_z = \frac{\partial w}{\partial z}, \\
&\varepsilon_{xy} = \frac{1}{2}\left(\frac{\partial v}{\partial x} + \frac{\partial u}{\partial y}\right), \varepsilon_{zx} = \frac{1}{2}\left(\frac{\partial w}{\partial x} + \frac{\partial u}{\partial z}\right), \varepsilon_{yz} = \frac{1}{2}\left(\frac{\partial w}{\partial y} + \frac{\partial v}{\partial z}\right)
\end{aligned}. \tag{3.5}$$

Here $\varepsilon_{xy}, \varepsilon_{yz}$, and ε_{zx} are the shear strain components, which are equal to half of the engineering shear strain. u, v, and w denote the displacement in the x, y, z direction, respectively. The strain components in Eq. (3.5) form the infinitesimal strain tensor, which can be transformed when the original Cartesian coordinate system rotates in the 3D space.

To characterize a deformation process, the first step is to adopt a reference configuration. In the undeformed reference configuration, we can use (X, Y, Z) to denote the position vector of a given material point at time zero. At time t, deformation happens, so this material point moves, and the position vector becomes (x, y, z). Strictly speaking, many relationships in solid mechanics (e.g., the differential equations of motion or equilibrium; the derivatives of the stress components) should be written in the deformed configuration. However, in linear elasticity we normally write these relationships relative to the undeformed configuration for simplicity. This is because when the stretch and rotation are small to allow for the use of infinitesimal strain tensor, the difference between the deformed and undeformed configurations can be neglected: $\frac{\partial}{\partial x} \approx \frac{\partial}{\partial X}$, $\frac{\partial}{\partial y} \approx \frac{\partial}{\partial Y}$, and $\frac{\partial}{\partial z} \approx \frac{\partial}{\partial Z}$. As a result, in Eq. (3.5) we do not distinguish if the reference configuration corresponds to the deformed or undeformed state.

During plastic deformation, solids undergo inelastic deformation that is significantly greater than elastic deformation. In some circumstances, the resulting stretch and rotation may be sufficiently small that many problems can still be solved using small-displacement kinematics. The definition in Eq. (3.5) thus remains widely used. However, when the stretch or rotation becomes sufficiently large (not small compared to unity), we must employ the theory of finite deformation in dealing with the kinematics (Bower, 2009; Nemat-Nasser et al., 2004).

In describing the *finite deformation kinematics*, it is convenient to use the vector and tensor notation. We use $\mathbf{X}$ and $\mathbf{x}$ to denote the position vector of a material point before and after deformation; $X_i (i = 1, 2, 3)$ and $x_i (i = 1, 2, 3)$ give the same meaning as (X, Y, Z) and (x, y, z). Imagine that there is an infinitesimal fiber going through the material point of consideration. Before deformation, this infinitesimal fiber is described by the vector $d\boldsymbol{X}$; after deformation, the fiber will be described by a new vector $d\boldsymbol{x}$. We can then define the *deformation gradient* $[F]$ in a matrix notation as:

$$F_{ij} = \partial x_i / \partial X_j. \tag{3.6}$$

Like stress and infinitesimal strain, the deformation gradient is a second rank tensor $\boldsymbol{F}$ and can be written as a 3×3 matrix $[F]$. It follows the same transformation rule when the Cartesian coordinate system rotates in the 3D space. The matrix components of $[F]$ must be nonsingular, with the Jacobian determinant $J = det[F] \neq 0$. The right Cauchy–Green tensor is defined as $\boldsymbol{C} = \boldsymbol{F}^T\boldsymbol{F}$. By definition, $\boldsymbol{C}$ is symmetric with $\boldsymbol{C} = \boldsymbol{C}^T$.

According to the polar decomposition theorem, a general deformation gradient $\boldsymbol{F}$ can be represented as a pure deformation $\boldsymbol{U}$ followed by a rigid rotation $\boldsymbol{R}$:

$$\boldsymbol{F} = \boldsymbol{R}\boldsymbol{U}. \tag{3.7}$$

$\boldsymbol{R}$ is a proper orthogonal tensor, with $\boldsymbol{R}^T = \boldsymbol{R}^{-1}$. $\boldsymbol{U}$ is symmetric with $\boldsymbol{U} = \boldsymbol{U}^T$ and $det[U] > 0$. Comparing $\boldsymbol{U}$ and $\boldsymbol{C}$, we can get $\boldsymbol{U} = \boldsymbol{C}^{1/2}$. For a given deformation, there will be three principal directions that are mutually orthogonal. For the infinitesimal fibers along these principal directions, they undergo extensional deformation without shearing between them. The three fibers in the deformed configuration remain orthogonal but are rotated by $\boldsymbol{R}$. The three principal stretches $\lambda_1, \lambda_2, \lambda_3$ are the eigenvalues of $\boldsymbol{U}$. $\lambda_1{}^2, \lambda_2{}^2, \lambda_3{}^2$ are the eigenvalues of $\boldsymbol{C}$. We have $\lambda_1\lambda_2\lambda_3 = 1$ because the volume stays constant in large plastic deformation.

Under conditions of finite deformation, the term "strain" may be defined in various ways:

1. A Lagrangian strain is a measure of how much $\boldsymbol{C}$ differs from $\boldsymbol{I}$, where $\boldsymbol{I}$ is the identity tensor. The Lagrangian strain tensor is defined as $\boldsymbol{E} = (\boldsymbol{C} - \boldsymbol{I})/2$, with the eigenvalues being $(\lambda_1{}^2 - 1)/2, (\lambda_2{}^2 - 1)/2, (\lambda_3{}^2 - 1)/2$.
2. The conventional strain tensor is $\boldsymbol{E_e} = \boldsymbol{U} - \boldsymbol{I}$, with the eigenvalues being $(\lambda_1 - 1), (\lambda_2 - 1), (\lambda_3 - 1)$.
3. The logarithmic strain tensor is $\boldsymbol{E_I} = ln\boldsymbol{U}$, with the eigenvalues being $\ln(\lambda_1), ln(\lambda_2), ln(\lambda_3)$.

3.2 Elements of fluid mechanics

Additive friction stir deposition is based on continuous plastic deformation of metals at temperatures below melting. That is, the material remains a crystalline solid yet deforms freely like a non-compressible fluid. A significant distinction between the two material types is that a solid can afford tension, compression, and shear, but a real fluid can only tolerate compression and shear. When dealing with the material flow behavior under the friction stirring condition, it is instructive to review basic concepts and conclusions from the fluid mechanics theory.

A fluid is defined as a substance that deforms continuously under shear stress of any magnitude. This is in stark contrast to a conventional solid, which initially deforms under a given stress state but does not necessarily deform continuously or flow. Historically, the investigation into fluid mechanics started with the concept of ideal fluid, which is defined as inviscid and incompressible (Schlichting and Gersten, 2017b). An ideal fluid has no internal resistance to shape change, and there are no tangential forces or shear stresses between adjacent layers. The only possible force is in the normal direction—it is compressive, corresponding to pressure. Without tangential forces, the contact condition between the fluid and solid boundary wall would be a *slip condition* due to their difference in velocity. In real fluid, however, tangential forces do exist between fluid layers or between the fluid and boundary walls. Such tangential forces are frictional forces in nature, originating from the viscosity of the fluid. The frictional forces in real fluid enable the fluid to adhere to the solid boundary wall, leading to the *no-slip condition*, which is widely observed, even in fluids with very low viscosity.

Consider the case of simple shear flow (or Couette flow) (White, 2015), in which a fluid flows between two parallel plates separated by a distance of h (Fig. 3.2). The bottom plate is stationary, and the top plate moves with a constant velocity $\tilde{U}$ in its own plane (along the X-direction). Under the no-slip condition, the top layer of fluid also has a velocity of $\tilde{U}$, while the bottom layer of fluid has a velocity of 0. In other words, there is a velocity gradient along the vertical direction (Y-direction). For water, the velocity is found to vary linearly with respect to the vertical location: $\tilde{u} = \tilde{U}y/h$. The velocity gradient is uniform: $d\tilde{u}/dy = \tilde{U}/h$. The symbol $\tilde{u}$ is used to denote the velocity in the x-direction, to distinguish it from the displacement in the x-direction, u.

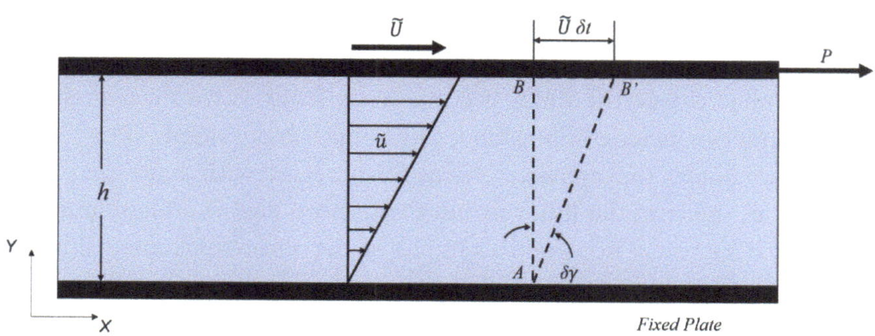

Figure 3.2 Schematic of the simple shear flow of a real fluid like water.

To maintain the velocity of the top plate, a force needs to be applied. Experimentally, it is found that the applied force is proportional to $\tilde{U}/h$. Since this force equals the product of the plate area and the frictional or shear stress τ at the fluid-plate interface, we have $\tau \propto \frac{\tilde{U}}{h} = \frac{d\tilde{u}}{dy}$. This gives rise to the elementary law of fluid friction:

$$\tau = \eta \frac{d\tilde{u}}{dy}. \tag{3.8}$$

The term η reflects the internal friction that resists the fluid flow, and it is defined as the *viscosity* of the fluid. The value of viscosity strongly depends on the temperature. As the temperature increases, the viscosity of gases generally increases, whereas the viscosity of liquids decreases.

For an infinitesimal time increment δt, the top layer of the fluid moves by a distance of $\tilde{U}\delta t$, but the bottom layer stays at the original location. The incremental engineering shear strain is thus $\delta\gamma = \tilde{U}\delta t/h$, and the shear strain rate is $\dot{\gamma} = \frac{\delta\gamma}{\delta t} = \tilde{U}/h = \frac{d\tilde{u}}{dy}$. In other words, the shear strain rate is characterized by the velocity gradient in the fluid. Eq. (3.8) then can be rewritten as

$$\tau = \eta\dot{\gamma}. \tag{3.9}$$

The relationship in Eq. (3.9) is based on a simple analysis of mechanics and experimental observation from water. This relationship indeed defines *Newtonian fluids*, in which the shear stress scales with the shear strain rate (Lai et al., 2010). Here, the viscosity is a function of temperature but is independent of the shear strain rate. By definition, water is a Newtonian fluid. Other Newtonian fluids include air, mineral oil, gasoline, alcohol, and most single-phase fluids made up of small molecules.

A fluid is defined as a non-Newtonian fluid if the shear stress is not linearly proportional to the shear rate. This corresponds to two scenarios: shear thinning and shear thickening (Amoo and Layi Fagbenle, 2020). For shear-thinning fluids, the apparent viscosity decreases with increasing shear rate. In other words, the fluids become less viscous as we shear them faster; examples include quicksand, ketchup, and blood. For shear-thickening fluids, the apparent viscosity increases with increasing shear rate, that is, shearing the fluids makes them more viscous. A suspension of corn starch in water is a good example of the shear-thickening fluid. In additive friction stir deposition, the material flows extensively under shear at elevated temperatures (but below the melting temperature). The shear stress does not scale with the shear rate but rather forms a complex relationship. In the context of fluid mechanics, therefore, the material under the friction stirring state may be viewed as a non-Newtonian fluid. For example, as explained in Chapter 2, it is a common practice to use the Sellars–Tegart relationship in modeling friction stirring of Al alloys (He et al., 2014). This relationship indicates a nonlinear constitutive model following a hyperbolic sine relationship between the effective stress σ_e and effective strain rate $\dot{\varepsilon}_e$. For such non-Newtonian fluids, the viscosity can be calculated as $\eta = \sigma_e / 3\dot{\varepsilon}_e$ (Nandan et al., 2007).

Depending on the viscosity, velocity, and density, the fluid may flow smoothly without lateral mixing between adjacent layers; alternatively, it may flow in a highly irregular and random manner with large amounts of mixing perpendicular to the flow direction. These two types of flow behavior are referred to as laminar flow and turbulent flow, respectively (Schlichting and Gersten, 2017a). Quantitatively, whether laminar flow or turbulent flow happens depends on the Reynolds number, which is a measure of the ratio between the inertial force and friction force. The Reynolds number is dimensionless and is mathematically defined as

$$Re = \frac{\rho V d}{\eta}, \tag{3.10}$$

where ρ is the fluid density, V is a characteristic velocity, and d is a characteristic dimension (e.g., the diameter of a pipe). A low Reynolds number corresponds to slow and viscous flow, which is more likely associated with laminar flow, while a high Reynolds number corresponds to fast and less viscous flow, which is more likely associated with turbulent flow. The critical Reynolds number that distinguishes these two scenarios depends on the geometry of the flow body; it is generally in the range of 10^3–10^5 (Kundu et al., 2016; Winslow et al., 2018).

As we have discussed in the beginning, the no-slip condition in real fluid means a transition from (1) zero velocity of the wall to (2) finite velocity of the fluid. This necessitates the existence of internal friction in the fluid, that is, viscosity. When the Reynolds number is large, this transition may happen within a thin fluid layer close to the wall, which is called the boundary layer or frictional layer. Along the flow direction, more and more fluid particles are caught up and retarded by the wall friction, so the thickness of the boundary layer increases. This physics picture is illustrated in Fig. 3.3.

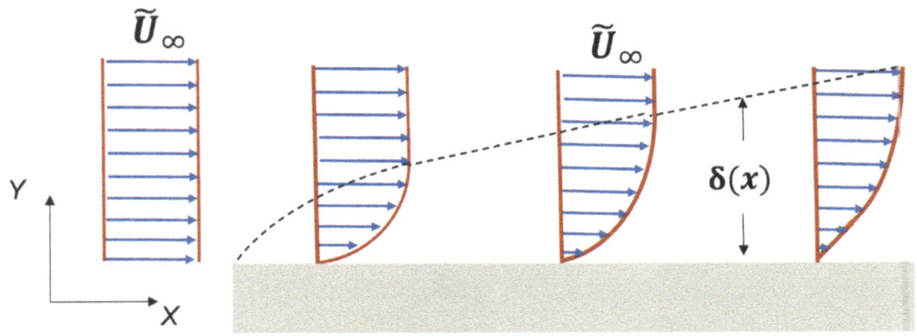

Figure 3.3 Schematic illustration of the boundary layer at a flat plate with zero incidence.

In the boundary layer, the flow can be either laminar or turbulent. For a laminar boundary layer, the thickness can be estimated as

$$\delta \sim \sqrt{\frac{\eta x}{\rho \tilde{U}_\infty}}, \tag{3.11}$$

where x denotes the spatial position along the flow direction, ρ is the fluid density, and $\tilde{U}_\infty$ is the far-field fluid velocity. Considering the flow past a flat plate with zero incidence angle, the exact fluid mechanics solution gives $\delta_{99}(x) = 5\sqrt{\frac{\eta x}{\rho \tilde{U}_\infty}}$ (Schlichting and Gersten, 2017a). Here, the subscript 99 means that at this "boundary," the velocity reaches 99% of the far-field velocity. If the plate length is l, the Reynolds number will be $Re = \frac{\rho \tilde{U}_\infty l}{\eta}$. Then, the boundary layer thickness can be written as $\delta_{99}(x) = 5\sqrt{xl}/\sqrt{Re}$. In other words, the boundary layer thickness decreases as the Reynolds number increases. In the limiting case of infinitely large Re, the boundary layer thickness reduces to zero, corresponding to the case of ideal fluids. For Newtonian fluids, using Eq. (3.8) we get an estimation of the wall shear stress

$$\tau = \eta \frac{d\tilde{u}}{dy} \sim \eta \frac{\tilde{U}_\infty}{\delta} \sim \eta \frac{\tilde{U}_\infty}{\sqrt{\frac{\eta x}{\rho \tilde{U}_\infty}}}, = \eta^{1/2} \rho^{1/2} {\tilde{U}_\infty}^{3/2} / x^{1/2}. \tag{3.12}$$

Note that here, we use the average velocity gradient across the boundary layer for estimation. This simple analysis suggests that the wall shear stress increases if the boundary layer is thinner; the wall shear stress also increases if the fluid is more viscous, has higher density, or flows more rapidly. Moreover, the wall shear stress is not a constant but decreases along the flow direction, because the boundary layer thickness increases along the flow direction.

3.3 Previous experimental studies on material flow in friction stir welding

In the previous two sections, we have reviewed the theoretical basis of plasticity and fluid dynamics. Owing to the intricate coupling of heat transfer and material flow as well as the poorly understood contact state between the print head and deposited material, theoretical efforts alone are insufficient to unravel the material flow phenomena in additive friction stir deposition. Instead, experiments are required to uncover any unknown underlying physics; when combined with theoretical studies, they can offer a complete physics picture of material flow. Since friction stir welding follows the same principle but is a much more mature field than additive friction stir deposition, we provide a brief review of the experimental studies on material flow in friction stir welding in this section.

Friction stir welding uses a non-consumable rotating tool, which includes a pin and a tool shoulder, to join two workpieces. Metallurgical bond formation is caused by high-temperature plastic deformation and material flow facilitated by frictional heating. To understand the material flow in friction stir welding, Seidel and Reynolds employed a marker insert technique, in which AA5454-H32 markers were embedded in the path of the spinning tool in welding of AA2195-T8 (Seidel and Reynolds, 2001). By comparing the initial and final positions of the markers, the material flow in the longitudinal, transverse, and vertical directions could be determined. Fig. 3.4A shows the locations of the six markers, while Fig. 3.4B demonstrates vertical mixing observed from the transverse cross-section. On the advancing side, the marker is pushed downward; on the retreating side, the marker is moved upward towards the top, showing a counterclockwise material transport path around the axis perpendicular to the paper. At the crown of the weld, the material is seen to transport from the retreating side to the advancing side.

Instead of using tracers or markers, the material flow phenomena can also be explored via dissimilar material welding. Li et al. examined the welding of AA2024 and AA6061 and visualized the flow patterns by leveraging the differential etch contrast of the two alloys (Li et al., 1999). Fig. 3.5A–C show the welding configuration; the mixing pattern observed on the transverse cross-section is shown in (D). The mixing seems to occur on multiple length scales. On the macroscale, the two alloys develop a tooth-like structure along the horizontal direction. On

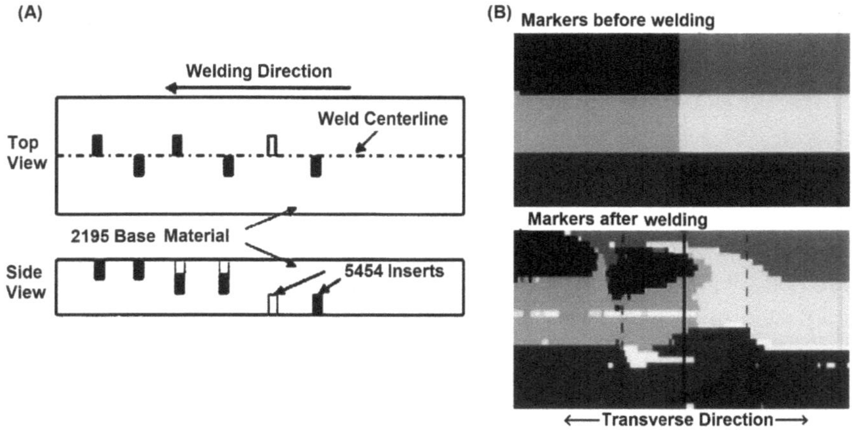

Figure 3.4 Marker insert technique to study material flow during friction stir welding of AA2195-T8. (A) Schematic drawings of the marker configuration, including the top view and side view. (B) Results of vertical mixing with a comparison between the markers (*top*) before and (*bottom*) after welding. *Reprinted with permission from Seidel, T.U., Reynolds, A.P. 2001. Visualization of the material flow in AA2195 friction-stir welds using a marker insert technique. Metallurgical and Materials Transactions A 32, 2879–2884.*

the mesoscale, secondary tooth-like patterns are seen to form on the primary teeth of AA2024. In addition, swirl patterns and "onion ring" features are observed. Overall, the mixing pattern may be described as intercalating lamellae with complex vortex, whorl, and swirl features typical of chaotic dynamic mixing. Microstructure characterization shows that both alloys are recrystallized after welding.

The above two examples use destructive 2D imaging to characterize the material flow pattern. Beyond that, 3D nondestructive characterization is better suited for material flow studies. Morisada et al. introduced a small spherical tungsten tracer in friction stir welding of pure aluminum plates (Morisada et al., 2011). As shown in Fig. 3.6A, two pairs of X-ray transmission real-time imaging facilities were used to monitor the movement of the tracer. Fig. 3.6B depicts the three-dimensional graph obtained using the tungsten tracer coordinates. Unexpectedly, the tracer was seen to flow around the probe seven times instead of one. Such information cannot be obtained by simply comparing the initial and final tracer positions, emphasizing the importance of in situ or time-resolved characterization. A few velocity trends are observed: (1) the tracer velocity increases along the radial direction; (2) the velocity of the tungsten tracer is lower than the peripheral velocity of the tool near the probe; (3) at the same distance

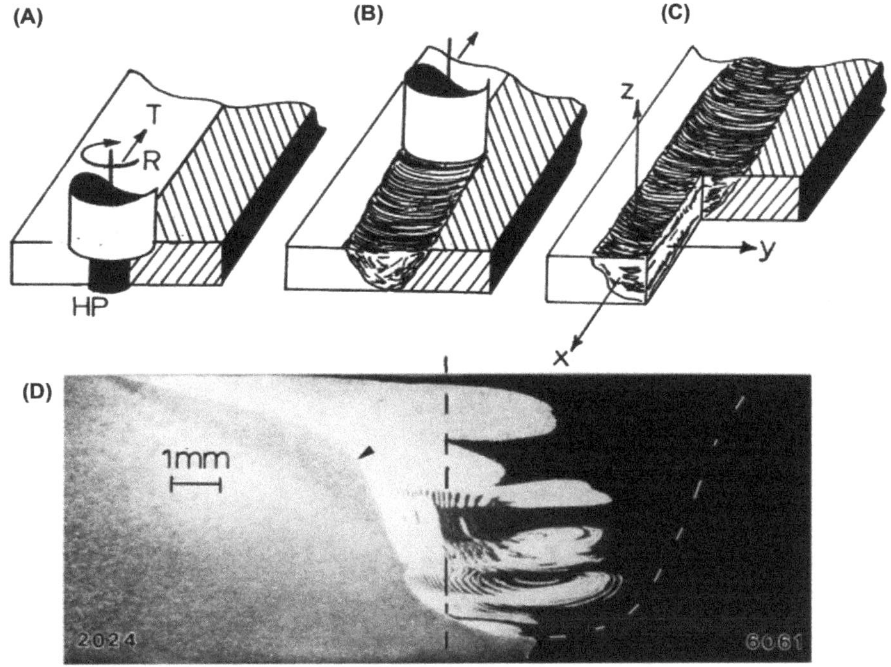

Figure 3.5 Dissimilar material friction stir welding between AA2024 and AA6061. (A), (B), and (C): The welding configuration. (D) Flow patterns observed on the transverse cross-section. *Reprinted with permission from Li, Y., Murr, L.E., Mcclure, J.C. 1999. Solid-state flow visualization in the friction-stir welding of 2024 Al to 6061 Al. Scripta Materialia 40, 1041–1046.*

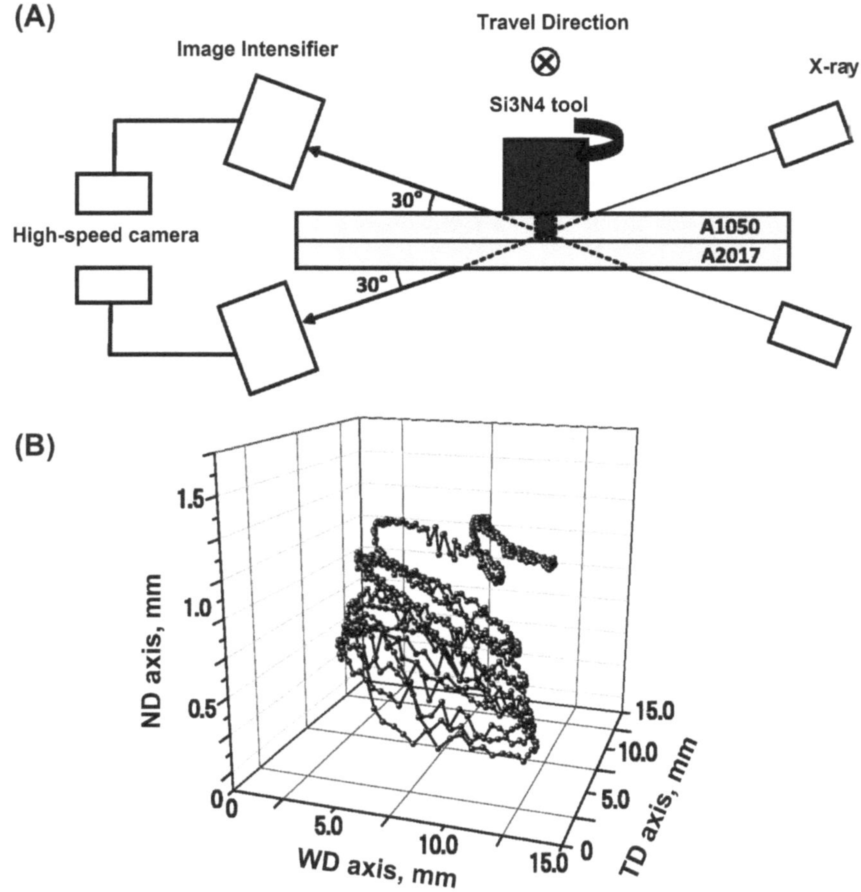

Figure 3.6 In situ X-ray radiography study on material flow in pure aluminum using a W tracer. (A) Configuration of the experimental set-up. (B) Trajectory of the W tracer during friction stir welding. *Reprinted with permission from Morisada, Y., Fujii, H., Kawahito, Y., Nakata, K., Tanaka, M. 2011. Three-dimensional visualization of material flow during friction stir welding by two pairs of X-ray transmission systems. Scripta Materialia 65, 1085–1088.*

from the center, the velocity on the retreating side is slightly higher than that on the advancing side.

While a single W particle tracer provides information about the material velocity in friction stir welding, more information regarding the flow field may be obtained by introducing a deformable tracer that can be distinguished from the base material in X-ray computed tomography. Schmidt et al. introduced a thin copper strip in the workpieces of AA2024-T3 in friction stir welding (Schmidt et al., 2006). The two- and three-dimensional computed tomography images were used in parallel with micrographs for visualization of the flow field. Fig. 3.7 shows the 3D characterization results for the scenario that the copper strip is placed in the transverse direction. As the tool is extracted, the center of the tool aligns with the plane of the strip. Placing the tracer strip in the transverse position allows for the characterization of the boundary between the highly distorted area close to the tool, which is often termed "shear layer," and the less deformed region. In the less deformed region, the originally planar copper strip gets twisted but remains intact and continuous. In the shear layer close to the probe, the copper strip is seen to break up into segments.

This observation supports the idea that a shear layer exists below the tool shoulder, but detailed deformation information of the shear layer is challenging to quantify due to the tracer fragmentation. This is because copper and AA2024-T2 differ considerably in thermomechanical properties. Replacing copper with another aluminum alloy exhibiting comparable mechanical properties to AA2024 may solve this problem (Mukherjee

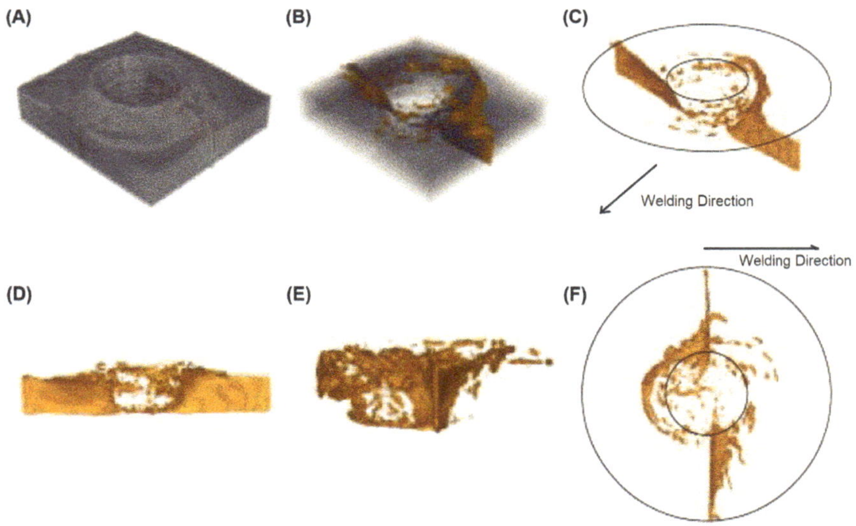

Figure 3.7 X-ray tomography results for the copper strip placed in the transverse position during friction stir welding of AA2024. (A) Isometric view without transparency. (B) Isometric view with half transparency. (C) Isometric view with full transparency, showing copper only. (D) Front view with full transparency. (E) Advancing side view with full transparency. (F) Top view with full transparency. *Reprinted with permission from Schmidt, H.N.B., Dickerson, T.L., Hattel, J.H. 2006. Material flow in butt friction stir welds in AA2024-T3. Acta Materialia 54, 1199–1209.*

and Ghosh, 2010). At the same time, the selected alloy should also be distinguishable from AA2024 in X-ray imaging.

3.4 Design of tracer experiments for material flow investigation in additive friction stir deposition

Inspired by the prior work on friction stir welding, tracer-based experiments have been employed to examine the material flow processes in additive friction stir deposition. In friction stir welding, a tracer material is usually embedded in the workpieces, interacting with the pin and shoulder as welding proceeds. In additive friction stir deposition, a tracer can be placed in the feed material, which is thermomechanically processed to become deposition tracks. Customized hybrid feed-rods are employed here, wherein a small cylinder tracer core made of AA2024 is embedded in the matrix material made of AA6061. These hybrid feed-rods are created as follows. First, drill a hole in an AA6061 bar (square cross-section of 9.525 × 9.525 mm). The tracer core is next inserted into the hole; the tracer is an 8.5 mm long AA2024 cylindrical rod with a 3.175 mm diameter. After that, another section of AA6061 bar is layered on top. As shown in Fig. 3.8, the tracer cylinder is either placed in the center or at

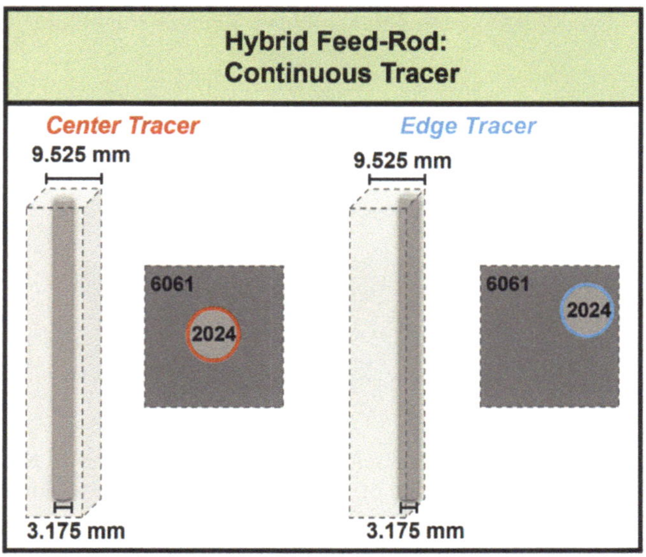

Figure 3.8 Feed-rod design to investigate the material flow phenomenon during additive friction stir deposition. AA2024 tracer is placed in the AA6061 matrix to make the heterogeneous feed material.

the edge of the feed-rod, in order to reveal the material flow information of the center and edge material voxels. AA6061 plates of 6.35 mm thick are used as substrates.

A material flow picture for additive friction stir deposition can be established by characterizing and evaluating how the tracer deforms, expands, or flattens at different stages of printing. The characterization method must offer 3D geometrical information while also being nondestructive. X-ray computed tomography is thus employed to characterize the tracer shape evolution. In this case, the tracer material is made of an Al-Cu alloy (AA2024) and the matrix material is made of an Al-Mg-Si alloy (AA6061). This pair of Al alloys are chosen because their composition differs by ~ 4%–5% in Cu (see Table 3.1), allowing them to be distinguished in X-ray imaging. In addition, they have comparable mechanical properties at elevated temperatures, which prevents significant tracer fragmentation during friction stirring (Li et al., 1999). The strain level can be quantified while the tracer is still intact.

Based on the tracer experiment and 3D characterization, the material flow paths of the center and edge tracers are presented in Sections 3.5 and 3.6, respectively. Our goal here is to understand the material flow fundamentals during both initial material feeding (with print head rotation only) and during steady-state deposition (with print head rotation and in-plane motion). During initial material feeding, the rotation rate is set at 300 RPM and the feed rate is 0.06 mm/s. The initial material feeding experiments are ended abruptly via "stop action" to keep the remaining feed-rod attached. During steady-state material deposition, the rotation rate stays the same, the feed rate is increased to 0.85 mm/s, and the in-plane velocity is 2.54 mm/s. These process parameter values are selected because they are proven to ensure good print quality. The current additive friction stir deposition facility is designed for large-scale additive manufacturing with a deposition layer thickness of ~1 mm and an average track width of 28.5 mm.

The 3D shape of the center or edge tracer is investigated using X-ray computed tomography at two steps: (1) after initial material feeding and (2) after steady-state material deposition. An X5500 3D X-ray inspection facility from North Star Imaging is used to scan the entire sample for global material shape information. The voxel size is between 33.1–43 μm, the voltage is between 120 and 180 kV, and the current is between 1.8 and 5.3 mA. A microwaterjet is used to make smaller specimens measuring 30 × 6 × 8 mm to explore local regions in greater detail.

Table 3.1 AA2024 vs. AA6061: alloy compositions by weight %.

	Al	Mg	Cr	Si	Cu	Mn	Fe	Ti	Zn	Ni	Zr	Other
AA6061	95.1–98.2	0.8–1.2	0.4–0.8	0.4–0.8	0.05–0.4	0.0–0.15	0.0–0.7	0.0–0.15	0.0–0.25	0.0–0.05	0.0–0.25	0.15
AA2024	90.75–94.7	1.2–1.8	0.0–0.1	0.0–0.5	3.8–4.9	0.3–0.9	0.0–0.5	0.0–0.15	0.0–0.25			0.0–0.15

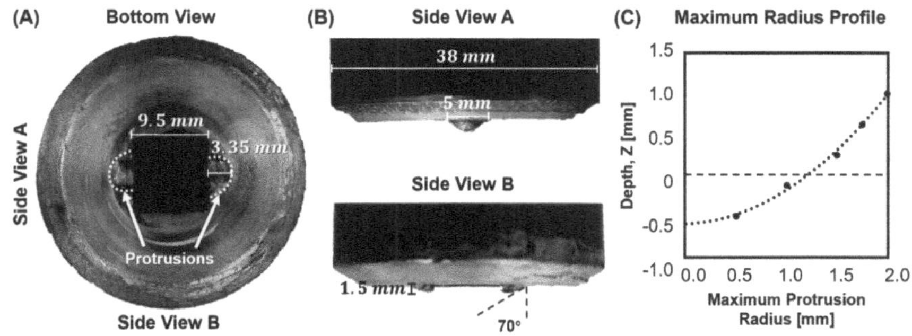

Figure 3.9 Geometry of the protrusions: (A) bottom view, (B) side views, and (C) plot of the protrusion profile. Here, the height of the substrate surface before printing is defined as $Z = 0$ mm. *Reprinted with permission from Perry, M.E.J., Rauch, H.A., Griffiths, R.J., Garcia, D., Sietins, J.M., Zhu, Y., et al. 2021. Tracing plastic deformation path and concurrent grain refinement during additive friction stir deposition. Materialia 18, 101159.*

Smaller samples are scanned at a higher resolution with a Zeiss Xradia 520 Versa Micro-CT facility. At a voltage of 60 kV and a current of 0.08 mA, the voxel size ranges from 3.57 to 9.85 μm. CTVox software is used to visualize the reconstructed volumes.

For all depositions, the bottom surface of the print head has two small protrusions near the edge of the channel opening. As shown in Fig. 3.9, the protrusions have a curved shape with the outermost radius increasing with the Z-height. From the side view, the profile of the protrusions forms an angle of 70 degrees with respect to the vertical direction. During printing, these surface protrusions are partially inserted into the substrate to generate additional heat and enhance material flow and mixing.

3.5 Flow path of the center volume of the feed material

3.5.1 Center tracer flow during initial material feeding

During initial material feeding, the rotating feed-rod is in contact with the substrate; the frictional heat arising from dynamic contact raises the material temperature. At the same time, the feed-rod is slowly pushed downwards. This is accompanied by complex evolution of the feeding force and torque, as shown in Chapter 2. To record the material flow information in the initial material feeding, the deposition is stopped once the steady-state force and torque are reached, with the feed-rod still attached to the substrate. The resultant sample is shown in Fig. 3.10. The

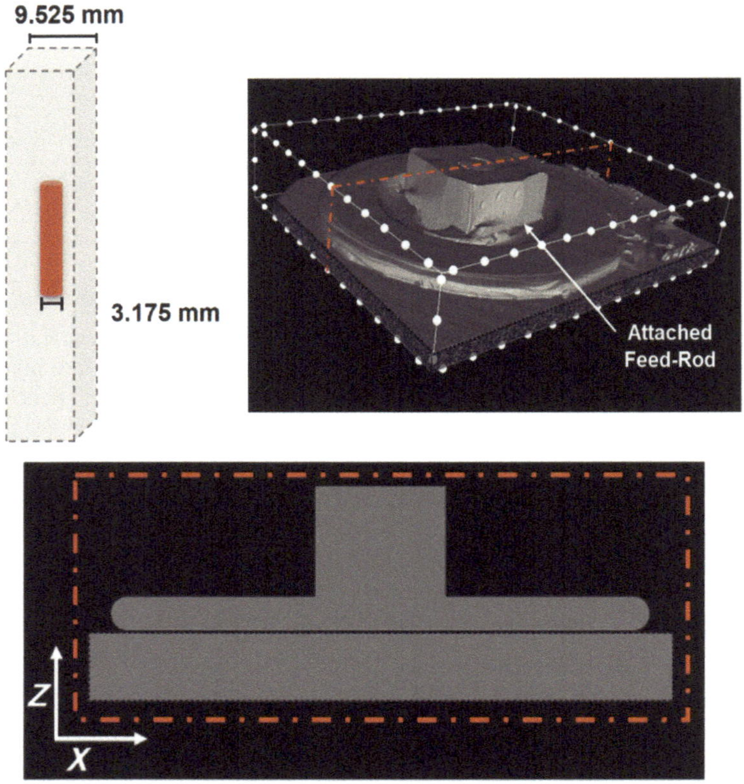

Figure 3.10 In additive friction stir deposition using center tracer-based feed materials, X-ray computed tomography and microstructure characterizations are performed after initial material feeding. The red dashed lines indicate the cross-section used for microstructure analysis; the distance between each pair of white dots is 2.5 mm. The *X*-direction is along the width of the deposition track; the *Y*-direction is along the print head traveling direction; the *Z*-direction refers to the vertical direction. *Reprinted with permission from Perry, M.E.J., Rauch, H.A., Griffiths, R.J., Garcia, D., Sietins, J.M., Zhu, Y., et al. 2021. Tracing plastic deformation path and concurrent grain refinement during additive friction stir deposition. Materialia 18, 101159.*

center tracer shape and microstructure are then characterized along the flow path accordingly.

The top image in Fig. 3.11 shows the X-ray computed tomography result of the center tracer, wherein $Z = 0$ mm and $Z = 1$ mm correspond to the height of the initial substrate surface and the bottom of the print head, respectively. The center tracer is seen to go deeply beneath the initial substrate level. This stems from the compression from the feeding apparatus, causing the surge of the substrate surface in surrounding

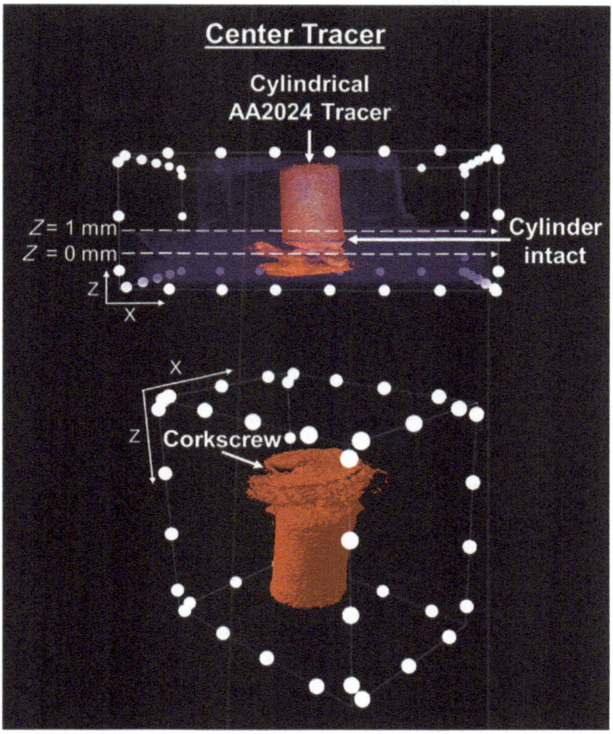

Figure 3.11 X-ray computed tomography of the center tracer after the initial material feeding. The AA2024 tracer is depicted in red and the surrounding AA6061 matrix is either in blue or transparent. The distance between each pair of white dots is 2.5 mm. *Reprinted with permission from Perry, M.E.J., Rauch, H.A., Griffiths, R.J., Garcia, D., Sietins, J.M., Zhu, Y., et al. 2021. Tracing plastic deformation path and concurrent grain refinement during additive friction stir deposition. Materialia 18, 101159.*

areas to form a non-planar interface. Clearly, the cylindrical shape of the tracer is maintained until Z $\sim$ 0.5 mm. After that, the center tracer develops into a corkscrew shape in the 3D space, whose upside-down view is shown as the bottom image in Fig. 3.11. The rotation rate of the center tracer within the print head channel is determined by the print head (300 RPM). However, the rotation rate drops to a very low level at the substrate surface level when the material co-deforms and mixes with the non-rotating substrate. As a result, the rotation rate of the deposited material decreases along the Z-direction. A helical-type shape should arise because of the combination of this rotation rate gradient and downward material feeding.

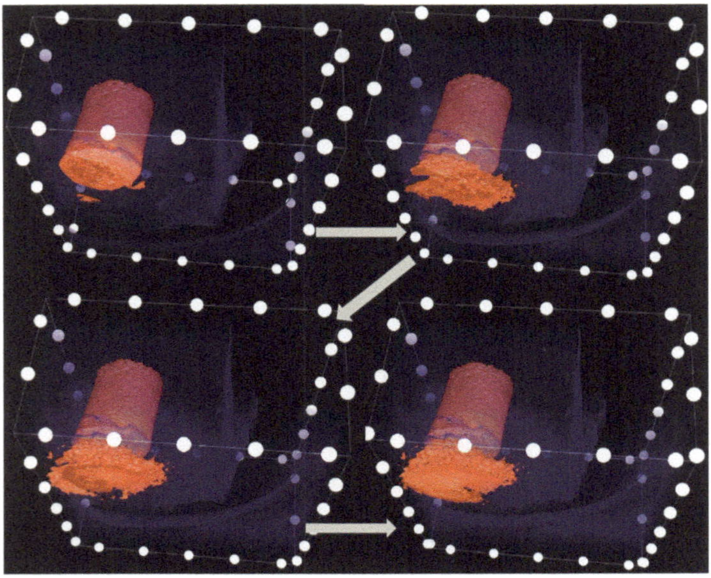

Figure 3.12 Highlight of the *X–Y* cross-section of the center tracer at different heights for Z between 0.5 and 1 mm. The arrows denote moving down along the Z-direction.

The rotating axis of the center tracer is the same as that of the print head, and the center tracer is mechanically constrained by the matrix material around it. The radius of the original tracer cylinder constrains the helical type, resulting in a tight corkscrew appearance. Moving down along the Z-direction, the $X-Y$ plane cross-sectional area is seen to increase due to extrusion of the feed material. This phenomenon is clearly seen in Fig. 3.12, in which the center tracer is truncated at given heights to highlight the corresponding $X-Y$ cross-section. The rotational pattern associated with the corkscrew appearance is also visible.

Generally, at least two strain components should be considered during initial material feeding. First, due to compression on the feed-rod, there is an extrusion strain that causes the cross-sectional area change. Second, the rotation rate gradient creates torsional strain, which results in the creation of a helical-type shape formation. On the global scale, extrusion happens when the feed material yields and spreads underneath the print head to form a thin disc. The diameter of this disc is measured as 28.5 mm in this work, which is the same as the average track width. The corresponding cross-sectional area is $A_{extrusion} = \frac{1}{4}\pi(28.5)^2 = 637.94\ \text{mm}^2$. The feed-rod has a square cross-section, whose area is $A_{feed-rod} = 9.525 \times 9.525 = 90.73\ \text{mm}^2$.

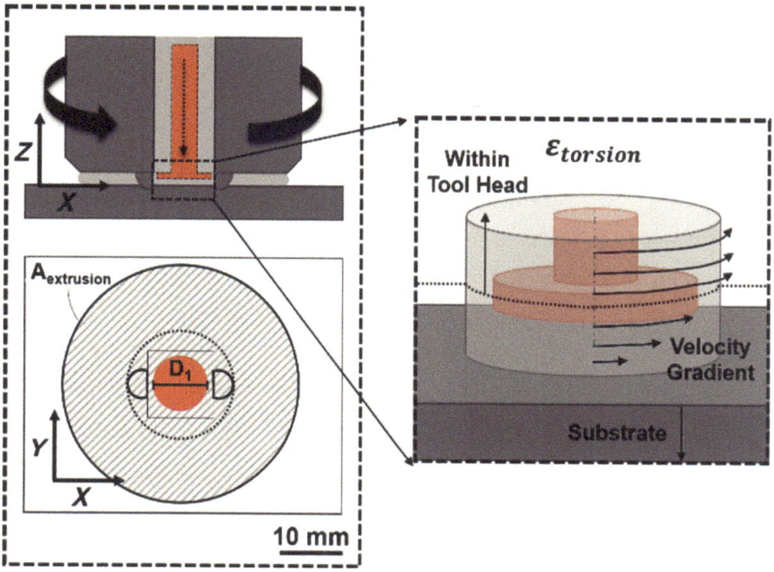

Figure 3.13 Strain caused by extrusion and torsion during initial material feeding. Because of material extrusion under the print head, the in-plane dimensions of the feed-rod and the cylindrical tracer both increase. Torsion is present because of the rotation rate gradient from the print head to the substrate. The red color denotes the AA2024 tracer, while gray denotes the AA6061 matrix. *Reprinted with permission from Perry, M.E.J., Rauch, H.A., Griffiths, R.J., Garcia, D., Sietins, J.M., Zhu, Y., et al. 2021. Tracing plastic deformation path and concurrent grain refinement during additive friction stir deposition. Materialia 18, 101159.*

As illustrated in Fig. 3.13, the engineering strain along the Z-direction in this extrusion step can be calculated as $\varepsilon_z = \frac{A_{feed-rod}}{A_{extrusion}} - 1 = -0.86$. Given that the volume stays constant during plastic deformation, the transverse strain ε_x and ε_y can be calculated based on $(1+\varepsilon_x)(1+\varepsilon_y)(1+\varepsilon_z) = 1$. With ε_x and ε_y being equal, we have $\varepsilon_x = \varepsilon_y = 1.6$. This extrusion step leads to dimensional changes and deformation of the center tracer. Before extrusion, the center tracer has an in-plane diameter of $D_0 = 3.175\text{mm}$. After extrusion, its diameter increases as: $D_1 = D_0\sqrt{\frac{A_{extrusion}}{A_{feed-rod}}} = 8.46\text{mm}$.

The center tracer also develops torsional strain during initial material feeding because of the rotation rate gradient from $Z = 1$ mm to $Z = 0$ mm. For a given point in the cylindrical tracer, one can take its radial direction as the x-direction and the circumferential direction as the y-direction to construct a Cartesian coordinate system. The corresponding shear strain can be calculated as $\gamma_{yz} = r\Delta\theta/h$. Here, $\Delta\theta$ is the total twisting angle in radians and h is the deposition thickness. For the scenario in Fig. 3.11, the total twisting angle of the corkscrew is measured as 3π, $h = 1\text{mm}$, and the outer edge radius is $r = \frac{3.175\text{mm}}{2} = 1.588\text{mm}$. This gives a minimum shear strain of 0 in the center ($r = 0$) and a maximum shear strain of $\gamma_{yz} = \frac{r\Delta\theta}{h} = \frac{1.588mm \times 3\pi}{1\text{mm}} = 14.96$ at the outer edge of the tracer.

When extrusion strain and torsional strain are both present in the tracer material, the equivalent strain ε_e can be calculated as

$$\varepsilon_e = \int \dot{\varepsilon}_e dt; \dot{\varepsilon}_e = \sqrt{\frac{2}{9}\left[(\dot{\varepsilon}_x - \dot{\varepsilon}_y)^2 + (\dot{\varepsilon}_y - \dot{\varepsilon}_z)^2 + (\dot{\varepsilon}_z - \dot{\varepsilon}_x)^2 + \frac{3}{2}\left(\dot{\gamma}_{xy}{}^2 + \dot{\gamma}_{yz}{}^2 + \dot{\gamma}_{zx}{}^2\right)\right]}.$$

Constant volume during plastic deformation leads to $d\varepsilon_x + d\varepsilon_y + d\varepsilon_z = 0$, so $d\varepsilon_x = d\varepsilon_y = -\frac{d\varepsilon_z}{2}$. The equation is then reduced to

$$d\varepsilon_e = \sqrt{\frac{2}{9}\left[\left(-\frac{1}{2}d\varepsilon_z - d\varepsilon_z\right)^2 + \left(-\frac{1}{2}d\varepsilon_z - d\varepsilon_z\right)^2 + \frac{3}{2}d\gamma_{yz}{}^2\right]}$$

$$= \sqrt{d\varepsilon_z{}^2 + \frac{1}{3}d\gamma_{yz}{}^2}.$$

Since the shear strain by torsion scales with radius, the equivalent strain reaches the maximum at the outer edge of the tracer (at $r = 1.588\text{mm}$). At this location, the shear component is dominant; considering it alone, we have $d\varepsilon_e > \frac{d\gamma_{yz}}{\sqrt{3}}$, and $\varepsilon_e > \frac{\gamma_{yz}}{\sqrt{3}} = 8.64$. In other words, the equivalent strain at the center tracer edge is larger than 8.64.

Fig. 3.14A shows an $X-Z$ cross-sectional slice along the centerline of the feed-rod during initial material feeding. The inverse pole figure map of the original AA2024 tracer material before printing is shown in Fig. 3.14B, and the color legend applies to all inverse pole figure maps in this investigation. The inverse pole figure maps for Areas 1–5 are shown in Fig. 3.14 C–G. Area 1 is located inside the print head (Z = 4.2 mm); Area 2 is located near the bottom of the print head (Z = 1.1 mm); Areas 3–5 are located in the deposited material below the print head (Z = 0.9, 0.26, and −0.2 mm, respectively). Areas 1 and 2 have grain sizes and shapes that are close to the original AA2024 tracer, implying that the original microstructure of the center tracer is preserved in the print head. Significant microstructure evolution happens after the center tracer leaves the print head. The grains in Area 3 are elongated and thinned to a width of 3–5 μm, bending horizontally. Fine, equiaxed grains emerge

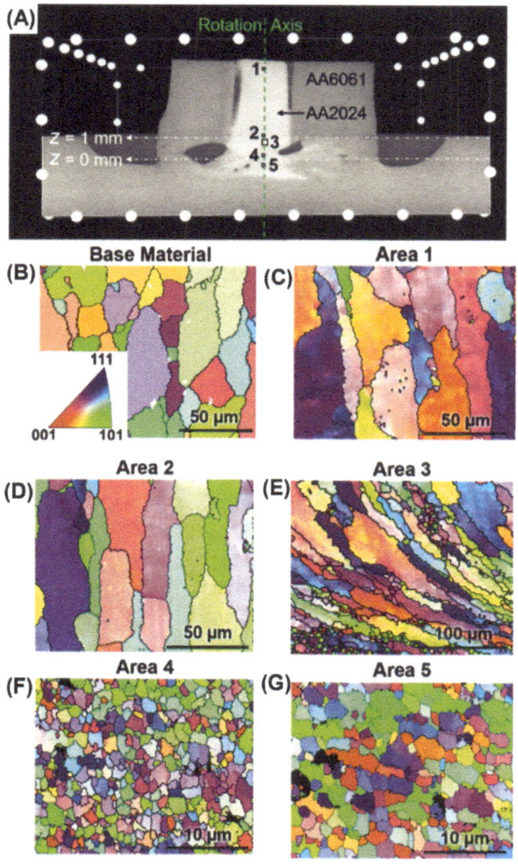

Figure 3.14 Microstructure mapping along the material flow path of the center tracer. (A) The cross-section used for microstructure characterization with areas of interest labeled. The distance between each pair of white dots is 2.5 mm. Inverse pole figure maps of the AA2024 tracer for (B) before printing, (C) Area 1 ($Z = 4.2$ mm), (D) Area 2 ($Z = 1.1$ mm), (E) Area 3 ($Z = 0.9$ mm), (F) Area 4 ($Z = 0.26$ mm), and (G) Area 5 ($Z = -0.2$ mm). *Reprinted with permission from Perry, M. E.J., Rauch, H.A., Griffiths, R.J., Garcia, D., Sietins, J.M., Zhu, Y., et al. 2021. Tracing plastic deformation path and concurrent grain refinement during additive friction stir deposition. Materialia 18, 101159.*

further down, such as in Areas 4 and 5. From $Z = 1.1$ mm (Area 2) to $Z = 0.26$ mm (Area 4), the microstructure of the AA2024 tracer changes dramatically, from vertically elongated, large grains of 25–30 μm to small, equiaxed grains of 1.36 μm. The related dynamic microstructure evolution mechanisms are elaborated in Chapter 4.

3.5.2 Center tracer flow during steady-state deposition

Fig. 3.15 shows the sample appearance after steady-state deposition, highlighting the cross-section used for microstructure characterization. Fig. 3.16 presents the high-resolution X-ray imaging results for several representative local volumes using a voxel size of 9.85 μm. Clearly, the initial cylindrical center tracer has been drastically transformed into a series of long and thin microribbons. The initial cylindrical tracer has a diameter of 3.175 mm, whereas the thickness of the microribbons is on the order of tens of microns. From the isometric view in Fig. 3.16A and $X-Y$ plane view in Fig. 3.16B, all the microribbons are curved, piling up along the Y-direction (i.e., the print head traveling direction). The microribbons extend both above ($Z > 0$ mm, in red) and below ($Z < 0$ mm, in blue)

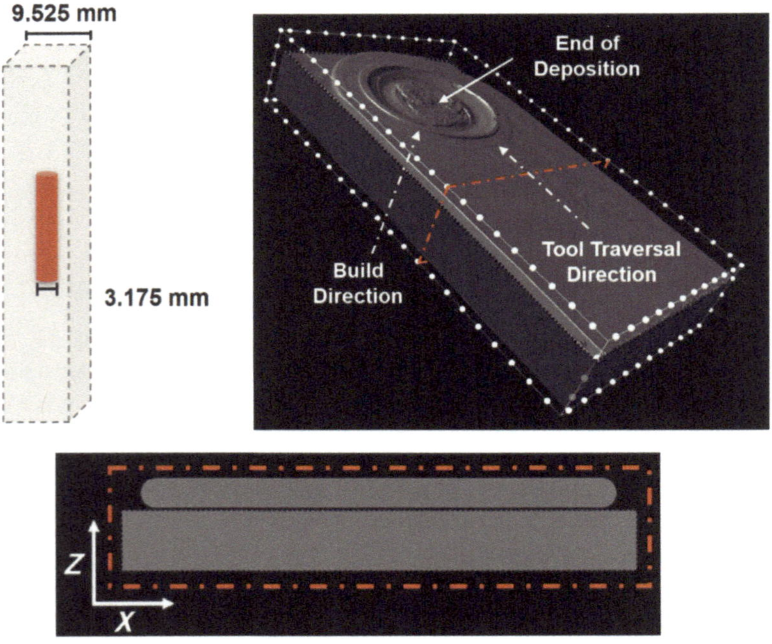

Figure 3.15 In additive friction stir deposition using center tracer-based feed materials, X-ray computed tomography and microstructure characterizations are performed after steady-state deposition. The red dashed lines indicate the cross-section used for microstructure analysis; the distance between each pair of white dots is 2.5 mm. *Reprinted with permission from Perry, M.E.J., Rauch, H.A., Griffiths, R.J., Garcia, D., Sietins, J.M., Zhu, Y., et al. 2021. Tracing plastic deformation path and concurrent grain refinement during additive friction stir deposition. Materialia 18, 101159.*

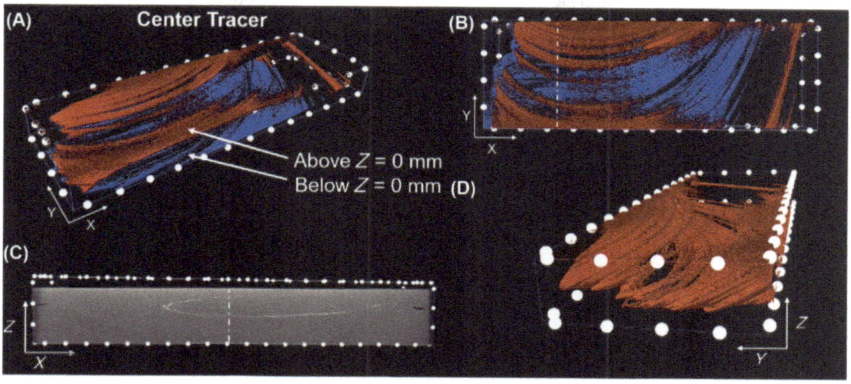

Figure 3.16 Local-scale X-ray computed tomography results after steady-state deposition. (A) Isometric view and (B) top view (*X–Y* plane), with red and blue representing the tracer material above and below $Z = 0$ mm, respectively. (C) Transverse view (*X–Z* plane) with AA6061 shown in gray and AA2024 tracer shown in white. (D) Longitudinal view (*Y–Z* plane). Distance between each pair of dots is ~ 1.25 mm.

the initial substrate surface level. When viewed in the $X-Z$ plane in (C), they appear as an incomplete ellipse with their center on the advancing side. When viewed in the $Y-Z$ plane in (D), the microribbons are seen to tilt ~ 70 degrees along the *Y*-direction and approximately resemble the surface shape of the tool protrusions shown in Fig. 3.9.

From X-ray imaging, the average thickness for the microribbons is measured to be 27.8 μm, suggesting a severe shear thinning process from the initial 3.175 mm wide cylinder. Severe material thinning by material-pin interactions has been commonly observed in friction stirring. For example, Mukherjee and Ghosh noticed a gradual thinning of the foil on the retreating side when the foil was sheared by the pin. This involves using a soft and ductile aluminum foil (AA5457) as the tracer in a workpiece (AA5083). The initial thickness of the foil is 290 μm; however, following friction stir welding, the thickness was reduced to 3 μm (Mukherjee and Ghosh, 2010). The dimension in the thickness direction was lowered by two orders of magnitude. Liu et al. investigated the strain development during friction stir welding of Cu using a 500 μm thick Cu-40Zn foil marker. The thinnest section of the foil was reduced to 11.5 μm after welding (Liu et al., 2019). This reduction was slightly less than two orders of magnitude. In the present work of additive friction stir deposition, no pin is used, but the two small protrusions on the bottom

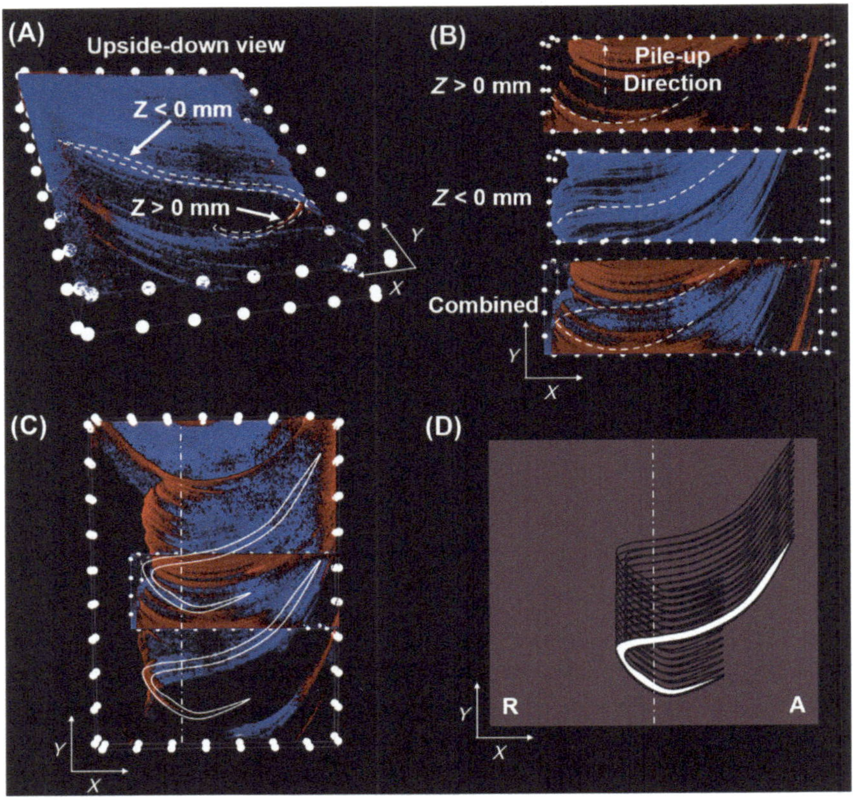

Figure 3.17 Global-scale characterization of the shapes of the microribbons after steady-state deposition. (A) Upside-down isometric view. (B) Top view, showing the top portion ($Z > 0$ mm), bottom portion ($Z < 0$ mm), and a combined image. (C) Top view with an extension along the Y-direction. (D) Illustration of the pile-up of microribbons along the deposition track. *Reprinted with permission from Perry, M.E.J., Rauch, H.A., Griffiths, R.J., Garcia, D., Sietins, J.M., Zhu, Y., et al. 2021. Tracing plastic deformation path and concurrent grain refinement during additive friction stir deposition. Materialia 18, 101159.*

surface can play a similar role in shear thinning. The reduction of the dimension from 3.175 mm to 27.8 μm is also two orders of magnitude.

To elucidate the full 3D shape of individual ribbons, Fig. 3.17 shows the X-ray computed tomography results on the global scale using a voxel size of 33.1 μm. Each microribbon can be best described by a stretched, mirrored "S" shape (i.e., "ƨ" shape). For each microribbon, the longer bottom portion extends in the forward direction and finally reaches the advancing edge, whereas the shorter top portion extends in the backward direction. The transition point between the two portions is on the retreating side and is approximately lined up with the circle defined by the rotating protrusions. When many of such mirrored "S"-shaped microribbons pile up next to each other, a coherent 3D shape forms, roughly resembling a flattened helix. This view is illustrated in Fig. 3.17D. On the advancing side, the footprint of the microribbons is outside of the outer radius of the protrusions, so microribbon formation is unlikely to be induced by a simple tear-off process. Instead, elongation and thinning must be caused by subsequent deformation via the rotating print head.

It is plausible to assume that the microribbons of the tracer material are generated by a multistep process that begins with an initial "tearing-off" event that separates a slice of the tracer. Following that, compression and shear forces from the rotating protrusions and print head flatten and lengthen the slice, resulting in shear thinning. Shear thinning can be caused by at least two different mechanisms. First, when the tracer material slice is on the side of the protrusions (red point in Fig. 3.18A), the rapid protrusion rotation can efficiently shear the material laterally, as seen in Fig. 3.18B. Second, when the tracer material slice is below the protrusions [blue point in (A)], the protrusions impose simultaneous compression and shear and smear the material along their path of motion. This is seen in Fig. 3.18C.

Shear thinning likely causes significant strain development, which is nevertheless difficult to measure without more information on the microribbon generation process. During the initial material feeding alone, the outer edge of the center tracer ($r = 1.588$ mm) has a shear strain of 14.96 and an equivalent plastic strain larger than 8.64. Adding the strain associated with microribbon generation, the total cumulative strain should be

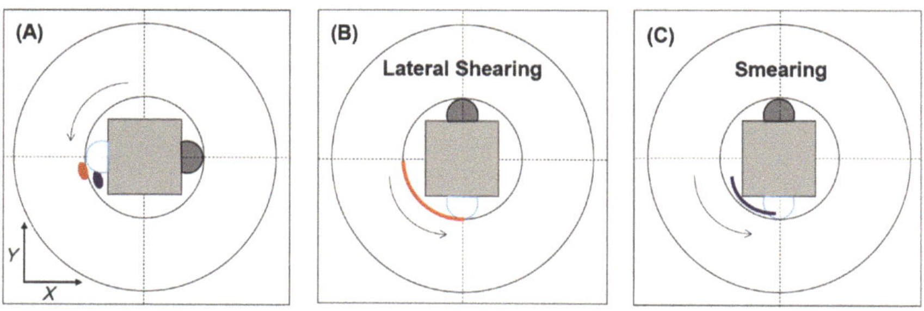

Figure 3.18 Possible mechanisms for material thinning during steady-state deposition. A small material volume in (A) can be thinned either by (B) lateral shearing or (C) smearing due to the rotating protrusions in additive friction stir deposition. *Reprinted with permission from Perry, M.E.J., Rauch, H.A., Griffiths, R.J., Garcia, D., Sietins, J.M., Zhu, Y., et al. 2021. Tracing plastic deformation path and concurrent grain refinement during additive friction stir deposition. Materialia 18, 101159.*

on the order of 10^1 or greater in additive friction stir deposition. This conclusion is consistent with prior microstructure characterization of AA6061 printed by additive friction stir deposition, which shows a strong C-type shear texture, suggesting a shear strain value larger than 10 (Griffiths et al., 2021).

The microstructure of the microribbon tracer is characterized in Fig. 3.19, showing the inverse pole figure and misorientation maps of four regions containing the AA2024 tracer. Although Areas i-iv correspond to characteristically different locations along the tracer oval in the transverse cross-section, their microstructures are all featuring fine, equiaxed grains with a small fraction of low angle boundaries. The grain size is ~ 2–3 μm, which is comparable to that in Areas 4 and 5 during the initial material feeding seen in Fig. 3.14. The shapes of the microribbons are 3D in nature, so Areas i-iv are not necessarily from the same microribbon. As a result, such characterization has been repeated for different *X-Z* cross-sections along the deposition track. The same conclusion

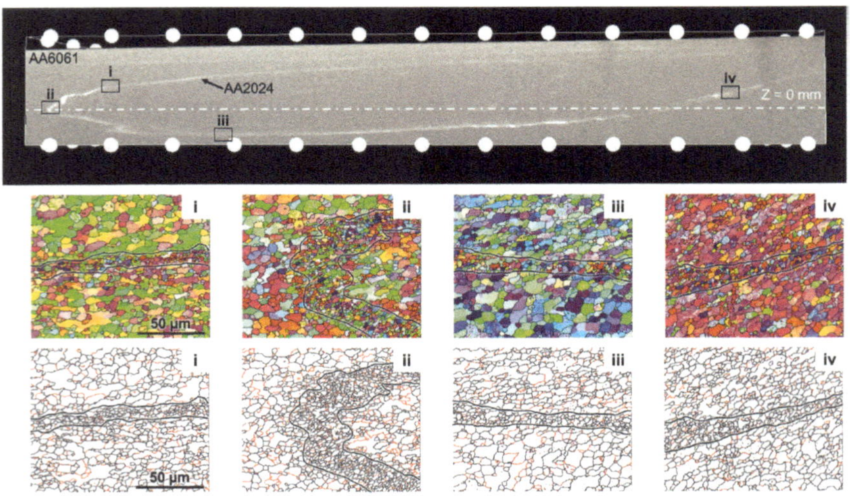

Figure 3.19 Microstructure of the center tracer after steady-state deposition. *Top*: An *X–Z* cross-sectional image based on X-ray computed tomography, showing the AA2024 as an incomplete oval with four areas of interest labeled. The distance between each pair of dots is 1.25 mm. Below that, *1st row*: inverse pole figure maps with the boundaries between the AA2024 tracer and AA6061 matrix shown as black lines. *2nd row*: Misorientation maps with low angle boundaries in red and high angle boundaries in black. *Reprinted with permission from Perry, M.E.J., Rauch, H.A., Griffiths, R.J., Garcia, D., Sietins, J.M., Zhu, Y., et al. 2021. Tracing plastic deformation path and concurrent grain refinement during additive friction stir deposition. Materialia 18, 101159.*

is reached: the microstructure of the AA2024 tracer is similar along the oval with a fine, equiaxed grain structure. After severe thinning, therefore, the tracer material completes the recrystallization process, showing a uniform recrystallized microstructure.

3.6 Flow path of the edge volume of the feed material

3.6.1 Edge tracer flow during initial material feeding

Because of the downward material feeding and rotation rate gradient along the Z-direction, the original edge tracer cylinder develops into a helix. As shown in Fig. 3.20, even when the edge tracer is still within the

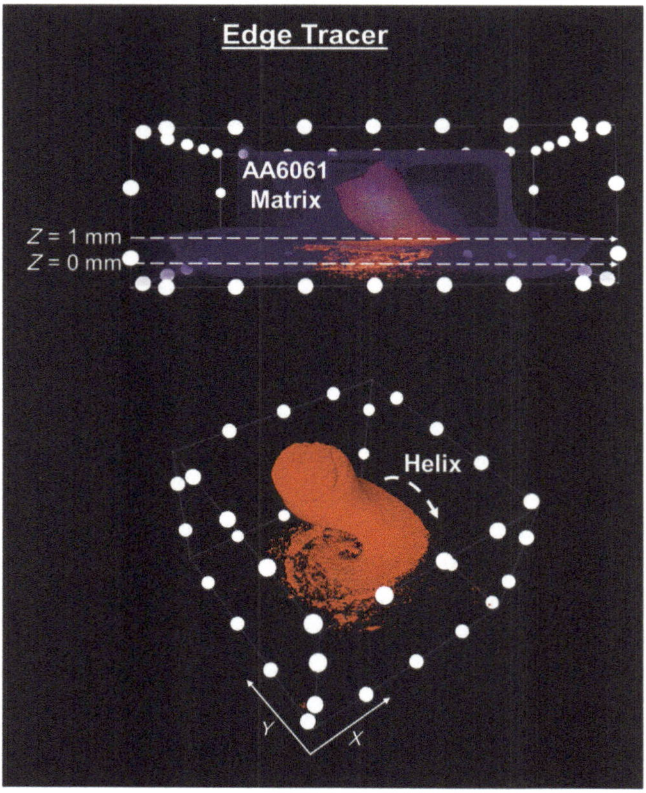

Figure 3.20 X-ray computed tomography results of the edge tracer after initial material feeding. The AA2024 tracer is depicted in red and the surrounding AA6061 matrix is either in blue or transparent. The distance between each pair of white dots is 2.5 mm. *Reprinted with permission from Perry, M.E.J., Rauch, H.A., Griffiths, R.J., Garcia, D., Sietins, J.M., Zhu, Y., et al. 2021. Tracing plastic deformation path and concurrent grain refinement during additive friction stir deposition. Materialia 18, 101159.*

print head ($Z > 1$ mm), it starts to tilt in the $X-Z$ plane and twist in the isometric view. This contrasts with the center tracer, which maintains the original geometry before it comes out of the print head channel. After leaving the print head, the edge tracer deforms significantly with its shape extending throughout the entire volume of the feed material. This again contrasts with the center tracer, which shows a tight corkscrew appearance. These distinctions stem from the differences in rotating axis and mechanical constraint between the center and edge tracers.

The microstructure evolution of the edge tracer is characterized along the flow path during initial material feeding. Fig. 3.21A shows an $X-Z$ cross-sectional slice along the centerline of the feed-rod. In Areas 1 and 2, the edge tracer is still constrained in the print head. Area 3 is directly beneath the print head channel, while Area 4 is under the print head. Areas 5 and 6 correspond to the thinner spiral region seen in Fig. 3.20. The inverse pole figure maps for Areas 1–6 are depicted in Fig. 3.21B–G. In Areas 3–6, both the AA2024 edge tracer and AA6061 matrix are present; their boundaries are highlighted on the inverse pole figure maps based on the composition scans using energy-dispersive X-ray spectroscopy (Wolfgong, 2016). The microstructure of Area 1 is similar to that of the feed-rod, which suggests that the microstructure of the edge tracer is preserved when it is far from being extruded out of the print head channel. Moving to Area 2 right before the tracer leaves the print head channel, the original vertically elongated grains $\sim$ 30 μm are seen to be transformed into smaller, approximately horizontally elongated grains $\sim$ 20 μm. Furthermore, more grains start to develop intragranular orientation gradients, indicating that these grains are being compressed with considerable deformation strain. This contrasts with the scenario for the center tracer, which shows almost no microstructure change at the same height.

As the edge tracer comes out of the print head, it quickly develops into a curved flat sheet beneath the print head as part of the helix. Area 3 in Fig. 3.21D is near the starting point for that flat sheet, whereas Area 4 in Fig. 3.21E represents a section of that sheet as it circles back into the characterized cross-section. Like Area 2, the grains in Area 3 and Area 4 are predominantly horizontal with intra-granular orientation gradients. The deformation levels in Areas 3 and 4 are considerably higher than Area 2, but the grain shape is similar, and the grain size is not significantly reduced. This suggests that the dynamic recrystallization is not extensive or near complete in these areas. This may be attributed to the relatively low temperature that leads to an insufficient dynamic recovery rate. In

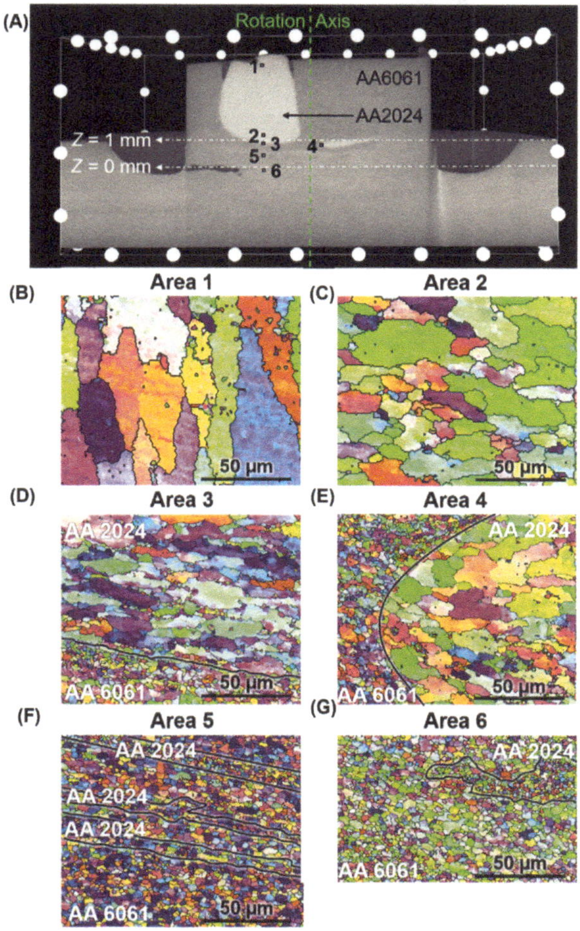

Figure 3.21 Microstructure mapping along the material flow path of the edge tracer. (A) The cross-section used for microstructure characterization with the areas of interest labeled. The distance between each pair of the white dots is 2.5 mm. Inverse pole figure maps are shown for (B) Area 1 ($Z = 3.6$ mm), (C) Area 2 ($Z = 1.1$ mm), (D) Area 3 ($Z = 0.87$ mm), (E) Area 4 ($Z = 0.58$ mm), (F) Area 5 ($Z = 0.39$ mm), and (G) Area 6 ($Z = -0.15$ mm). *Reprinted with permission from Perry, M.E.J., Rauch, H.A., Griffiths, R.J., Garcia, D., Sietins, J.M., Zhu, Y., et al. 2021. Tracing plastic deformation path and concurrent grain refinement during additive friction stir deposition. Materialia 18, 101159.*

Areas 5 and 6 (Fig. 3.21F and G), notably, the large, elongated grains are transformed into refined, equiaxed grains with a grain size of 1–2 µm. In other words, the transition from the flat sheet to the thinner, wispy spiral part causes significant microstructure changes.

3.6.2 Edge tracer flow during steady-state deposition

Like the center tracer, after steady-state deposition the edge tracer also appears as microribbons with a thickness of 20–30 μm, as shown in Fig. 3.22A. However, there are two major differences between the scenarios of the center tracer and edge tracer. The first difference is the shape of the microribbons. The center tracer shows a complex shape that may be described by a stretched, mirrored "S" shape, whereas the edge tracer is developed into a series of long arches piling up long the travel direction of the print head. This can be seen clearly in Fig. 3.22B. The second difference is the location of the microribbons. While a significant portion of the center tracer (in the form of microribbons) goes below $Z = 0$ mm, the edge tracer-based microribbons are all above $Z = 0$ mm. This stems from the extrusion nature of material feeding. The edge portion of the

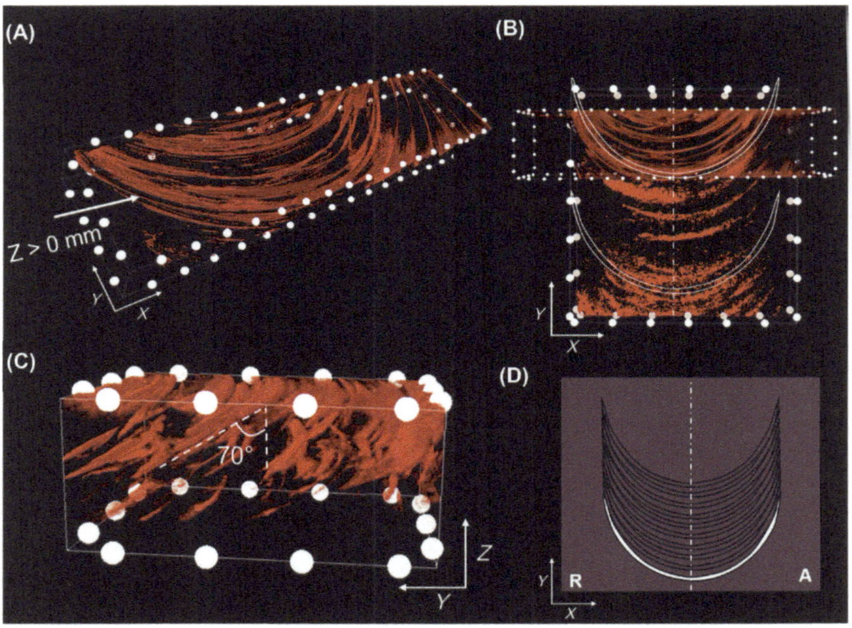

Figure 3.22 X-ray tomography results showing microribbon formation after steady-state deposition of the edge tracer. (A) Side view. (B) Top view of the X–Y plane, showing both results obtained from high-resolution and low-resolution X-ray imaging. (C) View of the longitudinal cross-section or the Y–Z plane. (D) Illustration of the pile-up of long arch microribbons along the deposition track. *Reprinted with permission from Perry, M.E.J., Rauch, H.A., Griffiths, R.J., Garcia, D., Sietins, J.M., Zhu, Y., et al. 2021. Tracing plastic deformation path and concurrent grain refinement during additive friction stir deposition. Materialia 18, 101159.*

feed material is extruded to the side, becoming the top portion of the deposition track; the center portion is compressed deeply along the vertical direction, reaching the level below the original substrate surface.

The edge tracer-based microribbons are also expected to form by shear thinning. Because the edge tracer never goes below the tool protrusions, the smearing mechanism described in Fig. 3.18 is unlikely to be important. Instead, the microribbons are mostly formed by lateral shearing through interactions with the tool protrusions. Notably, the radius of each long arch-shaped microribbon closely matches the circle defined by the outer edge of the rotating protrusions. From the $Y-Z$ plane view shown in Fig. 3.22C, these microribbons tilt $\sim$ 70 degrees along the Y-direction and approximately resemble the shape of the tool protrusions. These observations again suggest the key role played by the material–protrusion interactions in the shear thinning process.

The physics picture of microribbon formation in the edge tracer experiment may be described as follows:

1. At each time step, a small volume of the edge tracer is extruded out of the print head channel.
2. This small chunk of edge tracer is then sheared by the lateral side of the rotating protrusions.
3. The volume is thinned with a shape following the circular protrusion path. With the in-plane motion of the print head, this results in a long arch in the wake of the deposition.
4. When many of such long arch-shaped microribbons pile-up next to each other, the periodic appearance emerges, as illustrated in Fig. 3.22D.

When viewed in the $X-Z$ plane, the edge tracer-based microribbons appear as several concentric ovals, as shown in the top image of Fig. 3.23. Three areas of interest are labeled. Since the microribbons are thin, the $X-Z$ cross-section characterization in these areas includes both the edge tracer of AA2024 and the matrix of AA6061. The inverse pole figure and misorientation maps of these three locations are shown below, in which the boundaries between AA2024 and AA6061 are marked based on composition mapping. All three areas are characterized by fine, equiaxed AA2024 grains with a small percentage of low angle boundaries. Ranging from 1–3 μm, the grain size is comparable to that observed in the edge tracer after initial material feeding. Because Areas i, ii, and iii may not be from the same microribbon, characterization of several $X-Z$ cross-sections along the deposition track have been repeated. The same

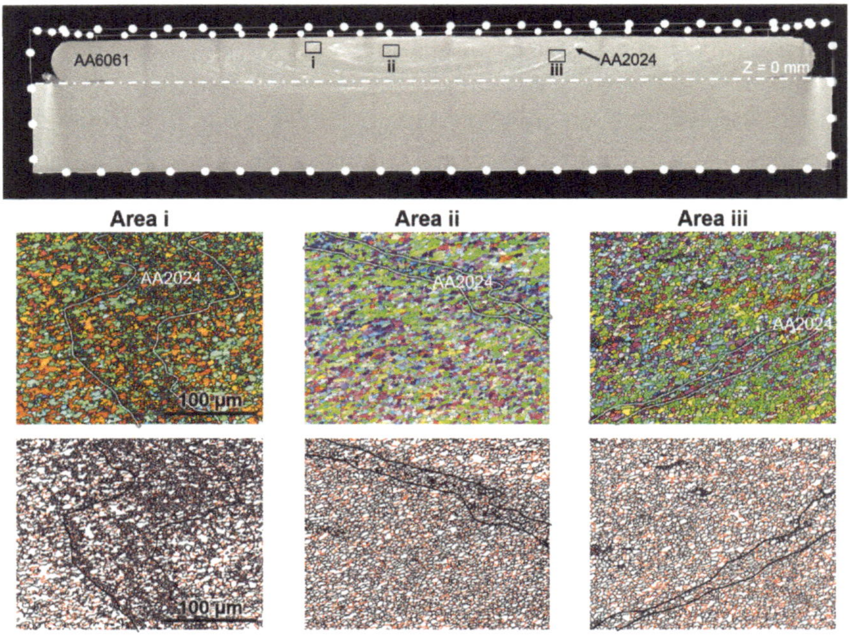

Figure 3.23 Microstructure of the edge tracer after steady-state deposition. *Top*: An *X–Z* cross-sectional image based on X-ray computed tomography, showing the AA2024 as bright rings with three areas of interest labeled. Below that, *1st row*: inverse pole figure maps, in which the boundaries between the AA2024 tracer and AA6061 matrix are highlighted. *2nd row*: misorientation maps with low angle boundaries depicted in red and high angle boundaries in black. *Reprinted with permission from Perry, M.E.J., Rauch, H.A., Griffiths, R.J., Garcia, D., Sietins, J.M., Zhu, Y., et al. 2021. Tracing plastic deformation path and concurrent grain refinement during additive friction stir deposition. Materialia 18, 101159.*

conclusion is reached: the dynamic recrystallization of the AA2024 edge tracer has been completed, resulting in a uniform, fine, equiaxed grain structure.

3.7 Material deformation and flow at the interface

Following the same logic as the tracer-based analysis of internal material flow, the interface flow phenomena and interface morphology formation are investigated via dissimilar material deposition in this section. This entails depositing a single layer of AA2024 onto an AA6061 substrate. For the results provided in this section, the average AA2024 deposition thickness is 1.65 mm, while the AA6061 substrate thickness is 6.35 mm. The print head geometry is the same as in earlier sections, showing two small protrusions. The protrusions are partially (0.5 mm deep) inserted into the substrate during deposition to improve vertical mixing between the deposited material and the substrate. The deposition is carried out at a print head rotation rate of $\Omega = 300$ RPM and an in-plane velocity of $V = 2.00$ mm/s. The material feed rate is $F = 0.85$ mm/s. The ratio of feed rate to in-plane velocity influences the layer thickness and track width.

3.7.1 Surface and interface morphology

Based on X-ray computed tomography, Fig. 3.24A shows the overall look of the AA2024 deposit from the top view. The rotating print head is elevated at the conclusion of deposition, and the feed-rod separates from the deposit, resulting in a rough circular region. The deposition surface exhibits a classic "onion skin" appearance, which has also been observed in friction stir welding (Chen and Cui, 2008; Cui et al., 2008). For printing using different sets of in-plane velocity and rotation rate, the measured onion skin spacing always matches the calculated advance per revolution for additive friction stir deposition. This indicates that the onion skin is caused by the interaction between the print head edge and the top surface of the deposited material during each revolution. As the print head moves forward, the top surface of the deposited material flows out of the deposition zone (i.e., the volume beneath the print head) from the trailing side.

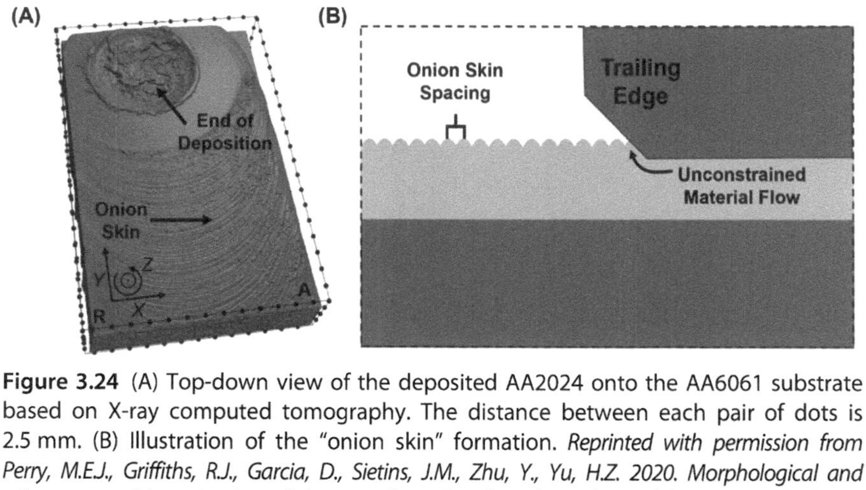

Figure 3.24 (A) Top-down view of the deposited AA2024 onto the AA6061 substrate based on X-ray computed tomography. The distance between each pair of dots is 2.5 mm. (B) Illustration of the "onion skin" formation. *Reprinted with permission from Perry, M.E.J., Griffiths, R.J., Garcia, D., Sietins, J.M., Zhu, Y., Yu, H.Z. 2020. Morphological and microstructural investigation of the non-planar interface formed in solid-state metal additive manufacturing by additive friction stir deposition. Additive Manufacturing 35, 101293.*

Then, the trailing edge of the print head mills an imprint into the deposited material. This effect is illustrated in Fig. 3.24B. The combination of print head movement and the imprint effect possibly leads to an onion skin pattern on the top surface.

To reveal the 3D interfacial morphology, the substrate material is set to be transparent in the reconstruction volumes. This allows for analysis of the underside of the deposited material. The result is shown in Fig. 3.25A, which is effectively a flipped picture of the bottom surface of

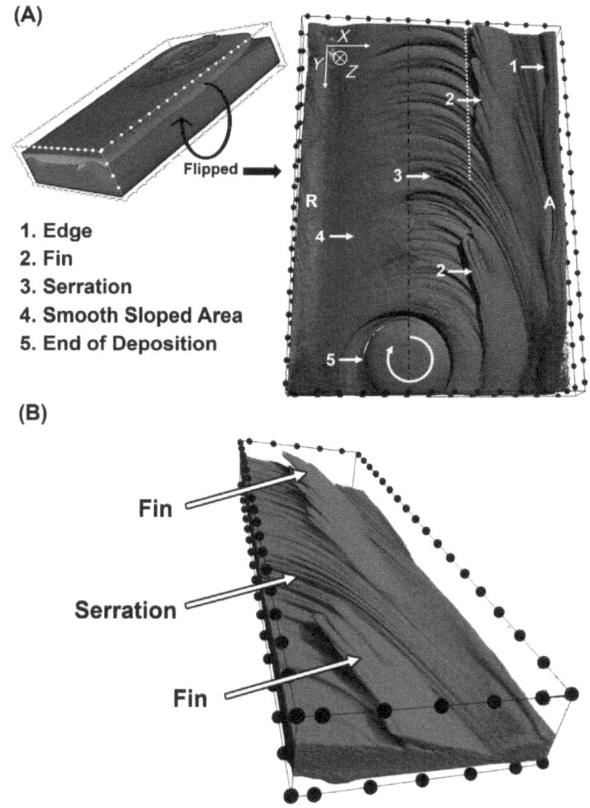

Figure 3.25 X-ray tomography of the interface. (A) Underside of the deposited AA2024, with a few prominent characteristics emphasized. This picture has been inverted so that the material flow is clockwise rather than counterclockwise. Each pair of dots is separated by 2.5 mm. (B) A slanted view of the deposited material on the advancing side to highlight the fin structures. *Reprinted with permission from Perry, M.E.J., Griffiths, R.J., Garcia, D., Sietins, J.M., Zhu, Y., Yu, H.Z. 2020. Morphological and microstructural investigation of the non-planar interface formed in solid-state metal additive manufacturing by additive friction stir deposition. Additive Manufacturing 35, 101293.*

deposition. The interface is not flat, but rather slopes from the midline to the two edges. On the advancing side, there are some distinct 3D features. For example, thin and flat strips are seen to accumulate near the advancing edge. A 3D image of these strips is shown in Fig. 3.25B, clearly showing that they are not in parallel with the X–Y plane but are slanted away from the remaining material. We refer to these strips as the 'fin' structures throughout this text because of their shape.

Fig. 3.25 also shows a sequence of periodic curves that begin at the midline and go toward the advancing edge. Such structures appear as a succession of jagged edges along the deposition track in the longitudinal cross-section seen in Fig. 3.26A. In this book, they are referred to as "serrations." The serrations are hidden by the fin structure at the advancing edge, but the high-resolution X-ray tomography results shown in Fig. 3.26B reveal that the serrations connect with the roots of the fins.

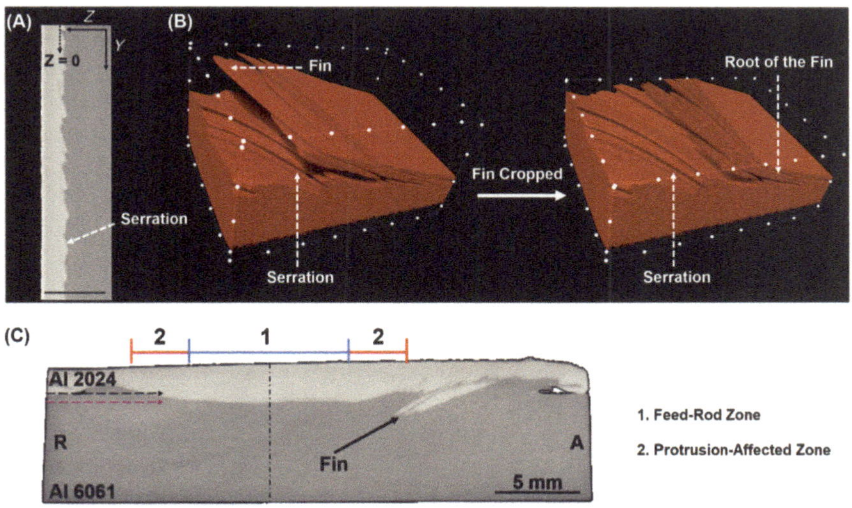

Figure 3.26 (A) Image of serrations in a longitudinal cross-section (*Y*–*Z* plane). Dark gray = AA6061; light gray = AA2024. The scale bar measures 5 mm. (B) High-resolution X-ray computed tomography data showing a full-volume image of the fins and serrations, as well as a cropped volume with the fins cropped. This demonstrates the continuity of serrations throughout the fin-dominated section. Each pair of dots is separated by 1 mm. (C) A representative transverse cross-section of the deposit and substrate highlighting the bowl-shaped interface. *Reprinted with permission from Perry, M.E.J., Griffiths, R.J., Garcia, D., Sietins, J.M., Zhu, Y., Yu, H.Z. 2020. Morphological and microstructural investigation of the non-planar interface formed in solid-state metal additive manufacturing by additive friction stir deposition. Additive Manufacturing 35, 101293.*

The retreating side of the interface, in contrast to the advancing side, lacks substantial topographic features, appearing as a smooth sloping surface. Fig. 3.26C shows a typical transverse cross-section of the deposition track on the substrate. We define the area directly below the feed-rod as the "feed-rod zone" and the area below the protrusions as the "protrusion-affected zone." In the feed-rod zone, the deposited AA2024 (light gray) plunges into the AA6061 substrate (dark gray) and reaches $Z = -0.5$ mm near the centerline. Outside the protrusion-affected zone, the substrate is driven above the original substrate surface level. The gradual slope of the deposited material away from the centerline creates a "bowl-like" shape. Between the protrusion-affected zone and the advancing edge, the interface features a tilted shelf structure deeply penetrating the substrate with a depth of -1.5 mm and a tilt angle of 20 degrees from the horizontal. This feature is indeed the projection of a fin structure on the transverse cross-section.

As shown in Fig. 3.27, the bowl shape of the interface is caused by the compression from the feeding apparatus, which forces the feed-rod to go deeply below the original substrate surface. Due to the dynamic friction at the interface, the temperature increases and the surface of the substrate becomes softened; yet, the bulk of the substrate remains stiff, resulting in strong mechanical constraint. The softened substrate surface must expand and surge in response to the feed-rod compression (i.e., via outward and

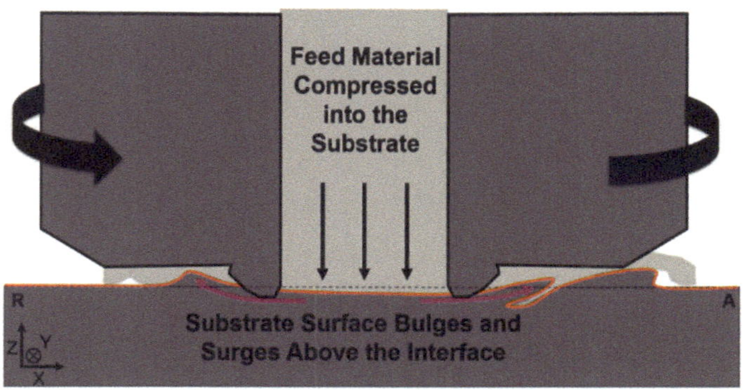

Figure 3.27 Illustration of interfacial material flow processes in additive friction stir deposition, showing the transverse cross-section of the deposition. *Reprinted with permission from Perry, M.E.J., Griffiths, R.J., Garcia, D., Sietins, J.M., Zhu, Y., Yu, H.Z. 2020. Morphological and microstructural investigation of the non-planar interface formed in solid-state metal additive manufacturing by additive friction stir deposition. Additive Manufacturing 35, 101293.*

upward motion). Because of the mechanical constraint imposed by the tool protrusions, this motion occurs mostly outside of the protrusion-affected zone, where AA6061 is seen at $Z > 0$ mm. The downward motion of the AA2024 feed-rod in the feed-rod zone, together with the upward motion of the AA6061 substrate outside the tool protrusion-affected zone, produces strong material mixing across the original interface level.

Why do the fin structures and serrations form on the advancing side rather than the retreating side? This is possibly caused by the interactions between the print head with in-plane material expansion. Changing from a tall cylinder or bar to a thin disk, the feed material is extruded beneath the print head as it is pressed downward and plasticized by frictional heating. As illustrated in Fig. 3.28A, this expansion favors the trailing side over the leading side when the print head travels in-plane. The rotating print head smears the deposited material through compression and shear, causing the deposited material to flow counterclockwise (in the direction of print head rotation). As shown in Fig. 3.28B, the influx of material from the trailing edge to the advancing side causes a broader accumulation of material on the advancing side. In comparison, the influx of the deposited material flowing from the advancing side to the retreating side has

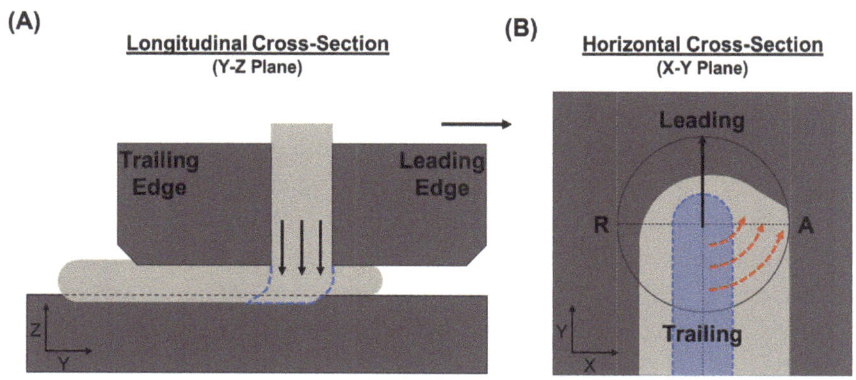

Figure 3.28 (A) Longitudinal cross-section demonstrating the process for feed-rod bending and shearing caused by in-plane motion. (B) A top-down perspective of the deposition process, with the light gray area corresponding to the deposition track. *Reprinted with permission from Perry, M.E.J., Griffiths, R.J., Garcia, D., Sietins, J.M., Zhu, Y., Yu, H.Z. 2020. Morphological and microstructural investigation of the non-planar interface formed in solid-state metal additive manufacturing by additive friction stir deposition. Additive Manufacturing 35, 101293.*

low in-plane constraint with empty space beneath the print head. As a result, less material is built-up on the retreating side.

The rotating protrusions play a critical role in determining the interface morphology. First, they push away the deposited material, which will further accumulate outside the protrusion-affected zone. Because of the severe in-plane restriction imposed by the excess material on the advancing edge, the accumulated material will be forced to move synergistically with the substrate surface. Outside the protrusion-affected zone, the motion of the AA6061 substrate is outward and upward, so the motion of the accumulated AA2024 is inward and downward. This results in more vertical mixing and fin structure development on the advancing side. On the retreating side, no 3D structures such as fins are created due to less material accumulation. The rotating protrusions also mill temporary trenches into the raised substrate material. As a result, the deposited material near the protrusion-affected zone will migrate downhill to rapidly fill these trenches. This process is more effective in locations of high in-plane constraint, resulting in the serration structure formation on the advancing side. Furthermore, the slope of the serrations shown in the longitudinal cross-section is in parallel with the outer surface of the protrusions. This supports the idea that the serrations are formed by direct material interactions with the protrusions.

3.7.2 Interfacial mixing

The material distribution is measured and quantified based on over 1000 collected X-ray cross-sectional images. Along the transverse direction, Fig. 3.29A shows a few representative longitudinal cross-section ($Y-Z$) slices with varying positions along the deposition width (i.e., the X-direction). Fig. 3.29B shows the AA2024 area percent plotted along the X-direction, in which each data point represents the ratio between the measured AA2024 area and the total area of AA2024 and AA6061 in a specific Y-Z slice. The plot shows that more AA2024 is present in the centerline of the deposition than the edges, which is consistent with the observed bowl shape of the interface. In fact, the shape of the plot largely resembles the inverse of the interface profile seen in the transverse cross-section. At the retreating edge ($X = 3$ mm), the AA2024 deposit accounts for approximately 16% of the overall material. Because the substrate material surges above $Z = 0$ mm outside the protrusion-affected zone, the AA2024 portion reduces to around 12% at $X = 5$ mm. The

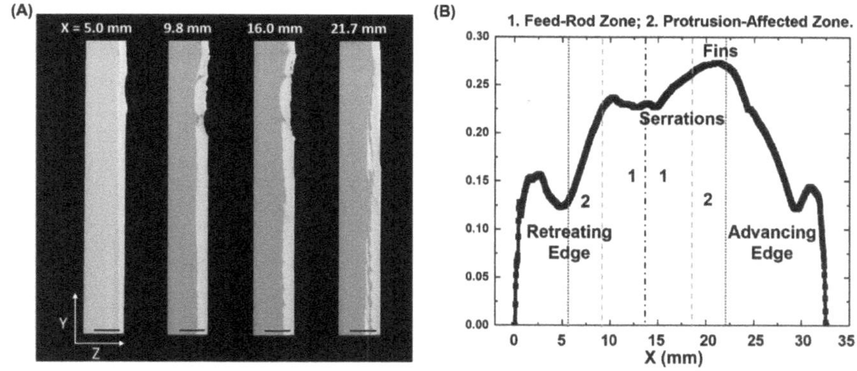

Figure 3.29 Quantified material distribution along the deposition width direction, X-direction. (A) Longitudinal cross-section (*Y*–*Z*) X-ray images along the *X*-direction. Dark gray = AA6061, light gray = AA2024. The scale bar is 5 mm. (B) Plot of the proportion of the deposited AA2024 along the deposition width, that is, *X*-direction. *Reprinted with permission from Perry, M.E.J., Griffiths, R.J., Garcia, D., Sietins, J.M., Zhu, Y., Yu, H.Z. 2020. Morphological and microstructural investigation of the non-planar interface formed in solid-state metal additive manufacturing by additive friction stir deposition. Additive Manufacturing 35, 101293.*

AA2024 proportion then increases towards the centerline due to the compression imposed on the feed-rod. The AA2024 proportion continues to increase with the presence of serrations and fins on the advancing side, reaching a maximum fraction of 27.5% at $X = 22$ mm. After that ($X > 22$ mm), the AA2024 proportion drops quickly towards the advancing edge.

The material distribution in the horizontal cross-sections ($X-Y$ plane) can be used to quantify macroscopic mixing over the original interface level. Fig. 3.30 shows a variety of X-ray imaging slices of the $X-Y$ planes, including those above, at, and below the initial interface level ($Z = 0$ mm). At $Z > 0.75$ mm, the deposit is a pure AA2024 coating. Below that, most of the material is still AA2024, with the substrate AA6061 being only visible around the margins (see Fig. 3.30A and B). There is about 60% AA2024 at the initial interface level, $Z = 0$ mm, which is inside a central band circumscribed by the protrusion-affected zone (see Fig. 3.30C). From the midline to the advancing side, the serrations appear as alternating curves of AA6061/AA2024. As Z decreases, the AA2024 percentage continues to decrease as shown in Fig. 3.30D. At $Z = -0.53$ mm, the deposited AA2024 is mostly found within the feed-rod zone (Fig. 3.30E). Below that, AA2024 is only seen as the end tip of fins on the advancing side. This is shown in Fig. 3.30F for $Z = -0.89$ mm.

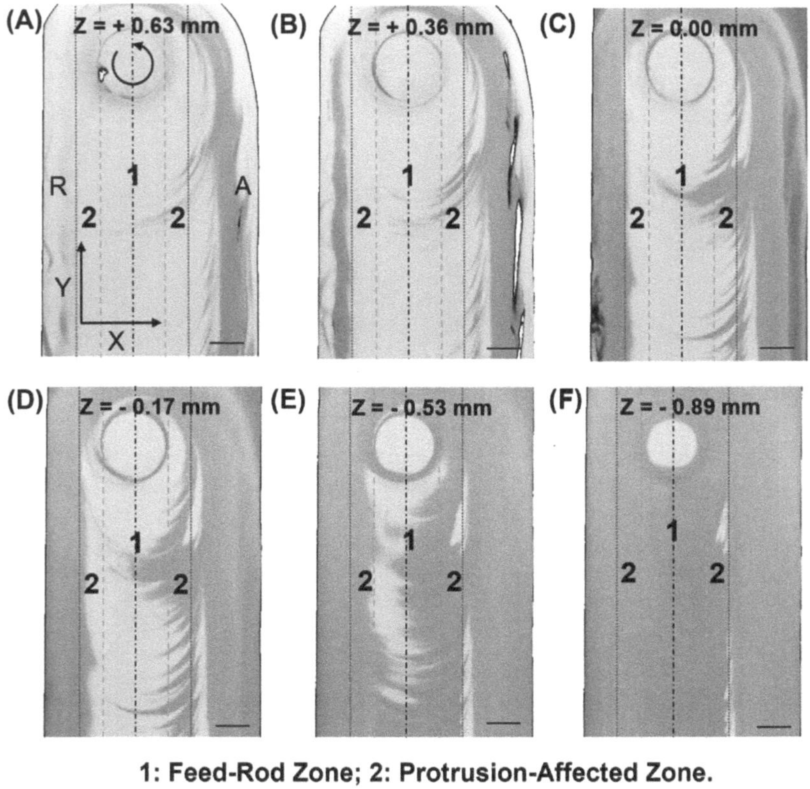

Figure 3.30 X-ray imaging of horizontal cross-sections. (A) $Z = 0.63$ mm, (B) $Z = 0.36$ mm, (C) $Z = 0$ mm, (D) $Z = -0.17$ mm, (E) $Z = -0.53$ mm, and (F) $Z = -0.89$ mm. The scale bar is 5 mm. *Reprinted with permission from Perry, M.E.J., Griffiths, R.J., Garcia, D., Sietins, J.M., Zhu, Y., Yu, H.Z. 2020. Morphological and microstructural investigation of the non-planar interface formed in solid-state metal additive manufacturing by additive friction stir deposition. Additive Manufacturing 35, 101293.*

Following the quantification approach for Fig. 3.29B, Fig. 3.31 plots the area percentage of AA2024 in the $X-Y$ plane along the Z-axis. Each data point represents the AA2024 area percentage in a specific $X-Y$ slice. The data points labeled as "A", "C", and "E" correspond to the images shown in Fig. 3.30 A, C, and E, respectively. The orange-shaded region indicates the quantity of AA6061 above $Z = 0$ mm, while the green-shaded area reflects the amount of AA2024 below $Z = 0$ mm. The existence of these shaded regions is the consequence of vertical material mixing on the macroscopic level. The green-shaded area is larger than the orange-shaded area because the protrusions penetrate the substrate during

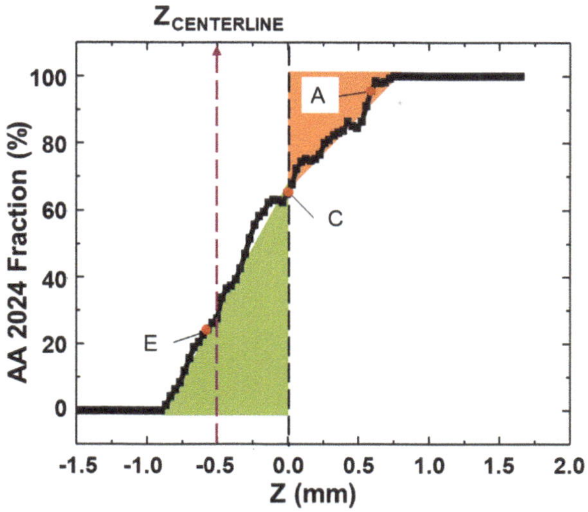

Figure 3.31 Plot of the deposited AA2024 volume fraction along the *Z*-axis (Perry et al. 2020). *Reprinted with permission from Perry, M.E.J., Griffiths, R.J., Garcia, D., Sietins, J.M., Zhu, Y., Yu, H.Z. 2020. Morphological and microstructural investigation of the non-planar interface formed in solid-state metal additive manufacturing by additive friction stir deposition. Additive Manufacturing 35, 101293.*

deposition. In other words, more AA2024 is deposited beneath the original interface level than the AA6061 surging above that.

The non-planar interface formed in additive friction stir deposition is further examined via microstructure mapping. Fig. 3.32 shows a confocal microscopy image of the transverse cross-section, in which microstructure characterization is performed for three interfacial areas of interest. Based on energy dispersive spectroscopy, AA2024 and AA6061 are distinguished according to the aluminum count in these areas. This reveals extensive mesoscopic material mixing in the middle and right areas, showing lamellar banding between AA2024 and AA6061. Considering the macroscopic mixing quantified by X-ray tomography, the interfacial mixing is concluded to be across multiple length scales. This is similar to what has been observed in dissimilar material friction stir welding between AA2024 and AA6061 (Li et al., 1999).

According to the related inverse pole figure maps, the deposited AA2024 consists of a fine, equiaxed microstructure with negligible color variation within individual grains in all locations. The microstructural homogeneity across different areas suggests that the material voxels in the feed-rod have comparable thermomechanical histories to render the

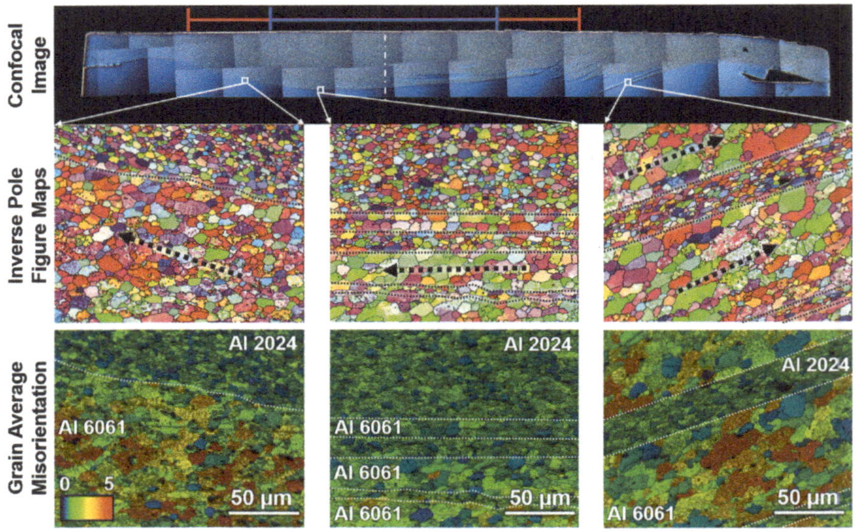

Figure 3.32 Microstructure characterization across the interface. *Top row*: A confocal image of a representative transverse cross-section. *Middle row*: Inverse pole figure maps of three interfacial areas of interest. *Bottom row*: GAM maps of the three areas. *Reprinted with permission from Perry, M.E.J., Griffiths, R.J., Garcia, D., Sietins, J.M., Zhu, Y., Yu, H.Z. 2020. Morphological and microstructural investigation of the non-planar interface formed in solid-state metal additive manufacturing by additive friction stir deposition. Additive Manufacturing 35, 101293.*

observed recrystallized microstructure. In the interfacial areas, the grain size of the substrate AA6061 is significantly greater than that of the deposited AA2024 (9.9 vs 4.9 μm), and the grain shape is typically more elongated. The long axis of these AA6061 grains, interestingly, frequently aligns with the local interface direction, as indicated by the black dashed arrows. This implies that the shear deformation associated with the material flow at the interface governs the dynamic microstructure development in AA6061. The grain average misorientation evaluates the intra-granular orientation gradients. Apparently, AA6061 has higher grain average misorientation values than AA2024, suggesting less complete dynamic recrystallization in the former than the latter.

3.8 Summary

This chapter focuses on the material flow phenomena in additive friction stir deposition. We first provide the basics of plasticity theory and fluid mechanics theory. After reviewing previous experimental material flow studies in friction stir welding, we present state-of-the-art experimental work on material flow in additive friction stir deposition. To understand the internal material flow pattern, tracer-based feed-rods are employed, involving placement of a AA2024 tracer into a AA6061 matrix. X-ray tomography is used to characterize the shape evolution of the tracers during initial material feeding and during steady-state deposition. To distinguish the material flow phenomena between the center and edge material voxels in the feed-rod, experiments based on center and edge tracers are performed separately.

During initial material feeding, the main deformation comes from compression and shear. Compression is from the feeding apparatus, resulting in an extrusion-like strain. Shear is caused by the rotation rate gradient along the thickness direction, resulting in a torsion-like strain. Both center and edge tracers undergo some sort of helical shape evolution, which is more obvious in the edge tracer. In the center tracer, the rotation axis is the same as the feed-rod, and the tracer is mechanically constrained by the surrounding matrix, resulting in a tight corkscrew appearance.

During steady-state deposition, the interactions between the protrusions and tracer lead to microribbon formation via shear thinning. This process transforms the original cylindrical center and edge tracers (dimensions of millimeter scale) to a series of microribbons (thickness ~ tens of microns), which pile up along the print head travel direction. The center tracer-based microribbons have a complex 3D shape extending both below and above the original interface level. The edge tracer-based microribbons are only present above the original interface level with a long arch shape. The total strain in additive friction stir deposition is estimated to be on the order of 10^1 or greater.

To understand the interfacial material flow and interface formation, dissimilar material printing is performed involving deposition of AA2024 onto AA6061 substrate. The surface is characterized by an onion skin structure, the spacing of which is consistent with the advance per revolution of the print head. The interface shows a bowl shape on the transverse cross-section, because of the compression force imposed on the feed-rod as well as the synergistic upward and outward flow of the substrate surface. As the material flows and rotates under the shear force imposed by the rotating print head, material accumulation more readily occurs on the advancing side. Together with the influence of the protrusions, intriguing 3D structures are formed on the advancing side, such as the fins and serrations. Both macroscopic and mesoscopic mixing occur in the interfacial regions.

References

Amoo, L.M., Layi Fagbenle, R., 2020. 14—Overview of non-Newtonian boundary layer flows and heat transfer. In: Fagbenle, R.O., Amoo, O.M., Aliu, S., Falana, A. (Eds.), Applications of Heat, Mass and Fluid Boundary Layers. Woodhead Publishing, pp. 413–435.

Assidi, M., Fourment, L., 2009. Accurate 3D friction stir welding simulation tool based on friction model calibration. International Journal of Material Forming 2, 327.

Bower, A.F., 2009. Applied Mechanics of Solids. CRC Press.

Bruhns, O.T., 2014. The Prandtl-Reuss equations revisited. ZAMM—Journal of Applied Mathematics and Mechanics / Zeitschrift für Angewandte Mathematik und Mechanik 94, 187–202.

Chen, Z.W., Cui, S., 2008. On the forming mechanism of banded structures in aluminum alloy friction stir welds. Scripta Materialia 58, 417–420.

Cui, G.R., Ma, Z.Y., Li, S.X., 2008. Periodical plastic flow pattern in friction stir processed Al–Mg alloy. Scripta Materialia 58, 1082–1085.

Dieter, G.E., 1976. Mechanical Metallurgy. McGraw-Hill.

Fung, Y.C. 1977. A First Course in Continuum Mechanics 2nd edition.

Griffiths, R.J., Garcia, D., Song, J., Vasudevan, V.K., Steiner, M.A., Cai, W., et al., 2021. Solid-state additive manufacturing of aluminum and copper using additive friction stir deposition: process-microstructure linkages. Materialia 15, 100967.

He, X., Gu, F., Ball, A., 2014. A review of numerical analysis of friction stir welding. Progress in Materials Science 65, 1–66.

Kundu, P.K., Cohen, I.M., Dowling, D.R., 2016. Chapter 11—Instability. In: Kundu, P. K., Cohen, I.M., Dowling, D.R. (Eds.), Fluid Mechanics, sixth ed. Academic Press, Boston, pp. 533–602.

Lai, W.M., Rubin, D., Krempl, E., 2010. CHAPTER 6—Newtonian viscous fluid. In: Lai, W.M., Rubin, D., Krempl, E. (Eds.), Introduction to Continuum Mechanics, fourth ed. Butterworth-Heinemann, Boston, pp. 353–410.

Li, Y., Murr, L.E., Mcclure, J.C., 1999. Solid-state flow visualization in the friction-stir welding of 2024 Al to 6061 Al. Scripta Materialia 40, 1041–1046.

Liu, X.C., Sun, Y.F., Nagira, T., Ushioda, K., Fujii, H., 2019. Experimental evaluation of strain and strain rate during rapid cooling friction stir welding of pure copper. Science and Technology of Welding and Joining 24, 352–359.

Lubliner, J., 2013. Plasticity Theory. Dover Publications.

Morisada, Y., Fujii, H., Kawahito, Y., Nakata, K., Tanaka, M., 2011. Three-dimensional visualization of material flow during friction stir welding by two pairs of x-ray transmission systems. Scripta Materialia 65, 1085–1088.

Mukherjee, S., Ghosh, A.K., 2010. Flow visualization and estimation of strain and strain-rate during friction stir process. Materials Science and Engineering: A 527, 5130–5135.

Nandan, R., Roy, G.G., Lienert, T.J., Debroy, T., 2007. Three-dimensional heat and material flow during friction stir welding of mild steel. Acta Materialia 55, 883–895.

Nemat-Nasser, S., Batchelor, G.K., Davis, S., Freud, L.B., Ablowitz, M.J., Leibovich, S., et al., 2004. Plasticity: A Treatise on Finite Deformation of Heterogeneous Inelastic Materials. Cambridge University Press.

Perry, M.E.J., Griffiths, R.J., Garcia, D., Sietins, J.M., Zhu, Y., Yu, H.Z., 2020. Morphological and microstructural investigation of the non-planar interface formed in solid-state metal additive manufacturing by additive friction stir deposition. Additive Manufacturing 35, 101293.

Perry, M.E.J., Rauch, H.A., Griffiths, R.J., Garcia, D., Sietins, J.M., Zhu, Y., et al., 2021. Tracing plastic deformation path and concurrent grain refinement during additive friction stir deposition. Materialia 18, 101159.

Sadd, M.H., 2019. Chapter 3—Kinematics of motion and deformation measures. In: Sadd, M.H. (Ed.), Continuum Mechanics Modeling of Material Behavior. Academic Press, pp. 57–109.
Schlichting, H., Gersten, K., 2017a. Fundamentals of Boundary–Layer Theory, Boundary-Layer Theory. Springer Berlin Heidelberg, Berlin, Heidelberg, pp. 29–49.
Schlichting, H., Gersten, K., 2017b. Some Features of Viscous Flows, Boundary-Layer Theory. Springer Berlin Heidelberg, Berlin, Heidelberg, pp. 3–27.
Schmidt, H.N.B., Dickerson, T.L., Hattel, J.H., 2006. Material flow in butt friction stir welds in AA2024-T3. Acta Materialia 54, 1199–1209.
Seidel, T.U., Reynolds, A.P., 2001. Visualization of the material flow in AA2195 friction-stir welds using a marker insert technique. Metallurgical and Materials Transactions A 32, 2879–2884.
Su, H., Wu, C.S., Bachmann, M., Rethmeier, M., 2015. Numerical modeling for the effect of pin profiles on thermal and material flow characteristics in friction stir welding. Materials & Design 77, 114–125.
White, F., 2015. Fluid Mechanics. McGraw-Hill Education.
Winslow, J., Otsuka, H., Govindarajan, B., Chopra, I., 2018. Basic understanding of airfoil characteristics at low Reynolds numbers (104–105). Journal of Aircraft 55, 1050–1061.
Wolfgong, W.J., 2016. Chapter 14—Chemical analysis techniques for failure analysis: part 1, common instrumental methods. In: Makhlouf, A.S.H., Aliofkhazraei, M. (Eds.), Handbook of Materials Failure Analysis with Case Studies from the Aerospace and Automotive Industries. Butterworth-Heinemann, Boston, pp. 279–307.

CHAPTER 4

Dynamic microstructure evolution

The only thing that is constant is change.

—Heraclitus.

During additive friction stir deposition, the feed material flows extensively at elevated temperatures. On the microscopic level, dislocations are created, accumulated, annihilated, and reorganized to form substructures. The misorientations between these substructures or subgrains gradually increase upon further deformation, so the subgrains will eventually become new grains. Alternatively, the dislocation annihilation and reorganization activities may be sluggish, and the dislocation density remains temporarily high. Then, new dislocation-free grains within the deformed structure can preferentially grow and consume neighboring deformed grains to reduce the stored energy in the system. In both scenarios, the grain structure is significantly changed during deposition, reducing the dislocation density and grain size. In the presence of deformation, these are termed dynamic microstructure evolution processes. After deposition, the material temperature decreases in the absence of deformation; static or metadynamic recrystallization possibly occurs during this cooling process.

With a thermomechanical processing nature, additive friction stir deposition falls in the category of metalworking. However, it differs significantly from conventional processes like forging, rolling, and extrusion (Lin et al., 2012). First, additive friction stir deposition is characterized by rapid and large shear deformation at elevated temperatures, in which the strain rate and total strain can be substantially higher than conventional metalworking. Second, at each time step, only a small volume of the feed material is being thoroughly and uniformly processed. Compared to ultrasonic and cold spray additive manufacturing, the plastic deformation in additive friction stir deposition has a global character, where every voxel in the feed material undergoes significant deformation, resulting in a more uniform microstructure. Fig. 4.1 compares the resulting microstructure among solid-state additive processes, including ultrasonic additive manufacturing, cold spray additive manufacturing, and additive friction stir deposition (Tuncer and Bose, 2020).

Additive Friction Stir Deposition
DOI: https://doi.org/10.1016/B978-0-12-824374-9.00008-3

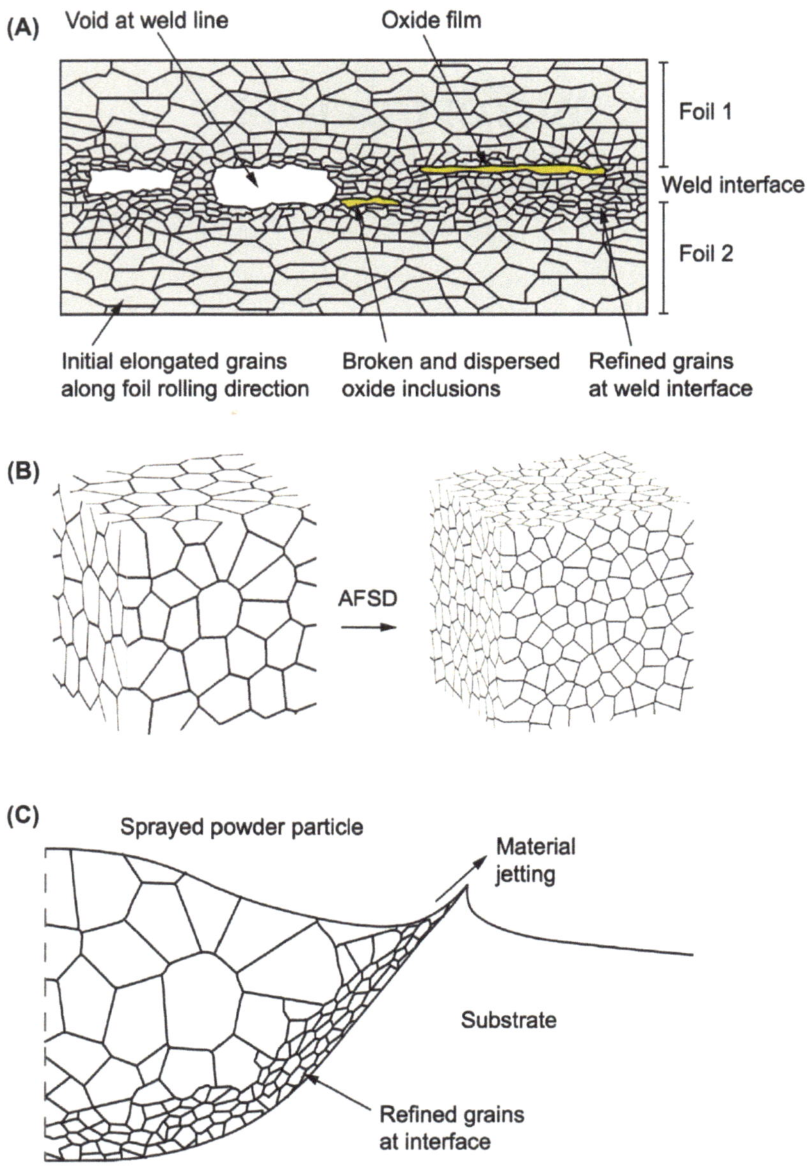

Figure 4.1 Schematic depiction of microstructures resulting from various solid-state additive manufacturing processes: (A) ultrasonic additive manufacturing, (B) additive friction stir deposition, and (C) cold spray additive manufacturing. *Reprinted with permission from Tuncer, N., Bose, A., 2020. Solid-state metal additive manufacturing: a review. JOM 72, 3090–3111.*

In this chapter, we focus on dynamic microstructure evolution during additive friction stir deposition of metals and alloys. We begin with the microstructure evolution basics, such as the concepts of grain boundary migration, recovery, recrystallization, and grain growth. This is followed by a review of the dynamic recrystallization mechanisms identified in hot deformation processes. We then discuss the thermomechanical stages in additive friction stir deposition and survey the characteristic microstructures observed in the as-printed alloys, such as Al, Mg, Ni, and steel. Based on the tracer experiments elaborated in Chapter 3, Material Flow Phenomena, we examine the dynamic microstructure evolution mechanisms along the flow path during deposition. After that, the process-microstructure linkage is explored with a comparison between Al and Cu, emphasizing the roles of the stacking fault energy and the material flow behavior in governing the as-printed microstructure. We finally discuss the dynamic and static phase evolution during deposition and subsequent cooling in the context of precipitation-strengthened alloys.

4.1 Elements of microstructure evolution

Grain boundaries separate crystalline regions of different crystallographic orientations. If the misorientation angle exceeds 15 degrees, the grain boundary is referred to as a high angle boundary; otherwise, the grain boundary is referred to as a low angle boundary. Grain boundaries tend to migrate to reduce the system energy, for example, the energy associated with excess grain boundary areas or the stored energy in worked metals. The velocity for grain boundary migration is

$$V = MP \tag{4.1}$$

where M is a kinetic term called grain boundary mobility and P is a pressure term denoting the thermodynamic driving force. Fig. 4.2 illustrates a possible mechanism for grain boundary migration, in which the atoms are constantly removed from one grain by thermal activation and migrate towards the disordered region of the boundary itself, followed by reattachment to the other grain. The energy barrier is associated with jumping across grain boundaries (Humphreys et al., 2017a).

The relationship in Eq. (4.1) is derived from the rate theory; it is valid if the mobility M is independent of the driving force P and $P \ll k_B T$,

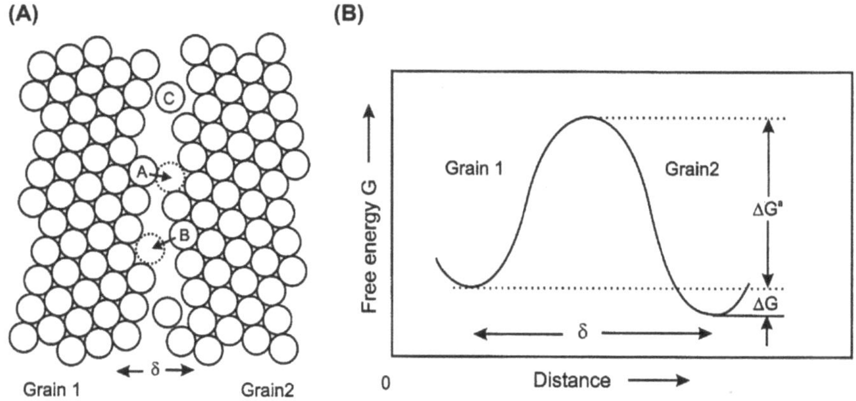

Figure 4.2 Grain boundary migration by atom jumping across the grain boundaries. (A) A schematic showing the mechanism. (B) The free energy landscape for the atom jump. *Reprinted with permission from Humphreys, J., Rohrer, G.S., Rollett, A., 2017a. Chapter 5—Mobility and migration of boundaries. In: Humphreys, J., Rohrer, G.S., Rollett, A. (Eds.), Recrystallization and Related Annealing Phenomena, third ed. Elsevier, Oxford, pp. 145–197.*

where k_B is the Boltzmann constant. In some other proposed boundary migration mechanisms such as boundary dislocation-controlled migration or rotation (Hirth and Balluffi, 1973; Wang et al., 2014), the mobility may depend on the driving force, so this equation becomes invalid. Generally, the mobility critically depends on temperature:

$$M = M_0 \exp\left(-\frac{Q}{RT}\right) \tag{4.2}$$

Here Q is the apparent activation energy, which may be related to atom-scale thermally activated processes. R is the gas constant: $R = k_B N_A$, with N_A being the Avogadro constant. The grain boundary mobility depends on the misorientation angle. It is high for boundaries of large misorientations ($>10-15$ degrees) and is quite low for subgrain boundaries formed during deformation and recovery.

If the stored energy is minimal (e.g., in annealed metals), the main driving force for grain boundary migration is the boundary curvature:

$$P_C = 2\gamma_{GB}/r \tag{4.3}$$

where γ_{GB} is the grain boundary energy per area and r is the grain radius. The velocity of grain boundary migration, therefore, scales with the boundary curvature, as validated by measurements and simulations of grain boundary migration in bicrystals (Gottstein et al., 2001). There is, however, evidence that neither mobility nor driving force is constant; rather, both are sensitive to boundary structure (Olmsted et al., 2009). The mobility may change with grain boundary complexion (Frolov, 2014); the grain boundary energy depends on the crystallography (Li et al., 2009). Recently, using three-dimensional orientation maps recorded by high-energy diffraction microscopy, the relationship between the grain boundary velocities and curvatures is investigated during annealing of a Ni polycrystal (Bhattacharya et al., 2021). Based on the data collected from 52,000-grain boundaries, the average grain boundary velocity is plotted against the curvature, showing no evident correlation. Instead, a correlation is found between the boundary velocity and the five macroscopic parameters specifying the grain boundary crystallography. This suggests the need for new models and understandings of grain boundary migration in polycrystals.

When the material is severely deformed, the dislocation density will increase dramatically to accommodate the large strain. The evolution of the dislocation density ρ with respect to the strain ε depends on two competing processes: dislocation storage and annihilation. It can be described by the Kocks–Mecking model (Estrin and Mecking, 1984):

$$\frac{d\rho}{d\varepsilon} = \mathrm{k}_1 \rho^{1/2} - \mathrm{k}_2 \rho \tag{4.4}$$

Here, k_1 describes the storage of dislocations and k_2 describes the dynamic recovery of dislocations, which proceeds via the annihilation of dislocation segments with opposite Burgers vectors. k_2 is dependent on strain rate and temperature, as well as the material intrinsic properties. For Cu, it is found that the dislocation density increases monotonically with the applied shear strain (Fig. 4.3A; Zehetbauer and Seumer, 1993). For Al, the dislocation density increases only marginally after the strain exceeds 0.3 due to the high dynamic recovery rate (Fig. 4.3B; Nes and Marthinsen, 2002).

In deformed metals, there are three important microstructure components: (i) statistically stored dislocations that are present in the grain interior; (ii) dislocation walls or geometrically necessary dislocations (Gao and Huang, 2003), which are either tangled to form cells or well-ordered to bound subgrains; (iii) high angle boundaries separating grains. For statistically stored dislocations with density of ρ, the stored energy per volume is

$$E_D = c\rho G b^2. \tag{4.5}$$

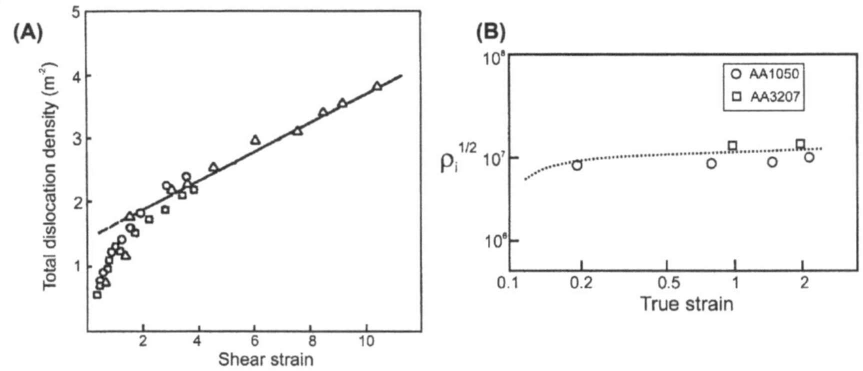

Figure 4.3 The measured dislocation density as a function of strain. (A) The total dislocation density of Cu. (B) The free dislocation density in Al alloys. *Reprinted with permission from Zehetbauer, M., Seumer, V., 1993. Cold work hardening in stages IV and V of F.C.C. metals—I. Experiments and interpretation. Acta Metallurgica et Materialia 41, 577–588. Reprinted with permission from Nes, E., Marthinsen, K., 2002. Modeling the evolution in microstructure and properties during plastic deformation of f.c. c.-metals and alloys – an approach towards a unified model. Materials Science and Engineering: A 322, 176–193.*

Here, c is a constant of ~0.5, G is the shear modulus, and b is the Burgers vector. For cell or subgrain structures, the stored energy per volume is

$$E_D = \alpha\gamma_s/D. \tag{4.6}$$

Here, α is a constant of ~1.5, D is the subgrain diameter, and γ_s is the specific energy of the subgrain boundaries. Note that the majority of the work done by plastic deformation is dissipated as heat; only 1%–10% is stored as microstructural and defect energy.

When the highly deformed material is subjected to heat treatment, there is a thermodynamic driving force to reduce the stored energy. At low temperatures (0.4–0.5 of the melting temperature T_m), recovery dominates, resulting in subgrain formation via individual or collective movement of dislocations. The recovery rate strongly depends on the stacking fault energy of the material. In high stacking fault energy materials like Al and BCC (body-centered cubic) Fe, recovery is effective because dislocation cross-slip and local annihilation are straightforward. In low stacking fault energy materials like austenitic steel and α-brass, recovery is more difficult (Raabe, 2014). Solute atoms can also reduce the dislocation mobility and slow down the recovery kinetics. Recovery may proceed by single dislocation annihilation based on diffusion of point defects or by mutual dislocation annihilation via dislocation glide, climb, and cross-slip. From that, dislocations are reorganized into dislocation walls, and complicated dislocation cell walls may be collapsed into neat subgrain boundaries. This will lead to the formation of cells or subgrain structures (Nes, 1995), within which the statistically stored dislocations have been eliminated. Subgrain growth can also occur, leading to the coalescence of dislocation walls.

At high temperatures ($> 0.7\ T_m$), recrystallization dominates, resulting in dislocation-free grains via high angle boundary migration. Recrystallization happens when new dislocation-free grains emerge inside the deformed or recovered structure, grow, and consume the neighboring grains with high stored energy (Doherty et al., 1997). There are two steps in recrystallization: "nucleation," or more precisely initiation, and growth. Differing from classical nucleation processes, here the recrystallized "nucleus" can be a large subgrain, a well-recovered region of a deformed grain, or a dislocation-free region within shear bands (especially in low stacking fault energy materials). If nonshearable particles are present, dislocations can be trapped around the particles, resulting in particle-deformed zones with large misorientation developments; recrystallization can initiate in these zones, referred to as particle stimulated nucleation (Robson et al., 2009).

Once the dislocation-free grains are formed, their grain boundaries will migrate, sweeping across the dislocation-heavy regions. Instead of the curvature-driven grain boundary migration described in Eq. (4.3), now the driving force is related to reduction of the stored energy. If the dislocations are inside the grains, then the driving force $P_{SE} = c\rho Gb^2$. If the dislocations are in the form of cell walls, then $P_{SE} = \alpha\gamma_s/D$. Now the curvature may play an opposite role to the conventional case: for a small new grain to consume larger grains, the Gibbs–Thomson relationship predicts a retarding pressure due to curvature (Perez, 2005). In other words, the curvature-driven pressure term is negative. Another notable retarding pressure term in recrystallization is the Zener drag or particle pinning (Nes et al., 1985): $P_z = 3F_v\gamma_{GB}/2r_{particle}$, where F_v and $r_{particle}$ are the volume fraction and radius of the second-phase particles. Solute drag and orientation pinning can also slow down the growth of recrystallized grains (Gottstein and Shvindlerman, 2009).

The annealing temperature has a significant impact on the recrystallization kinetics. Usually, the time for 50% recrystallization $t_{0.5}$ is a good measure of the recrystallization rate and can be expressed as $\frac{1}{t_{0.5}} \propto \exp\left(-\frac{Q}{RT}\right)$. Fig. 4.4 shows the experimental data for annealing of Fe-3.5% Si after deformation with a strain of 0.6 (Humphreys et al., 2017b). The $ln\left(\frac{1}{t_{0.5}}\right)$ *vs.* $1/T$ plot shows a straight line, from which the apparent activation energy Q is determined to be 290 kJ/mol. Like phase transformation, the recrystallization kinetics may be described by the JMAK (Johnson–Mehl–Avrami–Kolmogorov) equation (Fanfoni and Tomellini, 1998):

$$X_V = 1 - \exp(-Bt^n) \quad (4.7)$$

Here, X_V is the recrystallization fraction, B is a constant related to the nucleation rate and grain boundary velocity, t is the time, and n is the JMAK exponent with a theoretical value of 4 by assuming a constant nucleation rate. However, experiments reveal that the JMAK model prediction is too simplistic to quantitatively explain recrystallization, which is much more complicated than the model assumptions.

After recrystallization, curvature-driven grain coarsening can happen via normal grain growth or abnormal grain growth (Omori et al., 2013), but the driving force is roughly two orders of magnitude smaller than that of recrystallization. In normal grain growth, the grain size distribution nearly stays the same while the average grain size increases. In abnormal grain growth, a few grains preferentially grow and develop into a pinned grain structure (Hillert, 1965). The pinning is often caused by particles or

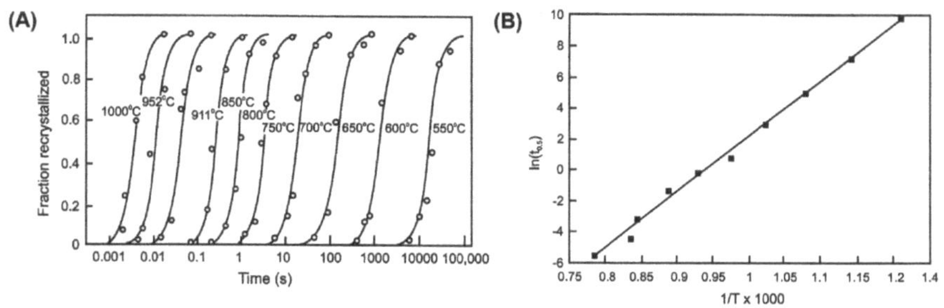

Figure 4.4 The measured recrystallization data for Fe-3.5% Si deformed at a strain of 0.6. (A) Effect of the annealing temperature on the recrystallization rate. (B) Arrhenius plot for 50% recrystallization time as a function of the inverse of temperature. *Reprinted with permission from Humphreys, J., Rohrer, G.S., Rollett, A., 2017b. Chapter 7—Recrystallization of single-phase alloys. In: Humphreys, J., Rohrer, G.S., Rollett, A. (Eds.), Recrystallization and Related Annealing Phenomena, third ed. Elsevier, Oxford, pp. 245–304.*

by low angle boundaries with low mobility. Both the grain size distribution and average grain size change in abnormal grain growth.

4.2 Dynamic recrystallization mechanisms

If large deformation occurs at high temperatures, salient microstructure changes like grain refinement and dislocation density reduction can be observed in situ. Such processes are generally termed dynamic recrystallization, which can be further divided into discontinuous and continuous types (Huang and Logé, 2016), based on whether high angle boundary migration plays an important role. Under a given thermomechanical condition, whether continuous or discontinuous dynamic recrystallization dominates depends on the intrinsic material properties, such as the stacking fault energy. For the same material, whether continuous or discontinuous dynamic recrystallization dominates depends on the thermomechanical condition, which is defined by the temperature T and strain rate $\dot{\varepsilon}$, or the Zener-Hollomon parameter (Mann et al., 2011):

$$Z = \dot{\varepsilon} \exp\left(\frac{Q}{RT}\right) \tag{4.8}$$

Here Q is an apparent activation energy to be determined from experiment. Generally, the value of Q is comparable to or above the activation energy of self-diffusion.

4.2.1 Discontinuous dynamic recrystallization

Discontinuous dynamic recrystallization occurs more prevalently in materials with low or medium stacking fault energy, such as Cu, Ni, and austenitic steel, and at temperatures above 0.5 T_m. This temperature is lower than what is typically required for static recrystallization, which occurs during heat treatment of worked metals in the absence of deformation. Characterized by "nucleation" and growth stages, the microstructure changes in discontinuous dynamic recrystallization rely on the migration of high angle boundaries. During hot deformation, new dislocation-free grains form in the deformed microstructure [e.g., by bulging of grain boundaries (Xie et al., 2019)] and grow to consume neighboring grains of high stored energy. To initiate discontinuous dynamic recrystallization, a critical deformation strain needs to be reached. The critical strain ε_c decreases by decreasing Z. As deformation continues, dislocations start to build up in these recrystallized grains, reducing the driving force for further growth and eventually halting said growth. Simultaneously, other dislocation-free grains will form and grow, leading to a "necklace" structure of grains around the original grain boundaries (Ponge and Gottstein, 1998). Fig. 4.5 illustrates the necklace structure formation process during discontinuous dynamic recrystallization.

Once dynamic equilibrium is reached, the mean grain size does not change as the recrystallization fraction increases, which is different from the case of static recrystallization. The steady-state grain size D_R increases monotonically with decreasing stress σ but has a weak temperature dependence. A universal relationship between the stress and grain size can be written as (Derby, 1991)

$$\frac{\sigma}{G}\left(\frac{D_R}{b}\right)^n = K_1 \tag{4.9}$$

where K_1 is a constant. This relationship is found to be valid for a wide variety of metals and geological materials, in which $n = 2/3$ and K_1 is in the range of 1–10. Eq. (4.9) is referred to as the Derby relationship for steady-state recrystallized grain size.

When discontinuous dynamic recrystallization occurs during hot deformation, the corresponding stress–strain curve is generally characterized by a broad peak. The new grain formation and growth will result in softening and decrease the work hardening rate until a stress peak is reached. Further increasing the strain, the flow stress will decrease until a dynamic equilibrium between strain hardening and recrystallization-induced strain softening is reached (Gao et al., 1999; Sakai and Jonas, 1984). As the value of Z decreases, the flow stress decreases, and the

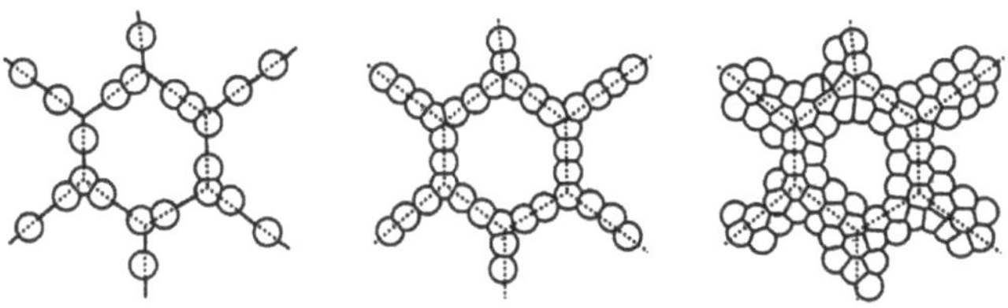

Figure 4.5 Illustration of necklace structure formation in discontinuous dynamic recrystallization, involving the formation of new dislocation-free grains at the grain boundaries. *Reprinted with permission from Humphreys, J., Rohrer, G.S., Rollett, A., 2017d. Chapter 13—Hot deformation and dynamic restoration. In: Humphreys, J., Rohrer, G.S., Rollett, A. (Eds.), Recrystallization and Related Annealing Phenomena, third ed. Elsevier, Oxford, pp. 469–508.*

stress–strain curve starts to develop multiple peaks (see Fig. 4.6A) (Cram et al., 2009). This is because, at low flow stresses, the material recrystallizes completely before the second cycle of recrystallization begins. The competition between strain hardening and strain softening in each cycle of recrystallization results in multiple peaks or oscillations in the stress-strain curve. At high stresses, the subsequent cycles start before the previous recrystallization cycle is complete; in other words, multiple peaks associated with each recrystallization cycle will overlap, resulting in a single broad peak.

A similar trend in the flow stress curve is observed with respect to the initial grain size (see Fig. 4.6B) (El Wahabi et al., 2005). As the initial grain size decreases, the stress–strain curve changes from a single broad peak to multiple peaks. It has been suggested that if the ratio of the initial grain size and recrystallized grain size is smaller than 2 (i.e., $\frac{D_0}{D_R} < 2$), all the new grains develop at the same time due to the high density of grain boundaries that act as "nucleation" sites. This will render complete recrystallization in one cycle. Upon further deformation, the fully recrystallized material hardens and recrystallizes again. Repeating the recrystallization cycle leads to an oscillatory stress–strain curve. In this case, grain coarsening, rather than grain refinement, takes place until the stable grain size is reached.

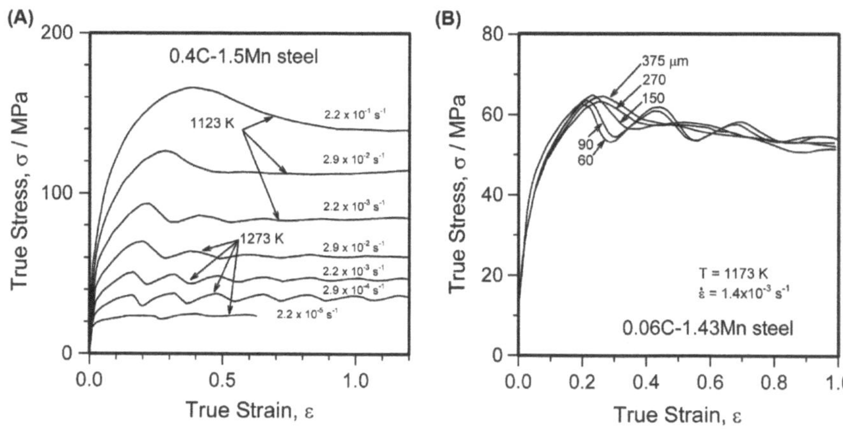

Figure 4.6 Stress–strain curves when discontinuous dynamic recrystallization occurs. (A) Effect of the Zener–Hollomon parameter. (B) Effect of the initial grain size. *Reprinted with permission from Sakai, T., Jonas, J.J., 1984. Overview no. 35 dynamic recrystallization: mechanical and microstructural considerations. Acta Metallurgica 32, 189–209.*

4.2.2 Continuous dynamic recrystallization

Continuous dynamic recrystallization typically occurs in materials of high stacking fault energy such as Al, Mg, and ferritic iron. In these materials, continuous dynamic recrystallization begins by dislocation annihilation and reorganization, forming subgrains via dynamic recovery; further deformation leads to strain-induced high angle boundary formation. In other words, the recrystallization is implemented by strain rather than by grain boundary migration (Gourdet and Montheillet, 2003). This is illustrated in Fig. 4.7A. Fig. 4.7B shows an example of continuous dynamic recrystallization observed in AA7475 during multi-directional forging. Regarding exactly how the subgrains are transformed into grains, two mechanisms are possible (Sakai et al., 2014).

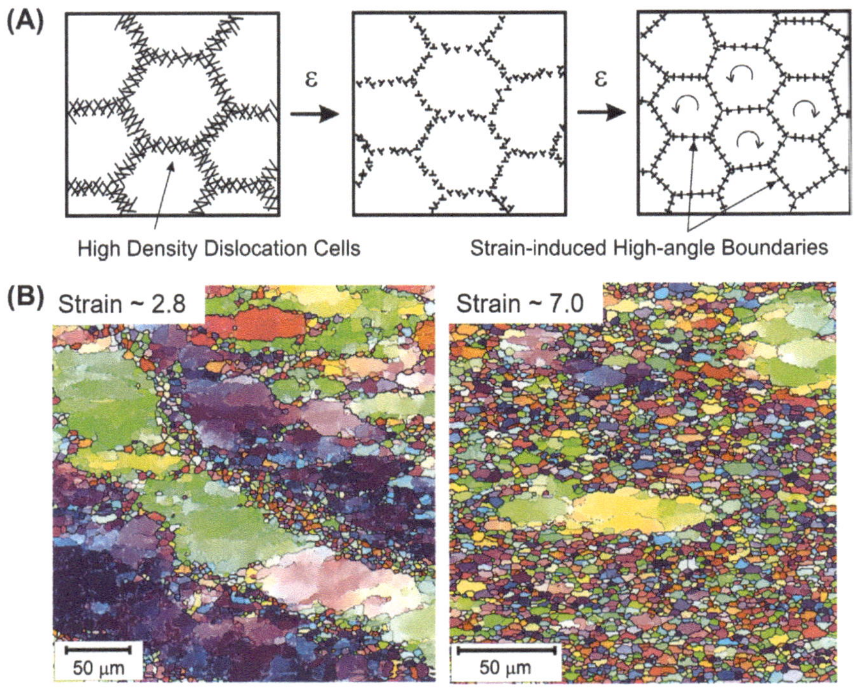

Figure 4.7 (A) Illustration of the continuous dynamic recrystallization process. (B) Orientation imaging microscopy results for AA7475 deformed at 673K at a strain rate of 0.0003 s^{-1}, with a comparison between two strain levels. *Reprinted with permission from Sakai, T., Belyakov, A., Kaibyshev, R., Miura, H., Jonas, J.J., 2014. Dynamic and post-dynamic recrystallization under hot, cold and severe plastic deformation conditions. Progress in Materials Science 60, 130–207.*

The first mechanism is progressive subgrain rotation (Poirier and Nicolas, 1975), which has been observed in geological materials and high solute metals. This mechanism is important especially if dislocation motion is inhibited by a lack of slip systems or by solute drag. During deformation, subgrains near grain boundaries may undergo substantially more rotation than those in the center. At sufficiently large strains, the misorientation between the subgrains near grain boundaries and the subgrains in the grain-interior may become more than 15 degrees, resulting in high angle grain boundary formation. This mechanism likely involves inhomogeneous plastic flow, accelerated dynamic recovery near the grain boundaries, and grain boundary sliding.

The second mechanism is geometric dynamic recrystallization (Kassner and Barrabes, 2005), which occurs when subgrains form and grain boundaries develop serrations during dynamic recovery. The serration wavelength is comparable to the subgrain size. Under large compression and shear strain, the cross-section area is significantly reduced, and the original grains become elongated and flattened. As the strain increases, the grain thickness decreases, eventually becoming comparable to the subgrain size and serration wavelength. Fig. 4.8A illustrates the geometric dynamic recrystallization process. Fig. 4.8B shows an example of serrated grain boundaries observed in high purity Al (Kassner, 1989). The grain thickness H can be described by the initial grain size D_0 through a geometric relationship:

$$H = D_0 \exp(-\varepsilon) \tag{4.10}$$

The pinch-off of serrated grain boundaries will then result in a microstructure of smaller-scale equiaxed grains, whose size is determined by the subgrain size. As an example, Fig. 4.8C plots the grain thickness as a function of strain based on measurement of the AA8006 alloy (Jazaeri and Humphreys, 2004). Setting the grain thickness H equal the subgrain size D, we get the critical strain for geometric dynamic recrystallization: $\varepsilon_{cr} \sim \ln\left(\frac{D_0}{D}\right)$.

In both mechanisms, the subgrain size D determines the recrystallized grain size. The subgrain size does not change as deformation proceeds but is determined by the strain hardening rate and dynamic recovery capability. For a given material, the subgrain size can be written as a function of the flow stress:

$$\frac{\sigma}{G}\frac{D}{b} = K \tag{4.11}$$

where K is a constant (Derby, 1991). This relationship is found to be valid for a wide variety of materials, in which K has a value of ~ 10 for FCC

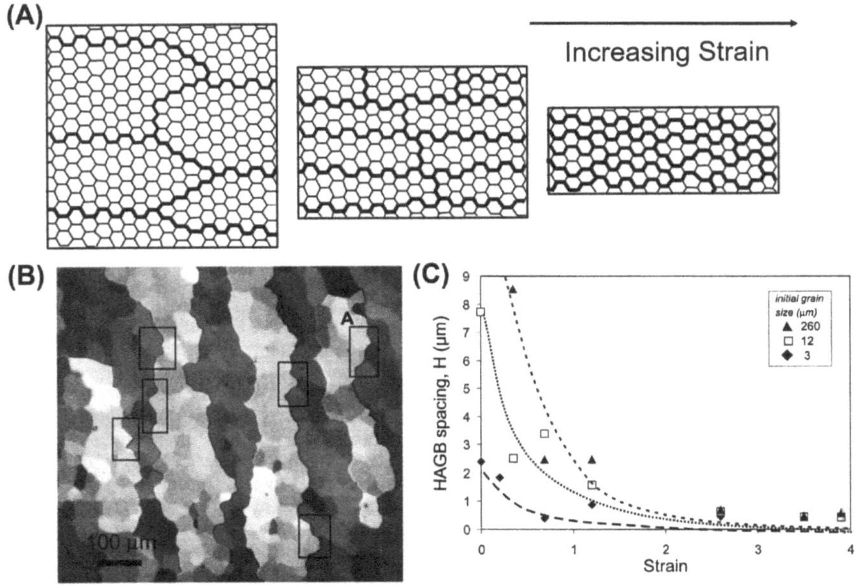

Figure 4.8 (A) Illustration of the geometric dynamic recrystallization mechanism. (B) Observation of serrated grain boundaries in high purity Al. (C) Plot of grain thickness as a function of strain during deformation of AA8006. *Reprinted with permission from Huang, K., Logé, R.E., 2016. A review of dynamic recrystallization phenomena in metallic materials. Materials & Design 111, 548–574. Reprinted with permission from Kassner, M. E., 1989. Large-strain deformation of aluminum single crystals at elevated temperature as a test of the geometric-dynamic-recrystallization concept. Metallurgical Transactions A 20, 2182–2185. Reprinted with permission from Jazaeri, H., Humphreys, F.J., 2004. The transition from discontinuous to continuous recrystallization in some aluminium alloys: II—annealing behaviour. Acta Materialia 52, 3251–3262.*

(face-centered cubic) metals and ~ 25–80 for ionic crystals of the NaCl structure. Eq. (4.11) can be referred to as the Derby relationship for subgrain size. Overall, in both discontinuous and continuous dynamic recrystallization, the steady-state grain size or the subgrain size is strongly dependent on the flow stress. The quantitative relationships are described by Eqs. (4.9) and (4.11), respectively. The experimental fitting of these equations is presented in Fig. 4.9 based on a wide variety of alloys and minerals.

The flow stress is governed by the thermomechanical condition or the Zener–Hollomon parameter. For example, the Sellars and Tegart model predicts $\sigma = \frac{1}{\alpha}\sinh^{-1}\left[\left(\frac{Z}{A}\right)^{\frac{1}{n}}\right]$, where α, n, and A are constants (Tello et al., 2010). In continuous dynamic recrystallization, therefore, the subgrain size

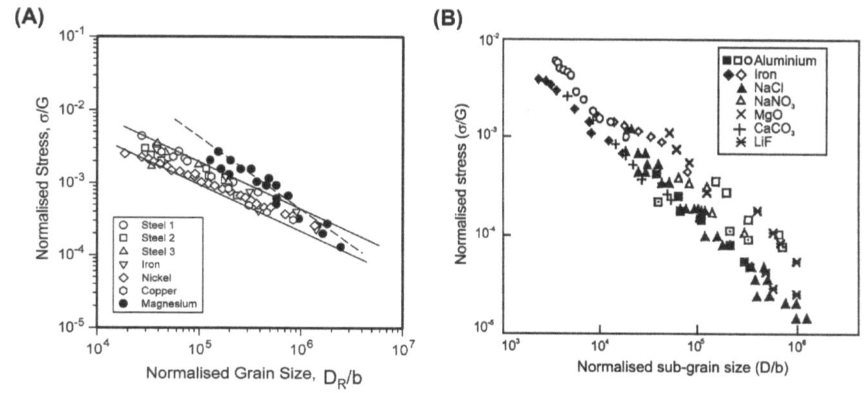

Figure 4.9 Derby's plots showing the correlation of (A) steady-state recrystallized grain size and flow stress and (B) subgrain size and flow stress. Both are based on data from a variety of material systems. *Reprinted with permission from Derby, B., 1991. The dependence of grain size on stress during dynamic recrystallisation. Acta Metallurgica et Materialia 39, 955–962.*

and the recrystallized grain size are determined by the temperature and strain rate. Since subgrains are equiaxed, complete continuous dynamic recrystallization generally leads to fine equiaxed microstructures.

When continuous dynamic recrystallization occurs during hot deformation, the corresponding stress–strain curve is characterized by a plateau after the initial rise. This plateau corresponds to the dynamic equilibrium between work hardening and dynamic recovery such that the dislocation density remains constant. However, this does not necessarily mean that a steady-state microstructure has been achieved. In fact, the grains continue to change shape during the deformation. Fig. 4.10 shows examples of stress–strain curves from hot deformation of Al and ferrite steel. In the Al alloy, the plateau is observed in all cases with different deformation rates. In the ferrite steel, the plateau is observed when the deformation is carried out at higher temperatures. When the temperature drops to 773K, no plateau is observed due to the decrease of the dynamic recovery rate.

Related to continuous dynamic recrystallization, a unique phenomenon occurs in low-temperature severe plastic deformation when the substructures separated by micro-shear bands undergo strain-induced rotation, eventually leading to grain fragmentation (Estrin and Vinogradov, 2013; Pippan et al., 2010; Valiev et al., 2000). This phenomenon has no need for high dynamic recovery rates. However, because there is no high angle boundary migration involved, this phenomenon may also be classified as "continuous"

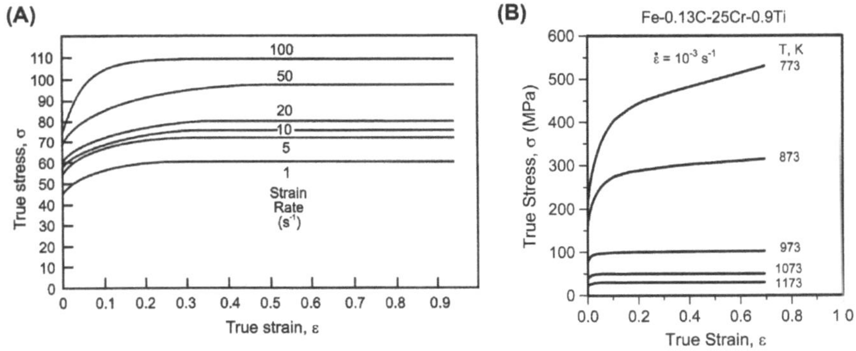

Figure 4.10 Stress–strain curves when continuous dynamic recrystallization occurs. (A) Al–1%Mg deformed at 400°C with various strain rates. (B) Ferrite steel deformed with strain rate of 0.001/s but at different temperatures. *Reprinted with permission from Humphreys, J., Rohrer, G.S., Rollett, A., 2017d. Chapter 13—Hot deformation and dynamic restoration. In: Humphreys, J., Rohrer, G.S., Rollett, A. (Eds.), Recrystallization and Related Annealing Phenomena, 3rd ed. Elsevier, Oxford, pp. 469–508. Reprinted with permission from Sakai, T., Belyakov, A., Kaibyshev, R., Miura, H., Jonas, J.J., 2014. Dynamic and post-dynamic recrystallization under hot, cold and severe plastic deformation conditions. Progress in Materials Science 60, 130–207.*

dynamic microstructure evolution. The two key factors are very large strain (usually > 10) and low temperature ($< 0.5\ T_m$), with which even materials of low or medium stacking fault energy can refine the grain size through continuous evolution.

During thermomechanical processing of a given material, whether continuous or discontinuous dynamic recrystallization happens will depend on the relationship between the dynamic recovery capability and grain boundary mobility. If dynamic recovery is effective but the grain boundary is pinned by second phase particles (like in many engineering alloys), continuous dynamic recrystallization will dominate in hot deformation. If the grain boundary mobility is high but dynamic recovery is inefficient, discontinuous dynamic recrystallization will dominate. If neither the grain boundary mobility nor dynamic recovery rate is high, the material may refine the grain size via grain fragmentation at very large strains.

4.3 Thermomechanical history in additive friction stir deposition

Consider the center material voxels in the feed-rod studied in Chapter 3, Material Flow Phenomena. When the feed material is still within the print head channel, the deformation is minimal, but the temperature is raised due to heat conduction from the end of the feed-rod, which is in dynamic contact with the substrate and generates frictional heat. As the center material voxels move out of the print head channel, they are under torsion and compression, undergoing rapid microstructure evolution in the deposition zone. The material continues to be deformed during steady-state deposition until it is behind the area covered by the moving print head. After that, the material cools in the absence of external strain. Fig. 4.11 schematically illustrates the temperature and strain rate profiles during additive friction stir deposition. The temperature profile is based on the in situ temperature measurement results in Chapter 2, Process Fundamentals. The strain rate profile is, at best, a notional estimation; the evolution of strain rate during additive friction stir deposition is difficult to experimentally measure and is not well understood. That being said, it is generally agreed that the timescale of heating and cooling is considerably larger than deformation.

Therefore, the whole thermomechanical history in additive friction stir deposition may be divided into three stages: preheating (Stage A), hot, rapid, and severe plastic deformation (Stage B), and cooling (Stage C) (Yu, 2022):

4.3.1 Stage A

In Stage A, the deposited material is heated-up within a moderate period (e.g., $\sim 10^1$ s). Phase transformation or processes with fast kinetics may occur. For example, BCC Fe can be transformed to the austenite phase if the peak temperature is above the transformation temperature.

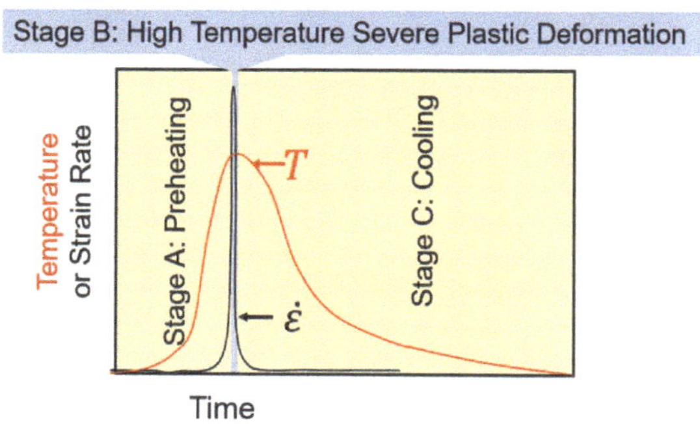

Figure 4.11 Illustration of the thermomechanical history during additive friction stir deposition, highlighting three stages, associated with preheating, high-temperature deformation, and cooling.

4.3.2 Stage B

In Stage B, the deposited material undergoes hot, rapid, and large shear deformation within several seconds. This time scale is several orders of magnitude higher than that of beam-based additive processes (King et al., 2014). Major dynamic microstructure evolution mechanisms include dynamic recovery, discontinuous dynamic recrystallization, and continuous dynamic recrystallization. Simultaneously, dynamic phase transformation may occur together with dislocation and grain structure evolution, such as dissolution or growth of precipitates (Mirihanage et al., 2021).

The maximum homologous temperature T_h in Stage B, which is defined as the peak temperature divided by the melting temperature of the material in Kelvin, can be as high as 0.9. The strain rate in additive friction stir deposition has not been reliably measured. Previous simulations of friction stir welding suggest the strain rate can be in the range of $10^0 - 10^2\ s^{-1}$ (Arora et al., 2009; Sharghi and Farzadi, 2018). The range of strain rate and temperature is close to hot working. From the material flow studies in Chapter 3, Material Flow Phenomena, the total strain is $\sim 10^1$ or above. This is a similar range to severe plastic deformation. Therefore, the dynamic microstructure evolution processes in hot working and severe plastic deformation can all be relevant in additive friction stir deposition. Fig. 4.12A summarizes the processing domains of various thermomechanical processes, including friction stirring, severe plastic deformation, and conventional hot working. Fig. 4.12B compares the temperature profiles of these processes.

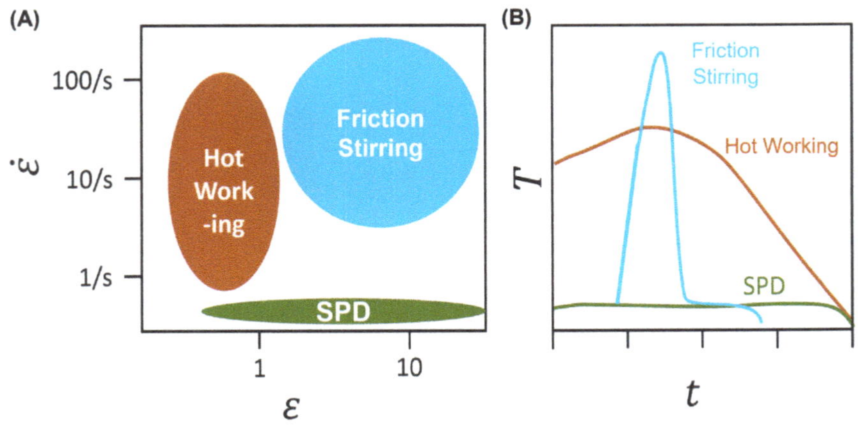

Figure 4.12 Comparison of the thermomechanical conditions among conventional hot working, friction stirring, and severe plastic deformation: (A) strain rate vs. strain; (B) temperature vs. time.

For materials with a high stacking fault energy, additive friction stir deposition likely results in continuous dynamic recrystallization. This is because a high stacking fault energy is associated with enhanced dislocation climb and cross-slip, increasing the dynamic recovery rate. As a result, subgrains form to initiate continuous dynamic recrystallization; at the same time, the dislocation density decreases, reducing the driving force for discontinuous dynamic recrystallization. For a material with lower stacking fault energy, additive friction stir deposition can lead to discontinuous dynamic recrystallization with a moderate strain. If the total strain is large, additive friction stir deposition may result in a refined grain structure via a continuous evolution mechanism like that in severe plastic deformation.

If the stacking fault energy is in the medium range, additive friction stir deposition may result in complex microstructure evolution mechanisms. High T and moderate to high $\dot{\varepsilon}$ typically result in discontinuous dynamic recrystallization, whereas a large ε likely results in continuous evolution. As a result, the extreme thermomechanical conditions of high T, moderate to high $\dot{\varepsilon}$, and large ε in additive friction stir deposition allow for multiple dynamic microstructure evolution regimes to exist, such as the discontinuous recrystallization-dominated regime, continuous recrystallization-dominated regime, or concurrent continuous and discontinuous recrystallization regime. This view has been supported by previous work on microstructure evolution during friction stir welding of pure copper (Mironov et al., 2015), wherein $T_h < 0.5$ leads to continuous dynamic recrystallization and $T_h > 0.5$ leads to discontinuous dynamic recrystallization.

The concurrent continuous and discontinuous recrystallization regime is interesting and underexplored. In this regime, the dislocation removal rate by dynamic recovery is sufficiently high to result in subgrain formation, while the "nucleation" rate in discontinuous dynamic recrystallization is also sufficiently high to produce stable "nuclei" near the high angle grain boundaries that are transformed from the subgrain boundaries. In this special regime, the two dynamic recrystallization mechanisms are interwoven. Fig. 4.13 illustrates the possible final microstructures arising from the competition of these two mechanisms. With an appropriate ratio of the discontinuous to continuous dynamic recrystallization rates, the grains resulting from the two recrystallization mechanisms may be similar in size to form uniform microstructures. Otherwise, we may obtain a hybrid microstructure consisting of a partially recrystallized microstructure from one of the two mechanisms.

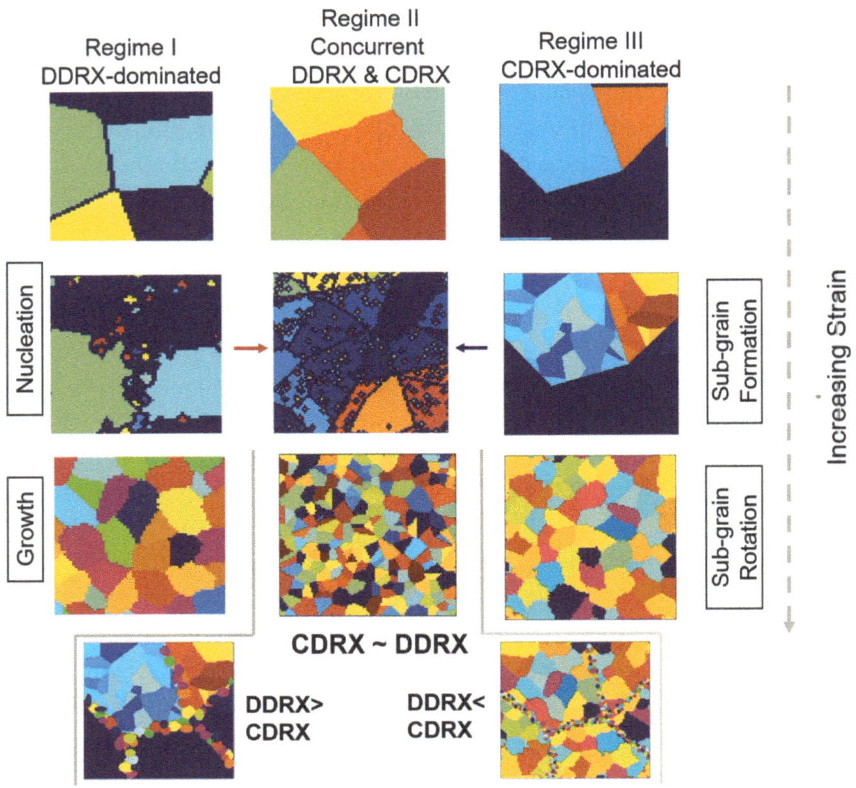

Figure 4.13 Possible microstructures arising from the competition of continuous dynamic recrystallization (CDRX in the figure) and discontinuous dynamic recrystallization (DDRX in the figure).

4.3.3 Stage C

In Stage C, the deposited material cools from elevated temperatures to room temperature within a relatively long period via natural cooling (e.g., $\sim 10^2$ s or above). Depending on the material system, the kinetic processes in this stage can include static recovery, static recrystallization, metadynamic recrystallization (Cho et al., 2001), and curvature-driven grain growth. Metadynamic recrystallization refers to the case of immediate formation of nuclei with small internal stored energy right after deposition, followed by quick growth of these nuclei into the deformed matrix without incubation time. This mechanism is especially possible if dynamic recrystallization is incomplete at the end of Stage B, leaving a high density of dislocations upon the start of Stage C.

Despite the possibility of curvature-driven grain growth, the final grain size is usually still smaller than that of the feed material. This is due to kinetic reasons: grain boundary mobility has an Arrhenius relationship with temperature, and the time spent at particularly high temperatures in Stage C is generally short. However, if the thermal conductivity of the material is low, grain growth can be substantial during cooling. Moreover, re-precipitation and precipitate coarsening may occur in Stage C, resulting in different precipitate attributes in the as-printed material than the feed material.

4.4 Characteristics of the resulting microstructures by additive friction stir deposition

4.4.1 High stacking fault energy materials: Al and Mg

In this section, we provide a survey of the representative microstructures achieved in additive friction stir deposition. We start from Al alloys, which are characterized by high stacking fault energy ($> 160\ mJ/m^2$) and excellent dynamic recovery capability. AA2219 is a wrought Al—Cu alloy frequently utilized in aerospace applications because of its excellent fracture toughness, weldability, and resistance to stress corrosion cracking (Srinivasan et al., 2010). Fig. 4.14 compares the microstructure of the feed material made of AA2219-T851 in (A) and the microstructure of the as-deposited material in (B) (Rivera et al., 2018). The former exhibits an average grain size of about 30 μm, whereas the latter shows a measured average grain size of roughly 2.5 μm, revealing significant grain refinement. The 3D EBSD (electron backscatter diffraction) representation of the feedstock material shows that the grains in the feed-rod are large and elongated in one direction, whereas the grains in the as-deposited AA2219 are equiaxed. In the as-printed material, a significant number of low angle boundaries are visible, implying that new high angle grain boundaries are generated from low angle boundaries as a result of severe deformation.

Fig. 4.15 shows the microstructure of an Mg alloy, WE43, before and after additive friction stir deposition (Williams et al., 2021). WE43 is a precipitation-hardenable Mg alloy with 3.7—4.3 weight percentage of Y, 2.4—4.4 weight percentage of Nd, and at least 0.4 weight percentage of Zr (Jahedi et al., 2018). Biocompatible and nonflammable, WE43 has recently attracted a lot of attention for temporary biomedical and armor applications (Cho et al., 2009). The feed material is made of WE43-T5,

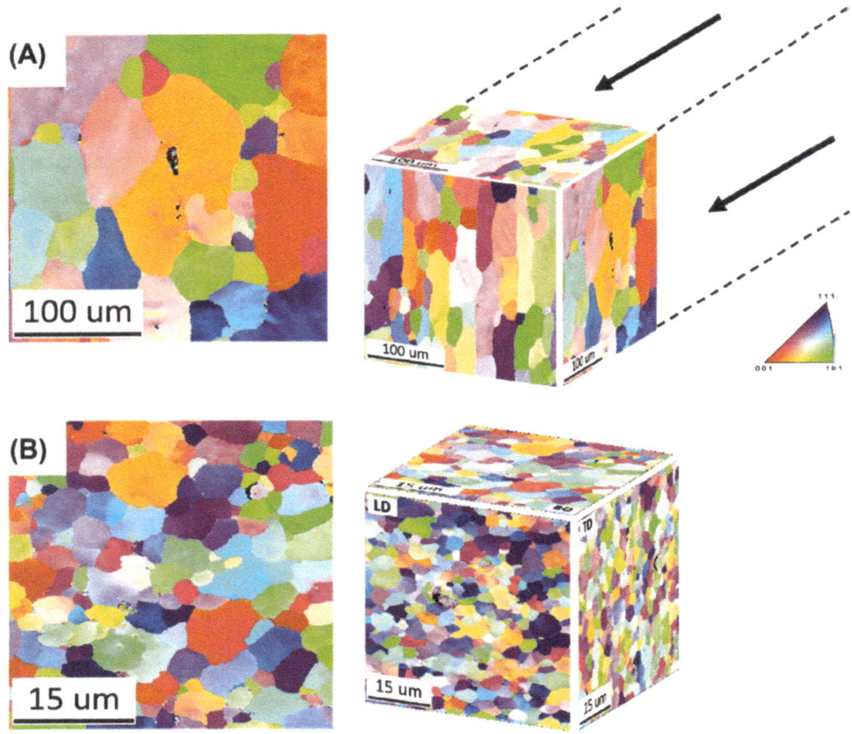

Figure 4.14 Inverse pole figure orientation maps of (A) a feedstock material made of AA2219-T851 and (B) the as-printed material. On the right side, the 3D EBSD results are presented for the two cases. *Reprinted with permission from Rivera, O.G., Allison, P.G., Brewer, L.N., Rodriguez, O.L., Jordon, J.B., Liu, T., et al., 2018. Influence of texture and grain refinement on the mechanical behavior of AA2219 fabricated by high shear solid state material deposition. Materials Science and Engineering: A 724, 547–558.*

showing a large grain size of 45 μm with a wide size distribution. After deposition, the grain size is more uniform, with the average grain size drastically reduced to 2.7 μm. The feed material shows a strong basal plane <0001> texture, whereas the as-printed material exhibits a random crystallographic orientation distribution. This microstructure change can be attributed to dynamic recrystallization through the severe plastic deformation caused by the tool.

With high stacking fault energy, Al and Mg alloys likely undergo continuous dynamic recrystallization to result in homogeneous refined grain structures—as observed in Figs. 4.14 and 4.15. Next, we examine the resulting microstructures of two materials with low or medium stacking fault energy, a Ni alloy, and an austenite steel.

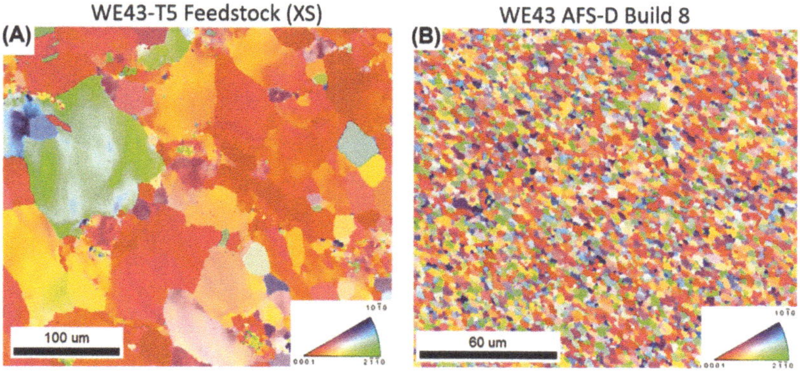

Figure 4.15 Representative EBSD maps of (A) the feed material made of WE43-T5 and (B) the as-printed material. *Reprinted with permission from Williams, M.B., Robinson, T.W., Williamson, C.J., Kinser, R.P., Ashmore, N.A., Allison, P.G., et al., 2021. Elucidating the effect of additive friction stir deposition on the resulting microstructure and mechanical properties of magnesium alloy WE43. Metals 11, 1739.*

4.4.2 Low (to medium) stacking fault energy materials: Inconel 625 and 316 L stainless steel

Inconel 625 is a Ni-based superalloy with good resistance against corrosion and oxidation. It is characterized by low (to medium) stacking fault energy (Gao et al., 2022). Fig. 4.16 presents the grain structure of the as-printed Inconel 625 by additive friction stir deposition, involving two locations along the build direction (Rivera et al., 2017). Fig. 4.16A and B correspond to the top portion of the deposition with a grain size on the order of 1 μm. Fig. 4.16C and D correspond to the middle portion of the deposition with a grain size below 300 nm. These numbers contrast with the feed material, which shows a large grain size of 30 μm. Remarkably, the middle portion of the deposition shows two orders of magnitude reduction in grain size, which is unlikely achieved via discontinuous recrystallization. The wide presence of low angle grain boundaries in the as-printed microstructure suggests that the recrystallization is related to substructure formation.

316 L stainless steel is a marine grade stainless steel, with primary alloying elements including chromium, nickel, molybdenum, and a small (<1%) quantity of silicon. Its stacking fault energy is low (Woo et al., 2020). Recently, additive friction stir deposition of 316 L stainless steel has been carried out (Beladi et al., 2021). The feed material consists of equiaxed grains with an average size of 32 $\pm$ 0.9 μm and a high fraction

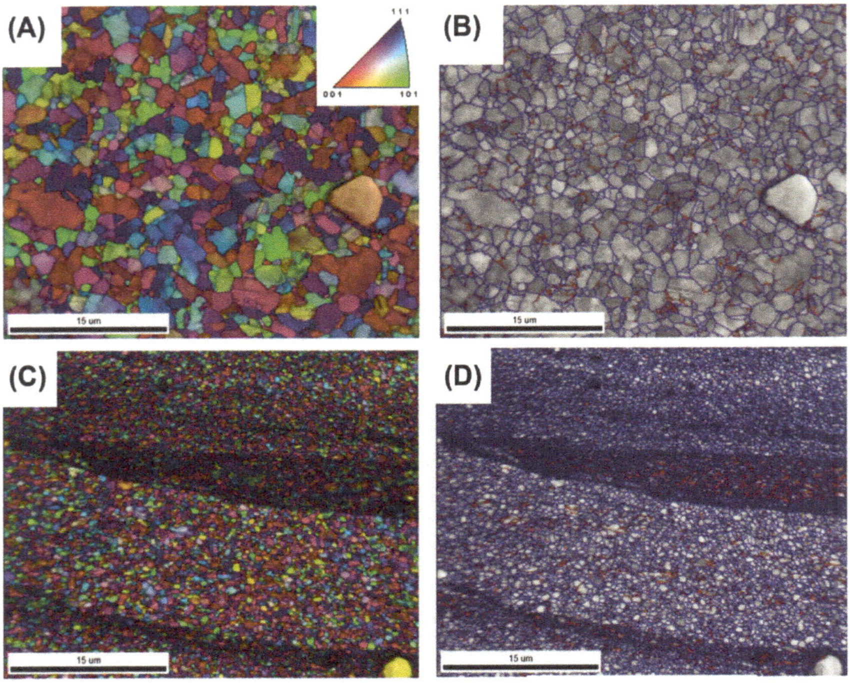

Figure 4.16 The grain structure of the as-printed Inconel 625, including EBSD maps in (A) and (C) and boundary rotation maps in (B) and (D). (A) and (B) are from the top portion of the deposition; (C) and (D) are from the middle portion of the deposition. *Reprinted with permission from Rivera, O.G., Allison, P.G., Jordon, J.B., Rodriguez, O. L., Brewer, L.N., McClelland, Z., et al., 2017. Microstructures and mechanical behavior of Inconel 625 fabricated by solid-state additive manufacturing. Materials Science and Engineering: A 694, 1–9.*

of annealing Σ3 twin boundaries. The length of Σ3 boundaries accounts for 46% of the total boundary length, including ~ 41% coherent Σ3 boundaries. The grain size in the deposition is drastically reduced from the feed-rod. In the last layer of the deposited 316 L stainless steel, the average grain size is 0.66 ± 0.05 μm. At the same time, there is a significant reduction in the relative areas of Σ3 annealing twin boundaries due to the distortion by severe plastic deformation. A number of incomplete high angle boundaries are observed to be frequently connected to low angle boundaries. This suggests that the high angle boundaries are formed through the increase of misorientation angle from low angle boundaries. In addition, the overall texture shows strong shear components. These observations all indicate that the tiny, equiaxed grains in the as-printed material are formed by continuous dynamic recrystallization through progressive rotation of subgrains.

Both Inconel 625 and 316 L stainless steel are characterized by low or medium stacking fault energy, for which dynamic recovery should not dominate. However, the microstructure characterization results suggest an important role played by continuous evolution and substructure formation. Considering the very large strain in deposition, the underlying microstructure evolution mechanism may be related to grain subdivision typically seen in severe plastic deformation (Hughes and Hansen, 1997), which is also a type of continuous evolution.

4.5 Dynamic microstructure evolution along the flow path of an Al–Cu alloy

In the previous section, we simply examine the final microstructure of the as-printed material and compare it with the feed-material. To better understand the dynamic microstructure evolution mechanisms, in this section we examine the microstructure evolution history of the printed material—by exploring the microstructures along the flow path during deposition. This is based on the tracer experiments elaborated in Chapter 3, Material Flow Phenomena, which unravel the flow paths of the center and edge voxels in the feed material.

As shown in Fig. 4.17, the feed-rod is still connected to the deposition track as a result of stop action. $Z = 0$ mm corresponds to the initial substrate surface level; $Z = 1$ mm corresponds to the bottom of the print head. Fig. 4.17A and B show the sites of microstructure analysis along the flow paths for the center and edge tracers during initial material feeding. So far, this type of experiment has been only carried out for Al alloys. In this case, the tracer is made of an Al-Cu alloy, AA2024; the matrix is made of an Al-Mg-Si alloy, AA6061. AA2024 and AA6061 have comparable mechanical properties at elevated temperatures, so they can be deposited simultaneously without significant tracer fragmentation. The compositions of the two alloys differ by ~ 4%–5% in Cu, allowing them to be distinguished in X-ray imaging.

4.5.1 Microstructure characterization along the flow path of the center tracer

For the center tracer, five sites are analyzed (see Fig. 4.17A). The basic inverse pole figure maps of these areas are already presented in Chapter 3,

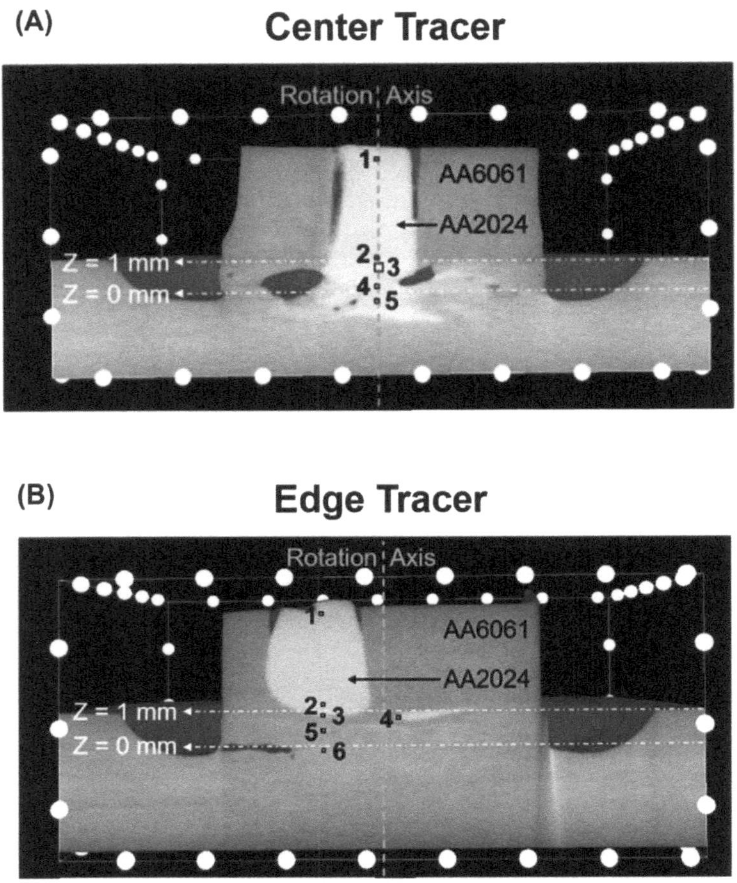

Figure 4.17 Sites of microstructure analysis along the flow path of (A) center tracer and (B) edge tracer. *Reprinted with permission from Perry, M.E.J., Rauch, H.A., Griffiths, R.J., Garcia, D., Sietins, J.M., Zhu, Y., et al., 2021. Tracing plastic deformation path and concurrent grain refinement during additive friction stir deposition. Materialia 18, 101159.*

Material Flow Phenomena, Section 3.5. Here, a detailed analysis of the microstructure evolution is performed along these sites. Fig. 4.18 compares the misorientation angle distributions, misorientation maps, and GAM (grain average misorientation) maps for Areas 1–5. The GAM value is a weighted average of the local misorientation in each neighborhood of EBSD pixels confined by specific grain boundaries. It allows for a quantitative comparison of the intragranular orientation gradients. This is in contrast to KAM (kernel average misorientation) maps, which depict the misalignment of each pixel with its immediate neighbors. The

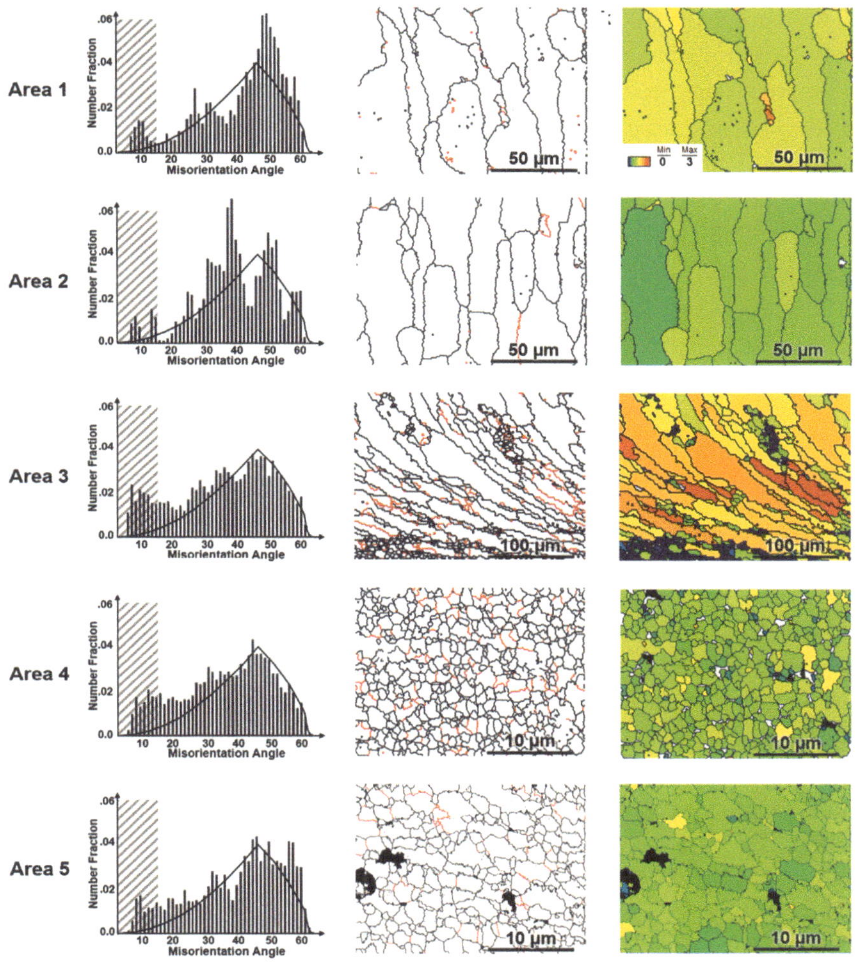

Figure 4.18 Microstructural characteristics of Areas 1–5 in the center tracer are compared in detail. Left column: misorientation angle distribution. The black line depicts the Mackenzie distribution for a randomly textured polycrystal. Center column: Misorientation maps with high angle boundaries in black and low angle boundaries in red. Right column: Grain average misorientation (0–3 degrees) maps. *Reprinted with permission from Perry, M.E.J., Rauch, H.A., Griffiths, R.J., Garcia, D., Sietins, J.M., Zhu, Y., et al., 2021. Tracing plastic deformation path and concurrent grain refinement during additive friction stir deposition. Materialia 18, 101159.*

microstructure of Areas 1 and 2 is close to the original AA2024 tracer. Significant microstructure evolution is observed immediately after the center tracer comes out of the print head channel: in Area 3, the grains become elongated and thinned with a width of 3–5 μm. In Areas 4 and

5, fine, equiaxed grains are formed. The distance from Area 2 to Area 4 is only 840 μm, but the microstructure changes dramatically, from vertically elongated, large grains with a size of 25–30 μm to small, equiaxed grains with a size of 1.36 μm.

Areas 1 and 2 have a small fraction of low angle boundaries, but the low angle boundary fraction increases significantly from Area 2 (5.9%) to Area 3 (14.2%). This is accompanied by a rise in GAM, indicating the emergence of substructures. In Area 4, there are still a substantial number of low angle boundaries, but the GAM is significantly reduced. Although both Area 4 to Area 5 are characterized by fine, equiaxed microstructures, the low angle boundary fraction reduces from 11.5% to 8.9%.

As discussed in earlier sections, recrystallization, which can be continuous or discontinuous, is the origin of the considerable grain size reduction observed here. Discontinuous recrystallization is characterized by "nucleation" and growth of dislocation-free grains, which often leads to heterogeneous microstructures prior to complete recrystallization. This is in opposition to what has been observed. Continuous dynamic recrystallization is characterized in part by progressive accumulation, annihilation, and reorganization of dislocations due to concurrent deformation and dynamic recovery, resulting in the creation of subgrains. Misorientation increases as deformation progresses; subgrain boundaries transform to low angle boundaries and then high angle boundaries. The observed microstructure evolution in the center tracer is consistent with this mechanism:

- With the slight deformation of the center tracer inside the print head, the microstructure of the original AA2024 tracer is largely preserved in Areas 1 and 2, showing most high angle boundaries.
- When the feed-rod is no longer constrained by the print head, shear deformation occurs, resulting in the development of low angle boundaries due to dynamic recovery in Area 3.
- With further deformation, the thin, elongated grains have evolved into small, equiaxed grains in Area 4. The low angle boundaries are seen to entirely cut across individual grains.
- As the misorientation increases to more than 15 degrees, these low angle boundaries transform themselves into high angle boundaries. The observation in Area 5 supports this view.

As illustrated in Fig. 4.19A, a closer look at the thin, curved grains in the vicinity of Area 3 reveals that the continuous dynamic recrystallization in the center tracer proceeds via the mechanism of geometric dynamic recrystallization. Geometric dynamic recrystallization is known for producing thin,

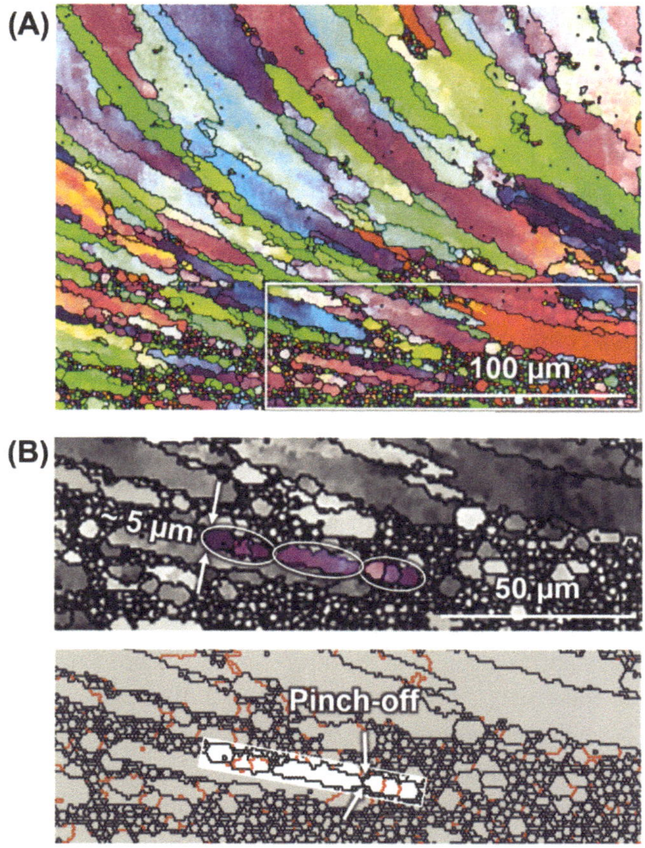

Figure 4.19 Direct evidence of geometric dynamic recrystallization observed in the center tracer. (A) An IPF map showing the transition of elongated grains to equiaxed grains. (B) The highlight of seven grains with similar crystallographic orientations that are formed by geometric dynamic recrystallization. *Reprinted with permission from Perry, M.E.J., Rauch, H.A., Griffiths, R.J., Garcia, D., Sietins, J.M., Zhu, Y., et al., 2021. Tracing plastic deformation path and concurrent grain refinement during additive friction stir deposition. Materialia 18, 101159.*

elongated grains with serrated high angle boundaries that pinch off to form equiaxed, subgrain-sized grains. In Fig. 4.19B, seven grains with similar crystallographic orientations are demonstrated to be from the same original elongated grain, which has been sheared to a thickness comparable to the subgrain size. There are three small, equiaxed grains on the left and three on the right. These grains are separated by low angle boundaries, and their size is similar to the Area 5 grains. The middle grain, on the other hand, is still elongated, with serrated high angle boundaries and a color gradient. With additional deformation, the middle grain is expected to pinch off into several equiaxed grains.

After Area 5, the center tracer is subjected to further deformation until it completely falls behind the moving print head. This stage is termed steady-state deposition. The final grain structure after steady-state deposition is found to be quite similar to that observed in Area 5, also featuring a fine, equiaxed grain structure with an average grain size below 3 μm. This suggests that the grain size is not further reduced during steady-state deposition. However, continuous dynamic recrystallization keeps occurring during steady-state deposition because the material is still subjected to deformation at elevated temperatures. This seeming discrepancy is caused by the dynamic equilibrium reached by various microstructure evolution mechanisms. As a result, the grain size does not show a salient change during steady-state deposition. This phenomenon is elaborated in Section 4.5.3.

4.5.2 Microstructure characterization along the flow path of the edge tracer

For the edge tracer, six sites are analyzed (recall Fig. 4.17B). Fig. 4.20 compares the misorientation angle distributions, misorientation maps, and GAM maps among these sites. From Area 1 to Area 2, there is an increase in the low angle boundary fraction from 2.6% to 9.7%, indicating the start of substructure formation inside the print head channel. Moving down along the flow path, the low angle boundary fraction maintains a relatively high level until Area 5 (17%), but substantially decreases in Area 6 (10.2%). From Area 1 to Area 4, moderate microstructure evolution is observed: the original vertically elongated grains gradually become horizontally elongated and the grain size is slightly reduced from ~ 30 to 15–20 μm. In Area 5 and Area 6, significant microstructure changes are observed, in which the large elongated grain structures have been replaced by fine, equiaxed microstructures with the grain size ~ 1–2 μm. These areas correspond to the thinner spiral regions seen in the X-ray computed tomography analysis in Chapter 3, Material Flow Phenomena. Considering these factors together, it can be concluded that the edge tracer of AA2024 also undergoes continuous dynamic recrystallization during additive friction stir deposition. In Area 6, the AA2024 tracer is close to being fully recrystallized.

After Area 6, the edge tracer is subjected to further deformation in steady-state deposition. Similar to the case of the center tracer, the grain structure of the edge tracer after steady-state deposition is found to be quite similar to that observed in Area 6, suggesting no further grain size

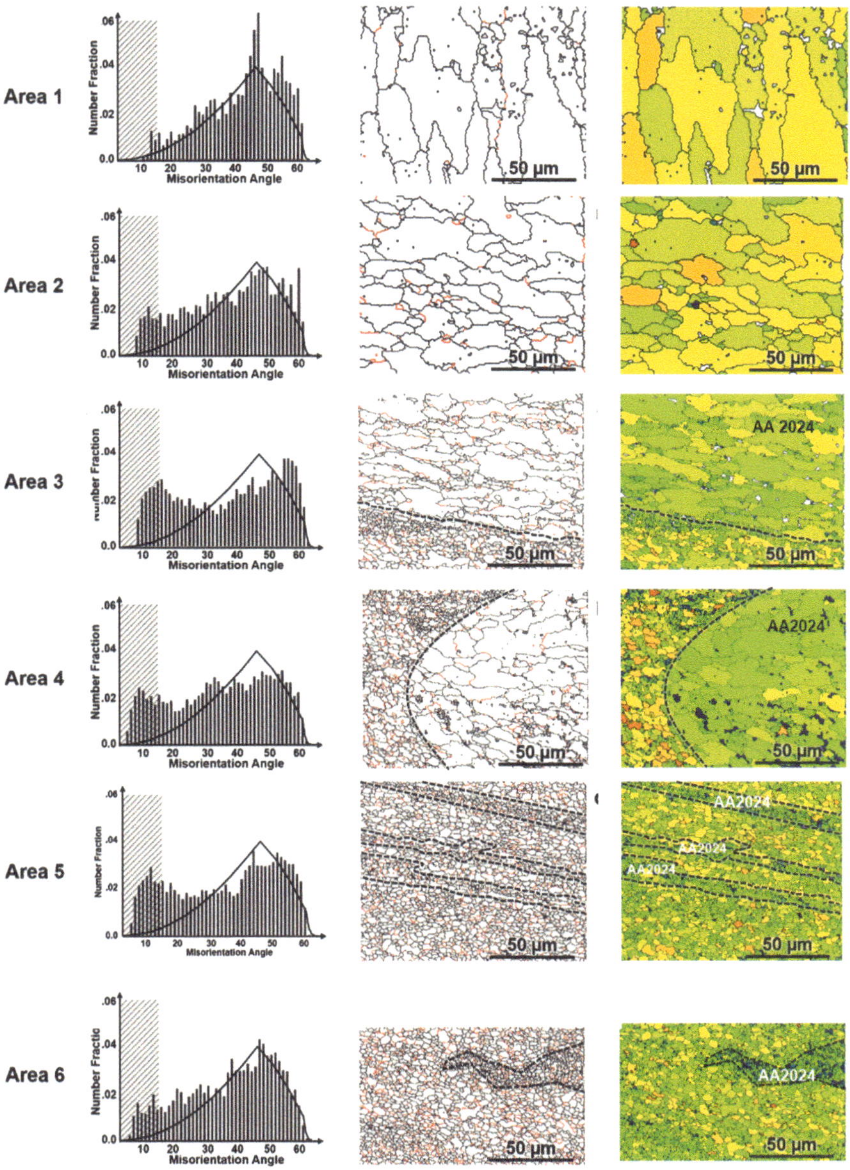

Figure 4.20 Microstructural characteristics of Areas 1–6 on the edge tracer are compared in detail. Left column: Misorientation angle distribution. The black solid line depicts the Mackenzie distribution for a randomly textured polycrystal. Center column: Misorientation maps with high angle boundaries in black and low angle boundaries in red. The black dashed lines in Areas 3–6 depict the boundaries between the AA2024 tracer and AA6061 matrix. Right column: Grain average misorientation (0–3 degrees) maps. *Reprinted with permission from Perry, M.E.J., Rauch, H.A., Griffiths, R.J., Garcia, D., Sietins, J.M., Zhu, Y., et al., 2021. Tracing plastic deformation path and concurrent grain refinement during additive friction stir deposition. Materialia 18, 101159.*

reduction despite deformation and continuous dynamic recrystallization. This is also a result of the dynamic equilibrium reached by multiple microstructure evolution mechanisms. By examining the misorientation maps (center column in Fig. 4.20) of Areas 2–4, low angle boundaries are seen to preferentially form near the grain boundaries compared to the grain interior. This indicates that under these deformation circumstances, subgrains are not uniformly distributed inside each grain. Instead, they are initially formed near grain boundaries, possibly via progressive lattice rotation as observed in the hot deformation of Al–Mg and Al–Zn alloys (Drury and Humphreys, 1986; Gardner and Grimes, 1979).

4.5.3 Quantification of the overall trend

Based on the results of both center and edge tracers, Fig. 4.21 summarizes the microstructure evolution trend along the flow path. Generally, the grain size is significantly reduced shortly after the material leaves the print head, which is accompanied by an increase of the low angle boundary fraction. This suggests the important role played by the continuous dynamic recrystallization that proceeds via substructure formation. After that, the grain size stays in the range of 1–3 μm, but the low angle boundary fraction eventually decreases as the recrystallization nears completion. This is caused by more low angle boundaries being transformed to high angle boundaries upon further deformation.

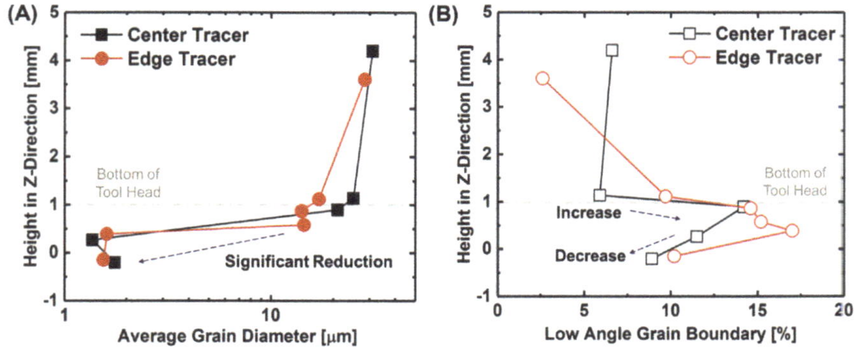

Figure 4.21 Plots of (A) grain size and (B) low angle boundary fraction against Z-height during initial material feeding, showing data from both the center and edge tracers. *Reprinted with permission from Perry, M.E.J., Rauch, H.A., Griffiths, R.J., Garcia, D., Sietins, J.M., Zhu, Y., et al., 2021. Tracing plastic deformation path and concurrent grain refinement during additive friction stir deposition. Materialia 18, 101159.*

By now, the microstructure characterization along the flow path has offered a complete physics picture of co-deformation and dynamic microstructure development. During initial material feeding, the grain size of the feed material rapidly decreases from tens of microns to a few microns. During steady-state deposition, the feed material develops a high degree of deformation as elaborated in Chapter 3, Material Flow Phenomena. Microstructure analysis, on the other hand, demonstrates no further grain refinement. By simultaneously evaluating the strain magnitude from material flow and microstructure information, Fig. 4.22A depicts the grain size for the center tracer as a function of the cumulative strain during initial material feeding (Perry et al., 2021). Overall, the grain size initially decreases rapidly with the cumulative strain, but it generally stays within the range of 1–3 μm without further reduction once a critical strain level of ~ 0.63 is reached.

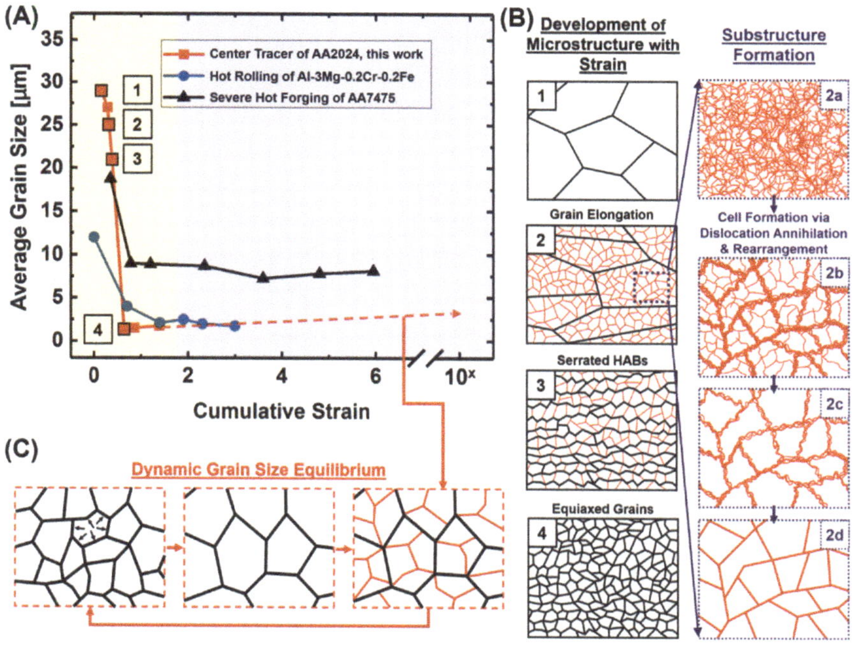

Figure 4.22 (A) Grain size vs. strain plots, including the results from this study for AA2024 by additive friction stir deposition, as well as the results from hot rolling of Al-3Mg-0.2Cr-0.2Fe and severe hot forging of AA7475. Dynamic microstructure evolution mechanisms in additive friction stir deposition of AA2024 are illustrated in (B) for initial material feeding and in (C) for steady-state deposition. *Reprinted with permission from Perry, M.E.J., Rauch, H.A., Griffiths, R.J., Garcia, D., Sietins, J.M., Zhu, Y., et al., 2021. Tracing plastic deformation path and concurrent grain refinement during additive friction stir deposition. Materialia 18, 101159.*

The observed linkage between the cumulative strain and grain size is consistent with the geometric dynamic recrystallization mechanism articulated earlier in the chapter. The increase of shear strain has two consequences. First, more dislocations are created, and subgrains form due to dynamic recovery. Second, the grains become elongated and thinner, and the grain size decreases. When a certain strain level is achieved, the high angle boundary spacing for the elongated grains approaches the subgrain size. This is followed by grain boundary pinch-off, yielding an approximately equiaxed microstructure. After the equiaxed microstructure is formed, further deformation will continue to cause dislocation generation and subgrain formation due to dynamic recovery. As the subgrain boundaries are transformed to low angle boundaries and ultimately to high angle boundaries, the grain size tends to be reduced. There is also a tendency for grain size increase due to migration of the high angle boundaries. When the grain size exceeds the subgrain size, substructure pinning limits the high angle boundary migration; when the grain size is less than the subgrain size, high angle boundary migration is accelerated. It is possible to eventually achieve a dynamic equilibrium between such opposing processes, maintaining the grain size at the subgrain size level. The subgrain size is a function of the temperature and strain rate rather than the total strain. This is why the grain size remains within a narrow range of a few microns despite the development of large cumulative strain during steady-state deposition. Such physics pictures are illustrated in Fig. 4.22B and C.

In addition to the AA2024 tracer printed by additive friction stir deposition, the grain size vs. strain is plotted for hot rolling of Al-3Mg-0.2Cr-0.2Fe (Gholinia et al., 2002) and severe hot forging of AA7475 (Sitdikov et al., 2005) in Fig. 4.22A. A similar pattern is observed: the grain size drops rapidly at first, followed by a plateau after the accumulated strain exceeds a threshold level. This is expected for hot rolling of Al-3Mg-0.2Cr-0.2Fe, wherein geometric dynamic recrystallization has also been discovered as a significant mechanism for grain size reduction. For severe hot forging of AA7475, the grain size is reduced through micro-shear band-based grain subdivision. This mechanism belongs to continuous recrystallization as it shares some similar features with geometric dynamic recrystallization: the final grain size is defined by the minimum spacing of micro-shear bands and is weakly dependent on the applied strain once a critical strain value is reached. Generally, the grain size trend in Fig. 4.22A can be viewed as a typical characteristic of dynamic microstructure evolution via continuous recrystallization.

4.6 Processing-microstructure linkages of Al-Mg-Si and Cu

The kinetics of microstructure evolution in additive friction stir deposition is dictated by the history of temperature and strain rate, which is controlled via the key processing parameters, such as the print head rotation rate Ω, in-plane travel velocity V, and material feed rate F. The heat generation and strain rate are normally controlled by the rotation rate, while the travel velocity defines the amount of time available for material deformation. The amount of deposited material for a given time is determined by F. The ratio Ω/V is related to the total strain development via print head-material interaction. Fig. 4.23 illustrates a process map for additive friction stir deposition, in which a feasible process window is defined by the process variable values to yield fully-dense, high-quality deposition (Griffiths et al., 2021). Intense heating beyond the feasible process window, such as at very high temperatures, can lead to distortion of the print head, cause feed material clogging inside the hollow channel, or melt the feed material in extreme situations. With a low rotation rate and high travel velocity, heat generation is insufficient to soften the material, preventing appropriate plastic

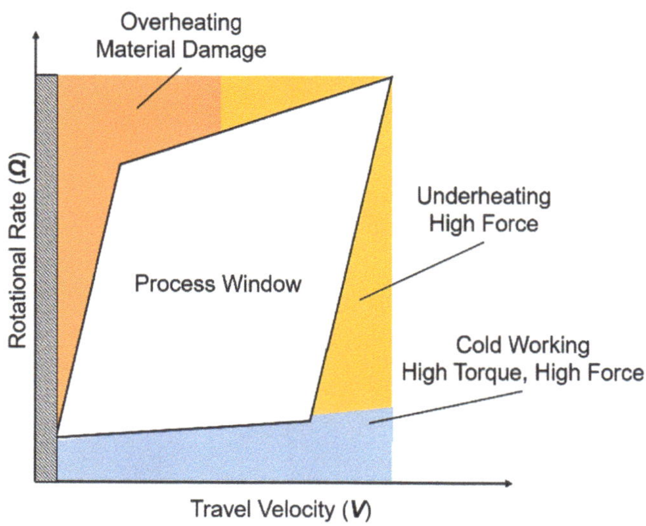

Figure 4.23 A schematic of the process map for additive friction stir deposition, showing a feasible process window defined by the rotation rate and travel velocity of the print head. *Reprinted with permission from Griffiths, R.J., Garcia, D., Song, J., Vasudevan, V.K., Steiner, M.A., Cai, W., et al., 2021. Solid-state additive manufacturing of aluminum and copper using additive friction stir deposition: Process-microstructure linkages. Materialia 15, 100967.*

deformation for deposition. With a low feed rate and high travel velocity, inadequate material is supplied for deposition, possibly resulting in poor surface quality and porosity development.

In practice, the feed rate to travel velocity ratio is often fixed at an optimal value to ensure good deposition quality. By controlling the print head rotation rate and travel velocity, the strain rate and temperature history can be robustly tuned, and various final microstructures may emerge. Another important process parameter to control is the layer thickness, which can be set by the initial gap between the print head and substrate. Layer thickness determines the gradient of the rotation rate through the deposition thickness, therefore significantly influencing the thermomechanical condition in printing.

4.6.1 Microstructure characterization of Al–Mg–Si printed at various conditions

Fig. 4.24 shows the microstructure of an Al-Mg-Si alloy (AA6061) deposited at three processing conditions, along with the microstructure of the feed material made of AA6061-T6. The rotation rate changes from 200 to 600 RPM, and the travel velocity changes from 1 to 3 mm/s. The feed

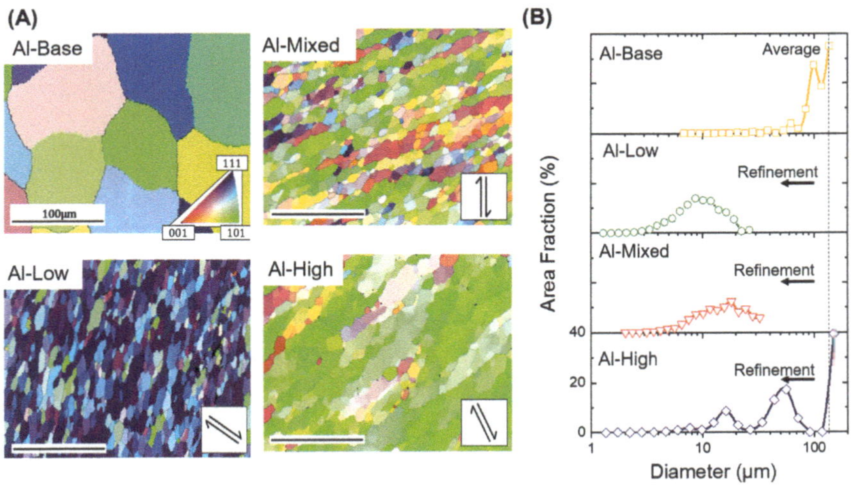

Figure 4.24 Microstructure characterization of Al–Mg–Si deposited under three conditions along with the feed material, including (A) inverse pole figure maps and (B) grain size distribution. *Reprinted with permission from Griffiths, R.J., Garcia, D., Song, J., Vasudevan, V.K., Steiner, M.A., Cai, W., et al., 2021. Solid-state additive manufacturing of aluminum and copper using additive friction stir deposition: Process-microstructure linkages. Materialia 15, 100967.*

rate is kept at 1/3 of the travel velocity for all cases. Table 4.1 details the processing parameters of the deposited Al–Mg–Si. For the sample notation, "Low" means that the print head rotation rate and travel velocity are both low. "High" means that the rotation rate and travel velocity are both high. "Mixed" means that the rotation rate is high, but the travel velocity is low.

Fig. 4.24A presents the inverse pole figure maps of the feed material along with the three deposits of Al–Mg–Si. For the latter, each map has been rotated to better present a particular shear texture, with the corresponding shear direction annotated on the bottom right. Because substructure creation is expected, the minimum misorientation establishing grain boundaries has been fixed at 2 degrees. For definition purposes, therefore, low angle boundaries are boundaries having misorientation angles in the range of 2–15 degrees; high angle boundaries are boundaries having misorientation greater than 15 degrees. Fig. 4.24B compares the grain size distribution of these samples. The feed material is originally processed through drawing with an additional step of T6 heat treatment, which effectively removes the local misorientations. As a result, the feed material exhibits large grains with no intragranular misorientation. The three Al–Mg–Si deposits all demonstrate grain refinement from the feed material, most notably in Al-Low and Al-Mixed, where the area-weighted average grain size decreases to 10 and 16 μm from 113 μm in the feed material. In Al-High, the inverse pole figure map clearly reveals significant deformation, with the large grains greatly flattened and elongated. Al-Mixed also exhibits flattened grains but at a much smaller scale, about the thickness of 1–2 subgrains. Substructure development is evident in all the Al–Mg–Si deposits. However, Al-Low has the smallest average value and standard deviation of the grain size.

According to the misorientation distribution in Fig. 4.25A, all of the Al–Mg–Si deposits contain a significant number of low angle boundaries, with the fraction being 32.2%, 42.7%, and 77.4% for Al-Low, Al-Mixed,

Table 4.1 Processing condition for additive friction stir deposition of Al–Mg–Si.

Sample	Rotational rate Ω (revolutions per minute)	Travel velocity V (mm/s)	Feed rate F (mm/s)
Al-Low	200	1	0.33
Al-Mixed	600	1	0.33
Al-High	600	3	1

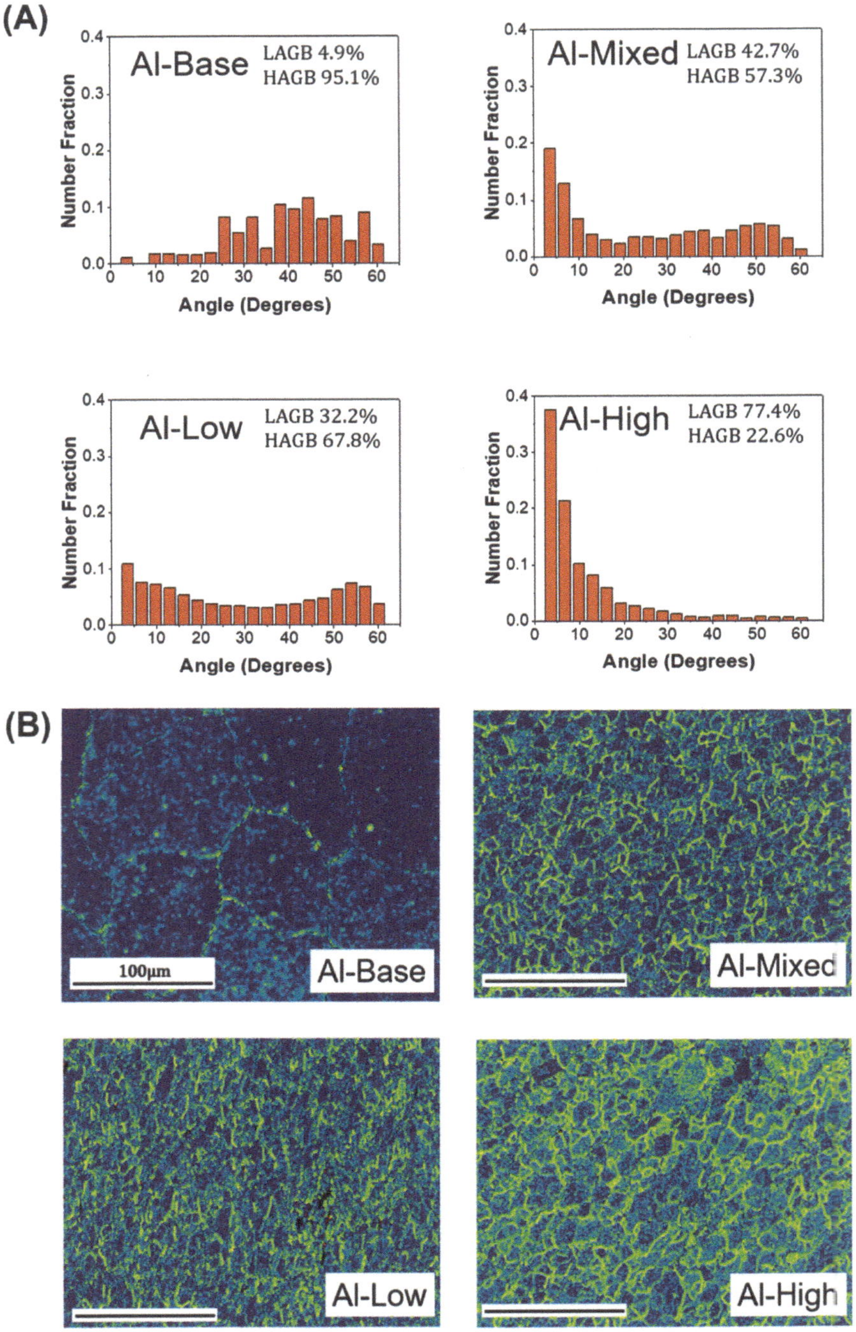

Figure 4.25 (A) Grain boundary misorientation distributions and (B) KAM maps of the three Al–Mg–Si deposits and the feed material. *Reprinted with permission from Griffiths, R.J., Garcia, D., Song, J., Vasudevan, V.K., Steiner, M.A., Cai, W., et al., 2021. Solid-state additive manufacturing of aluminum and copper using additive friction stir deposition: Process-microstructure linkages. Materialia 15, 100967.*

and Al-High, respectively. The relationship between grain refinement and low angle boundary fraction indicates that recovery and continuous dynamic recrystallization are responsible for microstructure evolution during additive friction stir deposition of Al–Mg–Si.

The observed recrystallization process involves subgrain creation and subsequent rotation. Al-Low has the most complete recrystallization, whereas Al-High has the least complete. The shear direction differs among Al-Low, Al-Mixed, and Al-High, although it always appears to be ~ 45 degrees from the elongation axis of the grains. Fig. 4.25B compares the KAM maps among different samples. Inside the grains of the Al–Mg–Si feed material, there is only a minor amount of intragranular misorientation. After printing, the KAM maps reveal a somewhat "organized" spatial distribution of local misorientation, which is particularly visible along low angle boundaries or subgrain boundaries formed by dislocation annihilation and rearrangement.

Fig. 4.26 shows the pole figures and orientation distribution functions for the as-printed Al–Mg–Si at three conditions, all of which show evidence of shear texture. The orientation of each pole figure map has been rotated to lock onto the C-type shear component. In deformation processing such as torsion (Montheillet et al., 1984) and friction stir welding or processing (Fonda et al., 2008), several types of shear texture have been

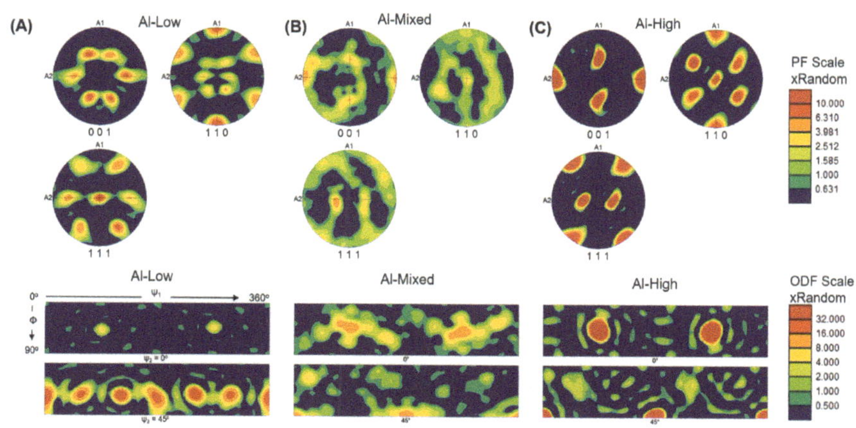

Figure 4.26 Texture of the deposited Al–Mg–Si represented by pole figures and orientation distribution functions, including (A) Al-Low, (B) Al-Mixed, and (C) Al-High. *Reprinted with permission from Griffiths, R.J., Garcia, D., Song, J., Vasudevan, V.K., Steiner, M.A., Cai, W., et al., 2021. Solid-state additive manufacturing of aluminum and copper using additive friction stir deposition: Process-microstructure linkages. Materialia 15, 100967.*

identified in FCC metals, such as Type A, Type B, and Type C. The ideal shear texture types for FCC metals and their coordinate systems can be found in Fonda and Knipling, 2011. Both Al-Mixed and Al-High have a strong C-type shear texture, whereas Al-Low has a weak C-type shear texture and a high $B/\bar{B}$ type. The type of shear texture is known to be related to the applied strain values, with A-type forming first (shear strain ~ 1), followed by B-type (shear strain $\sim 3-6$), and finally C-type (shear $\sim > 10$) (Canova et al., 1984). The observation thus suggests a lower total strain associated with Al-Low than the other two cases.

Overall, the dynamic microstructure evolution in the printed Al–Mg–Si is associated with continuous dynamic recrystallization, which can proceed by progressive subgrain rotation and geometric dynamic recrystallization (Humphreys et al., 2017d). From Fig. 4.27, both mechanisms have been observed. Fig. 4.27A shows pinch-off of elongated grains along serrated grain boundaries, which converts the subgrains to grains separated by high angle boundaries. This is direct evidence of geometric dynamic recrystallization in Al-Mixed. Fig. 4.27B highlights an example of progressive subgrain rotation in the Al-High case, wherein a subgrain at a serrated grain boundary is in the middle of rotation and could eventually become a new grain if the local strain is sufficient.

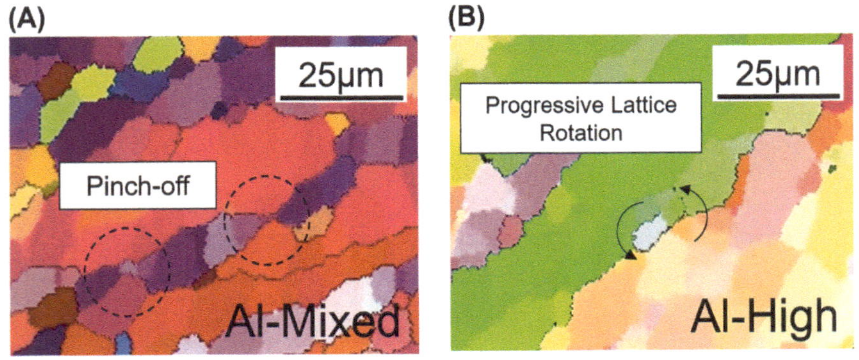

Figure 4.27 Continuous dynamic recrystallization observed in the printed Al–Mg–Si. (A) Geometric dynamic recrystallization in Al-Mixed. (B) Progressive lattice rotation in Al-High. *Reprinted with permission from Griffiths, R.J., Garcia, D., Song, J., Vasudevan, V.K., Steiner, M.A., Cai, W., et al., 2021. Solid-state additive manufacturing of aluminum and copper using additive friction stir deposition: Process-microstructure linkages. Materialia 15, 100967.*

4.6.2 Microstructure characterization of Cu printed at various conditions

Fig. 4.28 shows the microstructure of commercially pure Cu (Cu 110) printed at three processing conditions, along with the microstructure of the feed material, which is supplied in the H02 or "half-hard" tempered condition. The rotation rate changes from 300 to 600 RPM, and the travel velocity changes from 1 to 3 mm/s. The feed rate is kept at 1/3 of the travel velocity for all cases. Table 4.2 details the processing parameters of the deposited Cu. For the Cu feed material, the microstructure is characterized by an abundance of residual local misorientations due to its cold-worked state. For the as-printed Cu, all exhibit roughly equiaxed

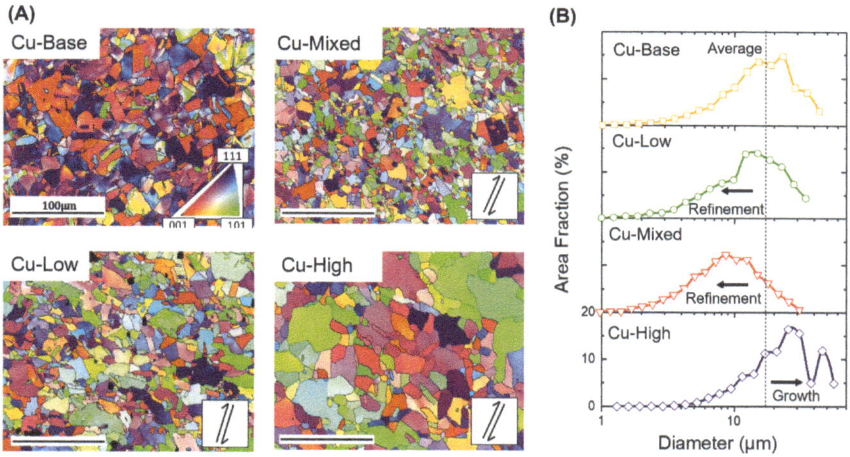

Figure 4.28 Microstructure characterization of Cu deposited under three conditions along with the feed material, including (A) inverse pole figure maps and (B) grain size distribution. *Reprinted with permission from Griffiths, R.J., Garcia, D., Song, J., Vasudevan, V.K., Steiner, M.A., Cai, W., et al., 2021. Solid-state additive manufacturing of aluminum and copper using additive friction stir deposition: Process-microstructure linkages. Materialia 15, 100967.*

Table 4.2 Processing condition for additive friction stir deposition of Cu.

Sample	Rotational rate Ω (revolutions per minute)	Travel velocity V (mm/s)	Feed rate F (mm/s)
Cu-Low	300	1	0.33
Cu-Mixed	600	1	0.33
Cu-High	600	3	1

grains and show less local misorientations. Both Cu-Low and Cu-Mixed have grain size reduction from the feed material, but with a heterogeneous grain structure consisting of larger grains intermixed with smaller ones. Cu-High shows a coarser grain structure than the feed material. The area-weighted average grain size of Cu-Base, Cu-Low, Cu-Mixed, and Cu-High is determined to be 18, 14, 10, and 27 μm, respectively. Despite variances in grain size and misorientation distribution, all the deposited Cu has roughly identical shear directions.

Given the heterogeneous nature of the resulting microstructure, grain refinement in Cu-Low and Cu-Mixed is most likely the consequence of several restoration events, with discontinuous recrystallization being one of the most important mechanisms. Cu-High also has a wide grain size distribution, but the average grain size is larger than the feed material. Furthermore, Cu-High has a lower LAB fraction (11%) than Cu-Low and Cu-Mixed (Fig. 4.29a), indicating less deformation or more high angle boundary migration.

The KAM maps in Fig. 4.29B show that nearly all grains in the feed material contain local misorientations, further highlighting the stored energy caused by its cold-worked state. After deposition, the stored energy in Cu has been significantly reduced. Compared to Al–Mg–Si that shows an organized structure, the local misorientations are more heterogeneously distributed in Cu; local misorientations are accumulated in some regions, while other regions are entirely free of them. This is especially prominent in Cu-High, which is comprised of primarily "clean" grains with little intragranular misorientation or substructures.

Fig. 4.30 shows the pole figures and orientation distribution functions for the as-printed Cu at three conditions, which show similar textures consisting of B/*B* and C components. Cu-Low also has an A1* component. Comparing the pole figures between Al–Mg–Si and Cu, a much stronger texture is observed in the former, noting the different scales of xRandom used for the two materials. This is partially because when new orientation-independent grains "nucleate" and develop in Cu during discontinuous recrystallization, the deformation-induced texture can be effectively diminished.

The printed Cu samples all exhibit a wide distribution of grain sizes. Fig. 4.31A only shows the small grains (grain size smaller than the mean grain size) in Cu, which appear in networks. Grain refinement in Cu-Low and Cu-Mixed, networking of small equiaxed grains, and a reduction in dislocation density all point to discontinuous dynamic recrystallization

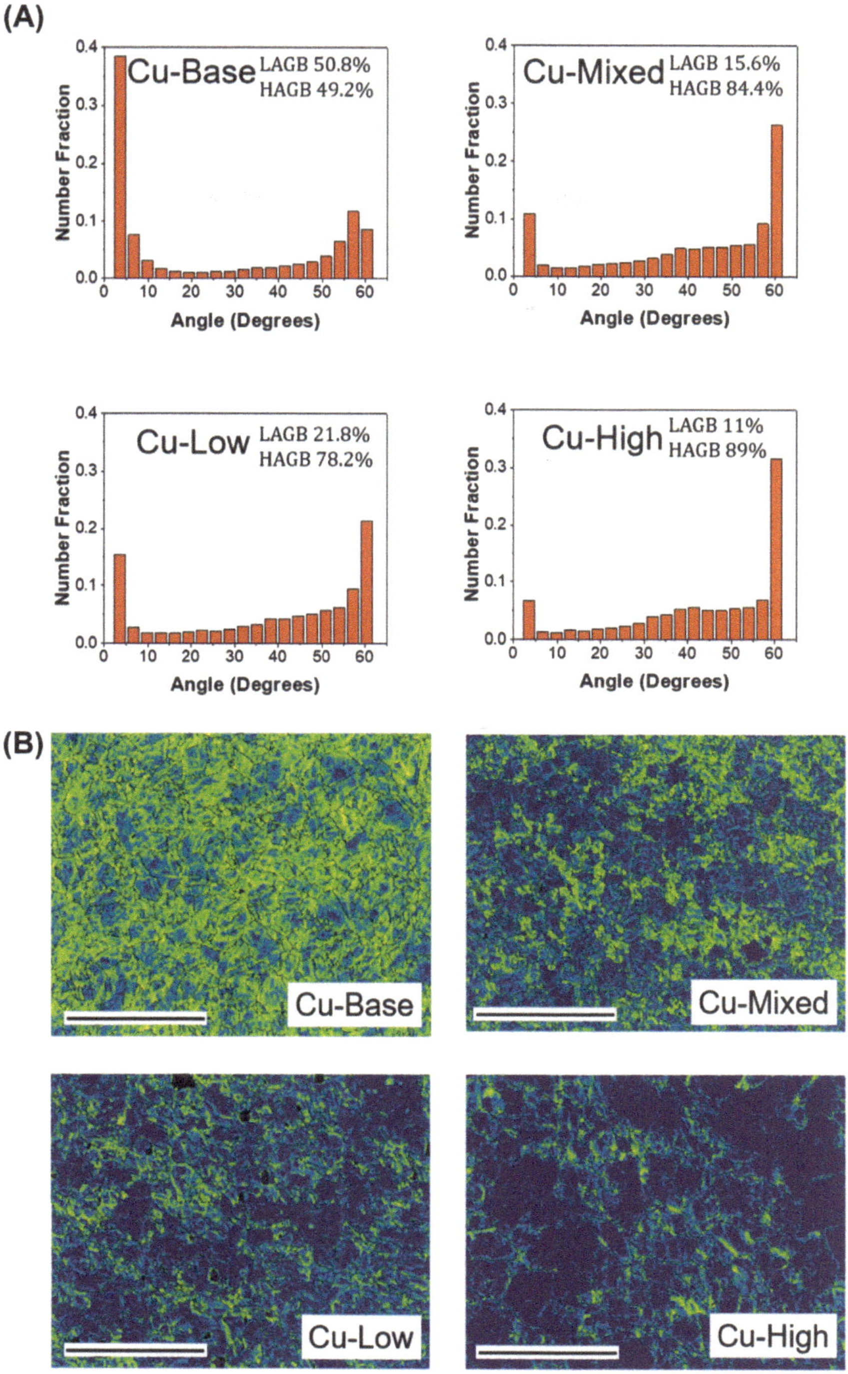

Figure 4.29 (A) Grain boundary misorientation distributions and (B) KAM maps of the three Cu deposits and the feed material. *Reprinted with permission from Griffiths, R.J., Garcia, D., Song, J., Vasudevan, V.K., Steiner, M.A., Cai, W., et al., 2021. Solid-state additive manufacturing of aluminum and copper using additive friction stir deposition: Process-microstructure linkages. Materialia 15, 100967.*

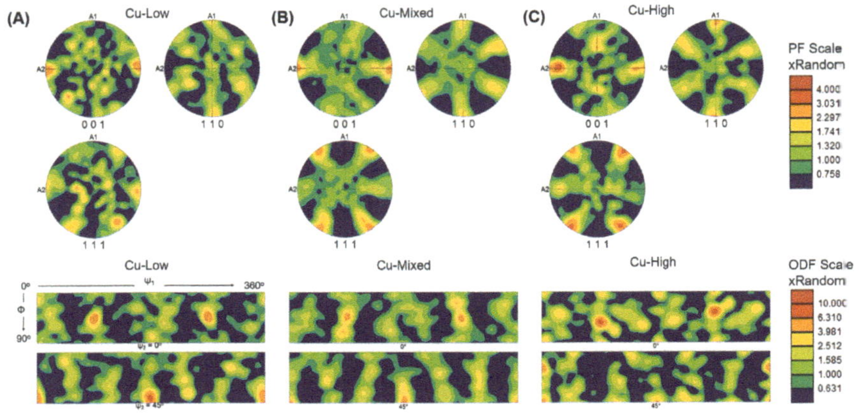

Figure 4.30 Texture of the deposited Cu represented by pole figures and orientation distribution functions, including (A) Cu-Low, (B) Cu-Mixed, and (C) Cu-High. *Reprinted with permission from Griffiths, R.J., Garcia, D., Song, J., Vasudevan, V.K., Steiner, M.A., Cai, W., et al., 2021. Solid-state additive manufacturing of aluminum and copper using additive friction stir deposition: Process-microstructure linkages. Materialia 15, 100967.*

as a fundamental mechanism in microstructure evolution. In Al–Mg–Si, continuous dynamic recrystallization progresses uniformly as deformation proceeds; in contrast, discontinuous dynamic recrystallization has a heterogeneous and cyclic nature.

In addition to discontinuous recrystallization, dynamic recovery happens in Cu-Low and Cu-Mixed. This can be seen from the substructure and tilt boundaries in Cu-Low and Cu-Mixed shown in Fig. 4.31B and C. The final microstructures in Cu-Low and Cu-Mixed are the result of both discontinuous dynamic recrystallization and dynamic recovery. Cu-High does not show obvious substructures; it is characterized by large "clean" grains with low dislocation content because of the migration of high angle grain boundaries during cooling. Therefore, the processing conditions for Cu-Low and Cu-Mixed result in Stage B-dominated (i.e., high temperature severe plastic deformation) microstructures, whereas the processing condition for Cu-High results in Stage C-dominated (i.e., cooling) microstructures.

4.6.3 Analysis of the microstructure evolution mechanisms and trends

4.6.3.1 Origin of the different microstructure evolution mechanisms

Given the above characterization results, additive friction stir deposition of Al–Mg–Si is characterized by continuous dynamic recrystallization,

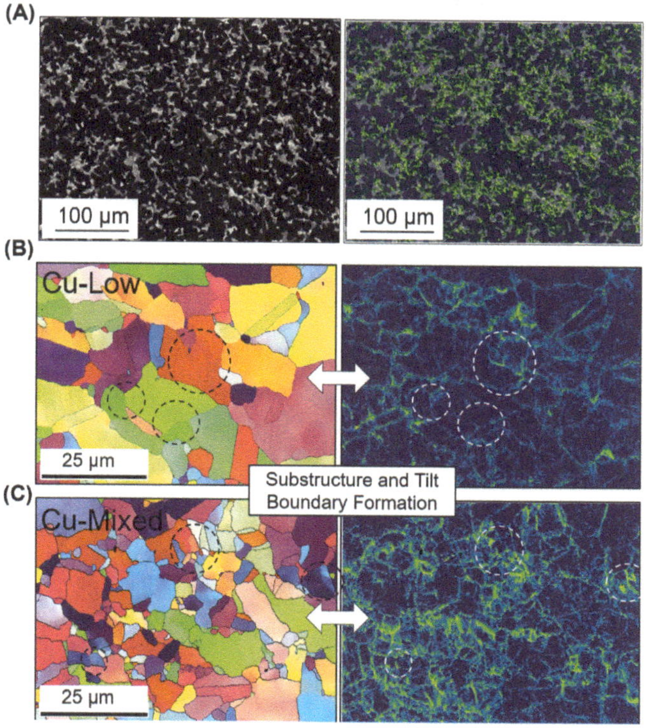

Figure 4.31 Microstructure evolution mechanisms in the printed Cu. (A) Highlight of the grains with a size smaller than the mean grain size. Inverse pole figure and KAM maps for (B) Cu-Low and (C) Cu-Mixed, showing substructures. *Reprinted with permission from Griffiths, R.J., Garcia, D., Song, J., Vasudevan, V.K., Steiner, M.A., Cai, W., et al., 2021. Solid-state additive manufacturing of aluminum and copper using additive friction stir deposition: Process-microstructure linkages. Materialia 15, 100967.*

whereas deposition of Cu is characterized by discontinuous recrystallization. The different microstructure evolution mechanisms between Al–Mg–Si and Cu stem from two aspects. First, from the in situ monitoring results in Chapter 2, Process Fundamentals, we know that Al–Mg–Si and Cu have drastically different material flow behaviors. Significant material rotation is seen in the deposition zone of Al–Mg–Si, whereas little rotational motion is detected in the Cu deposition zone. Therefore, the total strain in Al–Mg–Si is expected to be much higher than Cu. Second, dynamic recovery is more sufficient in Al–Mg–Si (high stacking fault energy $>160\ mJ/m^2$) than in Cu (medium stacking fault energy $\sim 80\ mJ/m^2$ (Heidarzadeh et al., 2019)). Overall, because of the large deformation and high dynamic recovery rate, continuous dynamic recrystallization is dominant in Al–Mg–Si; with moderate deformation and less sufficient dynamic recovery, discontinuous recrystallization is dominant in Cu.

4.6.3.2 Origin of the process-microstructure linkage in Al–Mg–Si

With extensive material rotation in the deposition zone, the strain level in Al–Mg–Si likely correlates with the ratio between the rotation rate and travel velocity, Ω/V. From the processing parameters listed in Table 4.1, Al-Low is expected to exhibit a lower total strain level than Al-Mixed. Based on the shear texture analysis earlier, Al-Low has a weak C-type texture, whereas Al-Mixed has a strong C-type texture. This also suggests a lower total strain level in Al-Low than Al-Mixed. On the other hand, the degree of recrystallization is dependent on the total strain in continuous dynamic recrystallization (Humphreys et al., 2017d). If the strain level were the only governing factor, Al-Low would have the least complete recrystallization, which is the opposite of the experimental observation.

This seeming discrepancy can be explained by the influence of precipitate evolution. Before deposition, the β'' precipitates exist in the feed material and their attributes lead to the peak strength. During additive friction stir deposition, the peak temperatures of Al-Low, Al-Mixed, and Al-High are ~706K, 824K, and 763K, respectively (Garcia et al., 2020). Al-Low corresponds to the lowest temperature condition, and the peak temperature is well below the precipitate solvus temperature (Davis et al., 1993). Therefore, Stage A preheating will result in the least dissolution of the β” phase (Gallais et al., 2008), and the remaining precipitates in Stage B can increase the dislocation density, rendering high-density tangled dislocations around the precipitates. This accelerates the subgrain development and the rate of continuous dynamic recrystallization at a given strain level (Huang et al., 2018).

To summarize, Al-Low is associated with the least deformation and lowest peak temperature among the three printing conditions. The deformation factor alone would lead to the least degree of recrystallization, whereas the temperature factor would accelerate the recrystallization. Overall, Al-Low shows the most complete recrystallization, suggesting that the temperature effect, which controls the remaining precipitates during recrystallization, is more dominant in governing the microstructure of Al–Mg–Si.

4.6.3.3 Origin of the process-microstructure linkage in Cu

Without material rotation in the deposition zone, the magnitude of strain is moderate in Cu. The print head rotation primarily applies frictional heating rather than imparting deformation in Cu, so the shear strain rate should then weakly depend on the rotation rate. As a result, the total strain is more reliant on the amount of time for deformation, which scales with the inverse of velocity, $1/V$. Because the travel velocity in Cu-High is three times that of Cu-Low and Cu-Mixed, the deformation level in Cu-High is projected to be significantly lower. Given the lowest strain level, the microstructure evolution in Cu-High is driven by post-dynamic recrystallization in Stage C and the resulting high angle boundary migration. Therefore, Cu-High has the least local misorientations and the largest portion of twin boundaries, featuring large "clean" grains with low dislocation concentration. In Cu-Low and Cu-Mixed, dynamic microstructure evolution in Stage B is dominant, resulting in a competition between discontinuous dynamic recrystallization and dynamic recovery (Mironov et al., 2015).

4.6.3.4 Origin of the texture differences

The combined impacts of recrystallization mechanisms and deformation levels lead to differences in the texture types and magnitudes between Al—Mg—Si and Cu. Cu has less deformation; it also has extensive high angle boundary migration due to discontinuous recrystallization. As a result, its deformation texture is significantly weaker than that of Al—Mg—Si. When the orientation of the produced shear textures is compared, the Cu samples are virtually identical, while the Al—Mg—Si deposits differ from one another. This implies that the imposed deformation changes minimally with process factors for Cu, but considerably for Al—Mg—Si. The evolution history of deformation texture also differs fundamentally between Cu and Al—Mg—Si: the texture in Cu may form primarily in the transition zone, whereas the texture in Al—Mg—Si develops in both the transition and deposition zones.

Finally, it is interesting to compare Cu, which is of medium stacking fault energy, with Inconel 625 and 316 L stainless steel, which are of low (to medium) stacking fault energy. The former is dominated by discontinuous recrystallization, but the microstructure in the latter arises from continuous recrystallization, as discussed in Section 4.4. This apparent discrepancy may be caused by the moderate strain level in Cu, which suppresses continuous dynamic recrystallization. Also, the presence of solutes and second-phase particles in Inconel 625 and 316 L stainless steel can slow down the high angle boundary migration via solute drag and Zener pinning. This will suppress discontinuous recrystallization and promote continuous recrystallization as the dominant mechanism. Moreover, Cu starts in a predeformed (cold-worked) state, which favors discontinuous recrystallization; as a result, the balance of recrystallization and recovery may be shifted.

4.7 Dynamic phase evolution

Many engineering alloys processed by additive friction stir deposition achieve their high strength via precipitation strengthening. Take 6xxx series Al—Mg—Si alloys as an example. The precipitation sequence is typically as follows: SSSS → atomic Si clustering → GP zones → β” → β’ → β-phase (Mg_2Si) (Edwards et al., 1998). SSSS is the supersaturated solid solution; the β”-phase is associated primarily with peak aged (T6) conditions; the β-phase is the most stable precipitate but contributes little to material strength and hardness (Simar et al., 2012). In 7xxx series Al—Zn—Cu—Mg alloys, the precipitation sequence is SSSS → GP zones → η' → stable η ($MgZn_2$) (Deshpande et al., 1998). In addition to the strengthening phases, other precipitates will form if aging occurs at higher temperatures and cooling is slow. These include S Phase (Al_2CuMg) and T Phase, which have been observed in as-deposited AA7050 via additive friction stir deposition (Mason et al., 2021). Achieving the optimal precipitate attributes in the as-printed alloys is extremely challenging in additive friction stir deposition due to the complex kinetics associated with precipitate dissolution, dynamic precipitation, dynamic recrystallization, dynamic recovery, particle pinning, precipitate coarsening, and re-precipitation. As a result, post-printing aging is necessary to restore wrought-like mechanical properties. More examples on this aspect are provided in Chapter 7, Mechanical Properties of the Printed Materials.

For precipitate-strengthened alloys such as Al and Mg alloys, precipitate-related phase evolution can occur in any stage of additive friction stir deposition. Precipitate dissolution will occur in Stage A if the peak temperature is above the solvus temperature for a long enough time. New precipitates may emerge in Stage B as a result of dynamic precipitation, which can interact with dynamic recrystallization in complex manners. In Stage C, some solute atoms will come out of the matrix, causing re-precipitation and precipitate coarsening.

Regarding the dynamic precipitation in Stage B and its interaction with dynamic recrystallization, there is a significant scientific gap with many questions remaining to be answered. It is instructive to review the static case, that is, the interaction of precipitation and recrystallization in post-deformation annealing. Hornbogen and coworkers (Hornbogen, 1979; Jones and Humphreys, 2003; Saji and Hornbogen, 1978) have carried out pioneering work for the static annealing case, showing that multiple process regimes could occur. These process regimes are divided by the characteristic temperatures that are dependent on recrystallization conditions and the TTT (time–temperature–transformation) curve of precipitation. As shown in Fig. 4.32, the kinetics of recrystallization and precipitation define two characteristic temperatures, T_A and T_B, with which three process regimes arise (Humphreys et al., 2017c):

- $T < T_B$: Precipitation before recrystallization.
- $T_B < T < T_A$: Precipitation and recrystallization occur concurrently.
- $T_A < T$: Recrystallization before precipitation.

In the interest of simplicity, the effect of prior strain level is not included in this picture, but it will have an important impact. Higher levels of cold work will lead to higher dislocation densities and stored strain energy, which will enhance (to different degrees) the precipitation and recrystallization kinetics.

Like the scenario described for static annealing, distinct regimes may exist in the dynamic case, regarding whether dynamic precipitation occurs before (Regime I), occurs concurrently with (Regime II), or occurs after (Regime III) dynamic recrystallization. Regime I could lead to particle

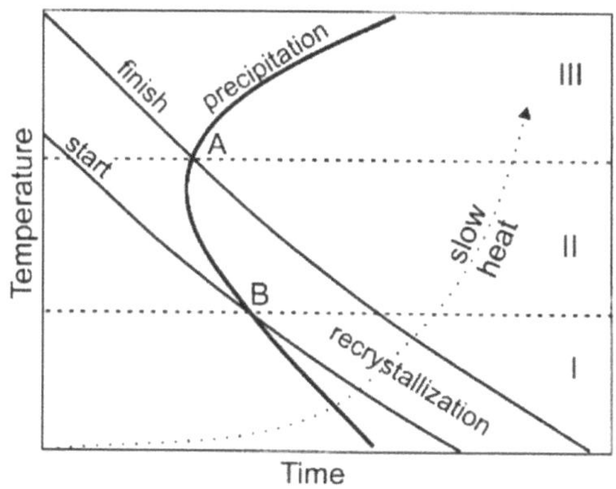

Figure 4.32 Schematic illustration of the competition between static precipitation and recrystallization of deformed alloys. *Reprinted with permission from Humphreys, J., Rohrer, G.S., Rollett, A., 2017c. Chapter 9—Recrystallization of two-phase alloys. In: Humphreys, J., Rohrer, G.S., Rollett, A. (Eds.), Recrystallization and Related Annealing Phenomena, third ed. Elsevier, Oxford, pp. 321–359.*

stimulated nucleation of dynamic recrystallization, which may have implications for texture evolution. Zener drag on the grain boundaries can also be notable. Regime III could lead to relatively large grain sizes because there is no Zener drag during dynamic recrystallization. Regime II promotes the interaction of dynamic recrystallization and dynamic precipitation, potentially causing simultaneous grain size reduction and effective precipitation. First, the precipitates may influence the dynamic recrystallization process once they are formed. For small particles ($< 1\ \mu m$), the pinning effect is stronger than particle stimulated nucleation of recrystallization, so dynamic recrystallization would be retarded, and the recrystallized grain size would be reduced. Second, the significant deformation will cause a high number density of suitable precipitation sites inside each grain, resulting in a uniform distribution of precipitates inside recrystallized grains. Regime II may thus promote fine grains containing a high density of precipitates.

Due to a dearth of in situ or time-resolved research, the dynamic evolution of precipitates is poorly understood in additive friction stir deposition. The majority of current research focuses on post-printing precipitate characterization and "hand-waving" explanations of what has happened during heating (Stage A), deposition (Stage B), and cooling (Stage C). Like tracer-based microstructure characterization in Section 4.5, precipitation characterization along the material flow path can provide new insights into the complex dynamic phase evolution. This could be a promising research area for additive friction stir deposition in the future.

4.8 Summary

Microstructure evolution in polycrystalline alloys typically proceeds by grain boundary migration, which may be driven by grain curvature or stored energy. The latter driving force is dominant during static annealing of a predeformed material or during deformation at high temperatures. Depending on the intrinsic material properties and thermomechanical condition defined by the temperature and strain rate, hot deformation can lead to discontinuous or continuous types of dynamic recrystallization. Discontinuous dynamic recrystallization proceeds in two steps: (i) "nucleation" (or initiation) of a new dislocation-free grain, which is typically a subgrain formed near prior grain boundaries, and (ii) growth of this grain with the boundary sweeping into heavily deformed regions. In the presence of deformation, the recrystallized grain will create new dislocations and increase the stored energy. Then, the "nucleation" and growth process will repeat cyclically. Continuous dynamic recrystallization proceeds via subgrain formation by dynamic recovery, followed by strain-induced high angle boundary formation. This can involve two mechanisms, progressive lattice rotation and geometric dynamic recrystallization.

Additive friction stir deposition is a thermomechanical process that can be divided into three stages. Stage A is preheating, in which precipitate dissolution or other phase transformation may happen. Stage B is high temperature severe plastic deformation, featuring multiple dynamic recrystallization and phase evolution processes. Stage C is cooling, possibly involving re-precipitation and static (or metadynamic) recrystallization. The thermomechanical condition of additive friction stir deposition is characterized by a high strain rate, high temperature, and very large strain. In conventional hot working, continuous dynamic recrystallization more readily occurs in high stacking fault energy materials, whereas discontinuous dynamic recrystallization is typically observed in low stacking fault energy materials. However, in additive friction stir deposition, even materials of low stacking fault energy, such as Ni alloys and austenitic steels, may undergo continuous recrystallization, resulting in extremely fine grains that cannot be obtained by discontinuous recrystallization. This is likely a result of the severe plastic deformation nature of additive friction stir deposition.

For an Al–Cu alloy (AA2024), microstructure characterization along the flow path reveals continuous dynamic recrystallization, no matter if the material is originally in the center or at the edge of the feed-rod. As the material comes out of the print head channel, the previous large grains are immediately deformed to become elongated grains with serrated grain boundaries. Evidence is found for geometric dynamic recrystallization, in which pinch-off of these serrated boundaries leads to new fine, equiaxed grains. The grain size is reduced to 1–3 μm by the end of the initial material feeding. There is no further grain size reduction in steady-state deposition despite large strain accumulation. This stems from the dynamic equilibrium between grain coarsening by high angle boundary migration and grain refinement by subgrain formation.

For an Al-Mg-Si alloy (AA6061), continuous dynamic recrystallization is also identified as the main mechanism leading to substantial grain refinement. Evidence of progressive lattice rotation and geometric dynamic recrystallization is found. The resulting microstructure shows a homogeneous grain size distribution and a relatively high fraction of low angle boundaries. Extensive material flow is observed, suggesting large total strain. The combination of continuous dynamic recrystallization and large deformation leads to a strong shear texture. Among various processing conditions, the one associated with the lowest peak temperature, rather than the one associated with the largest strain, results in the most complete recrystallization with the smallest grain size and lowest fraction of low angle boundaries.

For pure Cu, discontinuous dynamic recrystallization is identified as the main mechanism, which results in a wide distribution of grain size, network formation of the smaller grains, and a reduction of dislocation density in larger grains. The resulting microstructure shows a heterogeneous grain size distribution and a relatively low fraction of low angle boundaries. No material rotation is observed under the print head, suggesting a moderate strain level. The combination of discontinuous dynamic recrystallization and moderate deformation leads to a weak shear texture. Among various processing conditions, the one associated with the largest travel velocity and therefore the lowest strain level shows grain coarsening, featuring large "clean" grains with minimal local misorientations. This results from static or metadynamic recrystallization during cooling, which dominates the final microstructure.

Complex phase evolution can occur in precipitation-strengthened alloys during preheating, high-temperature severe plastic deformation, and cooling. Such phase evolution can interact with static and dynamic recrystallization, leading to new microstructures. This phenomenon can be explored via in situ characterization or time-resolved experiments in the future.

References

Arora, A., Zhang, Z., De, A., DebRoy, T., 2009. Strains and strain rates during friction stir welding. Scripta Materialia 61, 863–866.

Beladi, H., Farabi, E., Hodgson, P.D., Barnett, M.R., Rohrer, G.S., Fabijanic, D., 2021. Microstructure evolution of 316L stainless steel during solid-state additive friction stir deposition. Philosophical Magazine 1–16.

Bhattacharya, A., Shen, Y.-F., Hefferan, C.M., Li, S.F., Lind, J., Suter, R.M., et al., 2021. Grain boundary velocity and curvature are not correlated in Ni polycrystals. Science (New York, N.Y.) 374, 189–193.

Canova, G.R., Kocks, U.F., Jonas, J.J., 1984. Theory of torsion texture development. Acta Metallurgica 32, 211–226.

Cho, S.-H., Kang, K.-B., Jonas, J.J., 2001. The dynamic, static and metadynamic recrystallization of a Nb-microalloyed steel. ISIJ International 41, 63–69.

Cho, K.-J., Sano, T., Doherty, K.J., Yen, C.-F., Gazonas, G.A., Montgomery, J.S., et al., 2009. Magnesium technology and manufacturing for ultra lightweight armored ground vehicles.

Cram, D.G., Zurob, H.S., Brechet, Y.J.M., Hutchinson, C.R., 2009. Modelling discontinuous dynamic recrystallization using a physically based model for nucleation. Acta Materialia 57, 5218–5228.

Davis, J.R., Associates, J.R.D., Committee, A.S.M.I.H., 1993. Aluminum and Aluminum Alloys. ASM International.

Derby, B., 1991. The dependence of grain size on stress during dynamic recrystallisation. Acta Metallurgica et Materialia 39, 955–962.

Deshpande, N.U., Gokhale, A.M., Denzer, D.K., Liu, J., 1998. Relationship between fracture toughness, fracture path, and microstructure of 7050 aluminum alloy: Part I. Quantitative characterization. Metallurgical and Materials Transactions A 29, 1191–1201.

Doherty, R.D., Hughes, D.A., Humphreys, F.J., Jonas, J.J., Jensen, D.J., Kassner, M.E., et al., 1997. Current issues in recrystallization: a review. Materials Science and Engineering: A 238, 219–274.

Drury, M.R., Humphreys, F.J., 1986. The development of microstructure in Al-5% Mg during high temperature deformation. Acta Metallurgica 34, 2259–2271.

Edwards, G.A., Stiller, K., Dunlop, G.L., Couper, M.J., 1998. The precipitation sequence in Al–Mg–Si alloys. Acta Materialia 46, 3893–3904.

El Wahabi, M., Gavard, L., Montheillet, F., Cabrera, J.M., Prado, J.M., 2005. Effect of initial grain size on dynamic recrystallization in high purity austenitic stainless steels. Acta Materialia 53, 4605–4612.

Estrin, Y., Mecking, H., 1984. A unified phenomenological description of work hardening and creep based on one-parameter models. Acta Metallurgica 32, 57–70.

Estrin, Y., Vinogradov, A., 2013. Extreme grain refinement by severe plastic deformation: a wealth of challenging science. Acta Materialia 61, 782–817.

Fanfoni, M., Tomellini, M., 1998. The Johnson-Mehl- Avrami-Kohnogorov model: a brief review. Il Nuovo Cimento D 20, 1171–1182.

Fonda, R.W., Knipling, K.E., 2011. Texture development in friction stir welds. Science and Technology of Welding and Joining 16, 288–294.

Fonda, R.W., Knipling, K.E., Bingert, J.F., 2008. Microstructural evolution ahead of the tool in aluminum friction stir welds. Scripta Materialia 58, 343–348.

Frolov, T., 2014. Effect of interfacial structural phase transitions on the coupled motion of grain boundaries: a molecular dynamics study. Applied Physics Letters 104, 211905.

Gallais, C., Denquin, A., Bréchet, Y., Lapasset, G., 2008. Precipitation microstructures in an AA6056 aluminium alloy after friction stir welding: characterisation and modelling. Materials Science and Engineering: A 496, 77–89.

Gao, H., Huang, Y., 2003. Geometrically necessary dislocation and size-dependent plasticity. Scripta Materialia 48, 113–118.

Gao, W., Belyakov, A., Miura, H., Sakai, T., 1999. Dynamic recrystallization of copper polycrystals with different purities. Materials Science and Engineering: A 265, 233–239.

Gao, Y., Ding, Y., Ma, Y., Chen, J., Wang, X., Xu, J., 2022. Evolution of annealing twins in Inconel 625 alloy during tensile loading. Materials Science and Engineering: A 831, 142188.

Garcia, D., Hartley, W.D., Rauch, H.A., Griffiths, R.J., Wang, R., Kong, Z.J., et al., 2020. In situ investigation into temperature evolution and heat generation during additive friction stir deposition: a comparative study of Cu and Al-Mg-Si. Additive Manufacturing 34, 101386.

Gardner, K.J., Grimes, R., 1979. Recrystallization during hot deformation of aluminium alloys. Metal Science 13, 216–222.

Gholinia, A., Humphreys, F.J., Prangnell, P.B., 2002. Production of ultra-fine grain microstructures in Al−Mg alloys by coventional rolling. Acta Materialia 50, 4461−4476.

Gottstein, G., Shvindlerman, L.S., 2009. Chapter 3, Grain Boundary Migration in Metals: Thermodynamics, Kinetics, Applications, 2nd edn CRC Press.

Gottstein, G., Molodov, D.A., Shvindlerman, L.S., Srolovitz, D.J., Winning, M., 2001. Grain boundary migration: misorientation dependence. Current Opinion in Solid State and Materials Science 5, 9−14.

Gourdet, S., Montheillet, F., 2003. A model of continuous dynamic recrystallization. Acta Materialia 51, 2685−2699.

Griffiths, R.J., Garcia, D., Song, J., Vasudevan, V.K., Steiner, M.A., Cai, W., et al., 2021. Solid-state additive manufacturing of aluminum and copper using additive friction stir deposition: process-microstructure linkages. Materialia 15, 100967.

Heidarzadeh, A., Saeid, T., Klemm, V., Chabok, A., Pei, Y., 2019. Effect of stacking fault energy on the restoration mechanisms and mechanical properties of friction stir welded copper alloys. Materials & Design 162, 185−197.

Hillert, M., 1965. On the theory of normal and abnormal grain growth. Acta Metallurgica 13, 227−238.

Hirth, J.P., Balluffi, R.W., 1973. On grain boundary dislocations and ledges. Acta Metallurgica 21, 929−942.

Hornbogen, E., 1979. Combined reactions. Metallurgical Transactions A 10, 947−972.

Huang, K., Logé, R.E., 2016. A review of dynamic recrystallization phenomena in metallic materials. Materials & Design 111, 548−574.

Huang, K., Marthinsen, K., Zhao, Q., Logé, R.E., 2018. The double-edge effect of second-phase particles on the recrystallization behaviour and associated mechanical properties of metallic materials. Progress in Materials Science 92, 284−359.

Hughes, D.A., Hansen, N., 1997. High angle boundaries formed by grain subdivision mechanisms. Acta Materialia 45, 3871−3886.

Humphreys, J., Rohrer, G.S., Rollett, A., 2017a. Chapter 5—Mobility and migration of boundaries. In: Humphreys, J., Rohrer, G.S., Rollett, A. (Eds.), Recrystallization and Related Annealing Phenomena, 3rd edn Elsevier, Oxford, pp. 145−197.

Humphreys, J., Rohrer, G.S., Rollett, A., 2017b. Chapter 7—Recrystallization of single-phase alloys. In: Humphreys, J., Rohrer, G.S., Rollett, A. (Eds.), Recrystallization and Related Annealing Phenomena, 3rd edn Elsevier, Oxford, pp. 245−304.

Humphreys, J., Rohrer, G.S., Rollett, A., 2017c. Chapter 9—Recrystallization of two-phase alloys. In: Humphreys, J., Rohrer, G.S., Rollett, A. (Eds.), Recrystallization and Related Annealing Phenomena, 3rd edn Elsevier, Oxford, pp. 321−359.

Humphreys, J., Rohrer, G.S., Rollett, A., 2017d. Chapter 13—Hot deformation and dynamic restoration. In: Humphreys, J., Rohrer, G.S., Rollett, A. (Eds.), Recrystallization and Related Annealing Phenomena, 3rd edn Elsevier, Oxford, pp. 469−508.

Jahedi, M., McWilliams, B.A., Knezevic, M., 2018. Deformation and fracture mechanisms in WE43 magnesium-rare earth alloy fabricated by direct-chill casting and rolling. Materials Science and Engineering: A 726, 194−207.

Jazaeri, H., Humphreys, F.J., 2004. The transition from discontinuous to continuous recrystallization in some aluminium alloys: II—annealing behaviour. Acta Materialia 52, 3251−3262.

Jones, M.J., Humphreys, F.J., 2003. Interaction of recrystallization and precipitation: the effect of Al3Sc on the recrystallization behaviour of deformed aluminium. Acta Materialia 51, 2149−2159.

Kassner, M.E., 1989. Large-strain deformation of aluminum single crystals at elevated temperature as a test of the geometric-dynamic-recrystallization concept. Metallurgical Transactions A 20, 2182−2185.

Kassner, M.E., Barrabes, S.R., 2005. New developments in geometric dynamic recrystallization. Materials Science and Engineering: A 410–411, 152–155.
King, W.E., Barth, H.D., Castillo, V.M., Gallegos, G.F., Gibbs, J.W., Hahn, D.E., et al., 2014. Observation of keyhole-mode laser melting in laser powder-bed fusion additive manufacturing. Journal of Materials Processing Technology 214, 2915–2925.
Li, J., Dillon, S.J., Rohrer, G.S., 2009. Relative grain boundary area and energy distributions in nickel. Acta Materialia 57, 4304–4311.
Lin, J., Balint, D., Pietrzyk, M., 2012. Microstructure Evolution in Metal Forming Processes. Elsevier Science.
Mann, J.B., Chandrasekar, S., Compton, W.D., Trumble, K.P., Saldana, C., Swaminathan, S., et al., 2011. 7—Severe plastic deformation and the production of nanostructured alloys by machining. In: Whang, S.H. (Ed.), Nanostructured Metals and Alloys. Woodhead Publishing, pp. 178–210.
Mason, C.J.T., Rodriguez, R.I., Avery, D.Z., Phillips, B.J., Bernarding, B.P., Williams, M.B., et al., 2021. Process-structure-property relations for as-deposited solid-state additively manufactured high-strength aluminum alloy. Additive Manufacturing 40, 101879.
Mirihanage, W.U., Robson, J.D., Mishra, S., Hidalgo-Manrique, P., da Fonseca, J.Q., Daniel, C.S., et al., 2021. Direct observation of the dynamic evolution of precipitates in aluminium alloy 7021 at high strain rates via high energy synchrotron x-rays. Acta Materialia 205, 116532.
Mironov, S., Inagaki, K., Sato, Y.S., Kokawa, H., 2015. Microstructural evolution of pure copper during friction-stir welding. Philosophical Magazine 95, 367–381.
Montheillet, F., Cohen, M., Jonas, J.J., 1984. Axial stresses and texture development during the torsion testing of Al, Cu and α-Fe. Acta Metallurgica 32, 2077–2089.
Nes, E., 1995. Recovery revisited. Acta Metallurgica et Materialia 43, 2189–2207.
Nes, E., Marthinsen, K., 2002. Modeling the evolution in microstructure and properties during plastic deformation of f.c.c.-metals and alloys—an approach towards a unified model. Materials Science and Engineering: A 322, 176–193.
Nes, E., Ryum, N., Hunderi, O., 1985. On the Zener drag. Acta Metallurgica 33, 11–22.
Olmsted, D.L., Holm, E.A., Foiles, S.M., 2009. Survey of computed grain boundary properties in face-centered cubic metals—II: grain boundary mobility. Acta Materialia 57, 3704–3713.
Omori, T., Kusama, T., Kawata, S., Ohnuma, I., Sutou, Y., Araki, Y., et al., 2013. Abnormal grain growth induced by cyclic heat treatment. Science (New York, N.Y.) 341, 1500–1502.
Perez, M., 2005. Gibbs–Thomson effects in phase transformations, Scripta Materialia, 52. pp. 709–712.
Perry, M.E.J., Rauch, H.A., Griffiths, R.J., Garcia, D., Sietins, J.M., Zhu, Y., et al., 2021. Tracing plastic deformation path and concurrent grain refinement during additive friction stir deposition. Materialia 18, 101159.
Pippan, R., Scheriau, S., Taylor, A., Hafok, M., Hohenwarter, A., Bachmaier, A., 2010. Saturation of fragmentation during severe plastic deformation. Annual Review of Materials Research 40, 319–343.
Poirier, J.P., Nicolas, A., 1975. Deformation-induced recrystallization due to progressive misorientation of subgrains, with special reference to mantle peridotites. The Journal of Geology 83, 707–720.
Ponge, D., Gottstein, G., 1998. Necklace formation during dynamic recrystallization: mechanisms and impact on flow behavior. Acta Materialia 46, 69–80.
Raabe, D., 2014. 23—Recovery and recrystallization: phenomena, physics, models, simulation. In: Laughlin, D.E., Hono, K. (Eds.), Physical Metallurgy, 5th edn Elsevier, Oxford, pp. 2291–2397.

Rivera, O.G., Allison, P.G., Jordon, J.B., Rodriguez, O.L., Brewer, L.N., McClelland, Z., et al., 2017. Microstructures and mechanical behavior of Inconel 625 fabricated by solid-state additive manufacturing. Materials Science and Engineering: A 694, 1–9.
Rivera, O.G., Allison, P.G., Brewer, L.N., Rodriguez, O.L., Jordon, J.B., Liu, T., et al., 2018. Influence of texture and grain refinement on the mechanical behavior of AA2219 fabricated by high shear solid state material deposition. Materials Science and Engineering: A 724, 547–558.
Robson, J.D., Henry, D.T., Davis, B., 2009. Particle effects on recrystallization in magnesium–manganese alloys: Particle-stimulated nucleation. Acta Materialia 57, 2739–2747.
Saji, S., Hornbogen, E., 1978. Combined recrystallization and precipitation reactions in a Cu-4 Wt.% Ti-Alloy. International Journal of Materials Research 69, 741–746.
Sakai, T., Jonas, J.J., 1984. Overview no. 35 dynamic recrystallization: mechanical and microstructural considerations. Acta Metallurgica 32, 189–209.
Sakai, T., Belyakov, A., Kaibyshev, R., Miura, H., Jonas, J.J., 2014. Dynamic and post-dynamic recrystallization under hot, cold and severe plastic deformation conditions. Progress in Materials Science 60, 130–207.
Sharghi, E., Farzadi, A., 2018. Simulation of strain rate, material flow, and nugget shape during dissimilar friction stir welding of AA6061 aluminum alloy and Al-Mg2Si composite. Journal of Alloys and Compounds 748, 953–960.
Simar, A., Bréchet, Y., de Meester, B., Denquin, A., Gallais, C., Pardoen, T., 2012. Integrated modeling of friction stir welding of 6xxx series Al alloys: process, microstructure and properties. Progress in Materials Science 57, 95–183.
Sitdikov, O., Sakai*, T., Goloborodko, A., Miura, H., Kaibyshev, R., 2005. Grain refinement in coarse-grained 7475 Al alloy during severe hot forging. Philosophical Magazine 85, 1159–1175.
Srinivasan, P.B., Arora, K.S., Dietzel, W., Pandey, S., Schaper, M.K., 2010. Characterisation of microstructure, mechanical properties and corrosion behaviour of an AA2219 friction stir weldment. Journal of Alloys and Compounds 492, 631–637.
Tello, K.E., Gerlich, A.P., Mendez, P.F., 2010. Constants for hot deformation constitutive models for recent experimental data. Science and Technology of Welding and Joining 15, 260–266.
Tuncer, N., Bose, A., 2020. Solid-state metal additive manufacturing: a review. JOM 72, 3090–3111.
Valiev, R.Z., Islamgaliev, R.K., Alexandrov, I.V., 2000. Bulk nanostructured materials from severe plastic deformation. Progress in Materials Science 45, 103–189.
Wang, L., Teng, J., Liu, P., Hirata, A., Ma, E., Zhang, Z., et al., 2014. Grain rotation mediated by grain boundary dislocations in nanocrystalline platinum. Nature Communications 5, 4402.
Williams, M.B., Robinson, T.W., Williamson, C.J., Kinser, R.P., Ashmore, N.A., Allison, P.G., et al., 2021. Elucidating the effect of additive friction stir deposition on the resulting microstructure and mechanical properties of magnesium alloy WE43. Metals 11 (1739).
Woo, W., Jeong, J.S., Kim, D.K., Lee, C.M., Choi, S.H., Suh, J.Y., et al., 2020. Stacking fault energy analyses of additively manufactured stainless steel 316L and CrCoNi medium entropy alloy using in situ neutron diffraction. Scientific Reports 10, 1350.
Xie, B., Zhang, B., Ning, Y., Fu, M.W., 2019. Mechanisms of DRX nucleation with grain boundary bulging and subgrain rotation during the hot working of nickel-based superalloys with columnar grains. Journal of Alloys and Compounds 786, 636–647.
Yu, H.Z., 2022. Emerging processes—friction stir based. In: Caballero, F.G. (Ed.), Encyclopedia of Materials: Metals and Alloys. Elsevier, Oxford, pp. 153–161.
Zehetbauer, M., Seumer, V., 1993. Cold work hardening in stages IV and V of F.C.C. metals—I. Experiments and interpretation. Acta Metallurgica et Materialia 41, 577–588.

CHAPTER 5

Effects of tool geometry

Every tool carries with it the spirit by which it has been created.

—Werner Heisenberg.

Tools play a paramount role in deformation processing, such as rolling, forging, extrusion, and friction stir welding and processing. As a deformation-based additive manufacturing technology, the tool in additive friction stir deposition, which is usually referred to as the print head, is also of critical importance. Generally, the tool material needs to be stiffer and stronger than the deposited material, which will be severely deformed at elevated temperatures under the forces imposed by the tool. Moreover, the geometry of the print head can influence the material flow and heat generation, and therefore govern the interface quality and microstructure distribution. In additive friction stir deposition, the print head typically has a cylindrical shape containing a channel in the center for material feeding. Geometrical features can be added to the bottom surface, such as small surface protrusions, to promote vertical material flow and mixing.

This chapter focuses on the effects of tool geometry in additive friction stir deposition. Because tooling plays a similar role in other friction stir-derived technologies, we first provide a survey of the tool effects in jfriction stir welding. This is followed by an overview of the current tool types and geometries used in additive friction stir deposition, including those made of tool steels and tungsten (W)-based materials for printing low and high-temperature materials, respectively. We then compare the interface and microstructure that result from the print head with various geometries, including a flat bottom surface, a bottom surface with two small protrusions, and a bottom surface with four teardrop-shaped protrusions. The comparison focuses on dissimilar Al alloy deposition, for which the print head is made of tool steel. X-ray computed tomography is employed to discern the internal morphology of the interface in each scenario, while electron backscatter diffraction is used for characterizing the microstructure in the interfacial region.

Additive Friction Stir Deposition
DOI: https://doi.org/10.1016/B978-0-12-824374-9.00005-8

5.1 A survey of tool effects in friction stir welding

The tool in friction stir welding is made up of two parts: a rotating shoulder and a threaded cylindrical pin (Mishra et al., 2014). Both are in contact with the workpieces and rotate rapidly, generating frictional heat to soften the material. The pin inserts between the workpieces and moves the softened alloy around it to form the joint. The rotating tool shoulder imposes compression and shear onto the material to plastically deform the material and facilitate material flow. For joining softer metals, such as Al and Cu, steel or cobalt alloy tools are typically used. For joining high strength alloys, tungsten alloys, tungsten carbide composites, or polycrystalline cubic boron nitride tools are the options although they are more expensive (Rai et al., 2011). The reactivity of the tool with the workpiece is an important metric. High temperatures, intimate contact, and oxygen exposure can all result in the formation of brittle intermetallic phases that may worsen the welding quality.

As shown in Fig. 5.1, the shoulder can have a flat, convex, or concave shape. Convex shoulders typically have a positive or negative scroll, but they

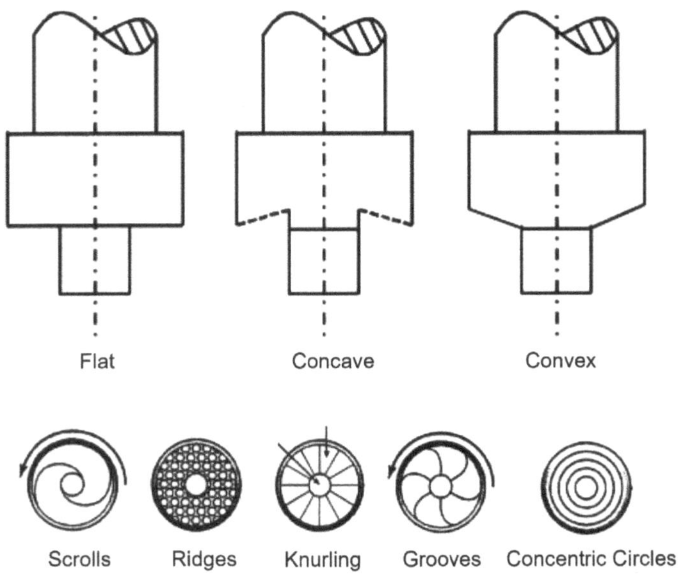

Figure 5.1 Illustrations of possible geometries of the tool shoulder in friction stir welding. From the side view, the shoulder profile can be flat, concave, or convex. From the bottom view, various features can be created, including, scrolls, ridges, knurling, grooves, and concentric circles. *Reprinted with permission from Singh, K., Singh, G., Singh, H., 2018. Review on friction stir welding of magnesium alloys. Journal of Magnesium and Alloys 6, 399–416.*

may also have ridges or knurling. Concave shoulders, on the other hand, can be featureless. To prevent the flash formation and to push material behind the pin, a small angle tilt is required for concave shoulders; this reduces the force placed on the pin. A convex shoulder does not require tool tilt because the surface features, particularly scrolls, act to push material inward, preventing the formation of volumetric defects by promoting the material filling (Singh et al., 2018). Another advantage of convex shoulders is their ability to function on curved surfaces as well as for workpieces of varying thicknesses, for which a featureless, flat shoulder would become unsuitable. Appropriate design of the shoulder geometry is critical because its large contact area with the workpiece surface dictates the thermal condition and material flow in friction stir welding.

The primary functions of the pin are to promote the vertical flow of material, prevent volumetric defects in the weld, and promote stirring. To optimize the material flow and filling, a variety of pin geometries have been used. Two of the most common designs are a cylindrical pin and a truncated cone pin. Surface features of the pins include threads or stepped spirals. The pin could also be featureless, which may compromise the mechanical properties of the weld due to insufficient vertical material flow. In this regard, threaded or spiral truncated conical pins are the superior geometry. Additionally, the incorporation of flats and flutes in the pin surface may result in increased intermixing of dissimilar metals and a reduction in weld defects. Fig. 5.2 shows several examples of pin geometries that have been used for joining dissimilar Al alloys in friction stir welding. These include the straight cylindrical, the threaded cylindrical, and the tapered cylindrical shapes (Ilangovan et al., 2015).

Along with the processing parameters, the combination of the shoulder and pin geometry ultimately determines the weld quality. In principle, the geometry design of the shoulder and pin should be considered simultaneously. Much research has been conducted to determine the optimal ratios of shoulder to pin diameters, wherein lower ratios are found to be preferable. For a specific study, a ratio of approximately 3:1 has the best weld quality and resulting mechanical properties (Khan et al., 2015). The effect of plunging depth of the tool has also been studied, which implies that a shallower plunge may produce welds with lower defect generation and better mechanical performance (Memon et al., 2021). Although optimization of the tool geometry should be conducted for specific material systems, it is widely accepted that tools with embossed geometries on the shoulder and pin outperform their featureless counterparts with fewer defective welds and improved mechanical performance.

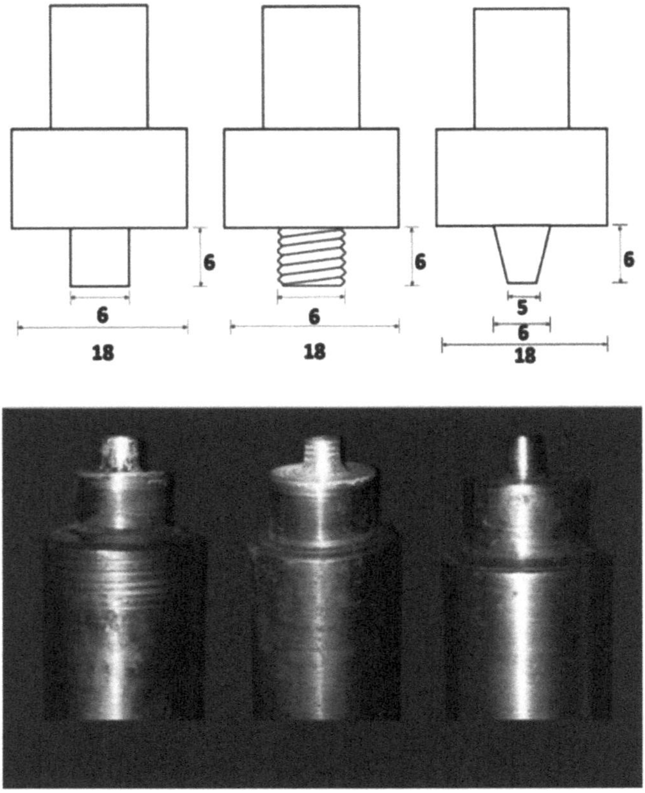

Figure 5.2 Illustrations and pictures of three pin geometries: straight cylindrical, threaded cylindrical, and tapered cylindrical. *Reprinted with permission from Ilangovan, M., Rajendra Boopathy, S., Balasubramanian, V., 2015. Effect of tool pin profile on microstructure and tensile properties of friction stir welded dissimilar AA 6061—AA 5086 aluminium alloy joints. Defence Technology 11, 174—184.*

5.2 Tool types and geometries for additive friction stir deposition

Compared to friction stir welding, there is no non-consumable pin in additive friction stir deposition. Instead, the feed-rod is delivered through the channel in the print head. By analogy, the print head acts as the shoulder, whereas the feed-rod acts like the pin—but it is a "consumable pin" (Yu et al., 2018). In friction stir welding, the shoulder and non-consumable pin are made of the same material, which is stiffer and stronger than the workpieces. In additive friction stir deposition, the print head material needs to be significantly stronger than the "consumable pin," i.e., the feed material. Therefore, the tool in this instance simply refers to the print head, which has been hollowed out for material feeding.

During printing, the deposited material increases its temperature to enable continuous plastic deformation. The peak temperature is typically in the range of 50–90% of the melting temperature of the feed material. Under such conditions, the print head material is preferred to have minimal permanent deformation. As a result, for printing low-temperature materials such as Al, Cu, and Mg alloys, as well as the corresponding metal matrix composites, tool steel (e.g., H13) is used to make the print head (Griffiths et al., 2021). So far, the majority of additive friction stir deposition experiments have been conducted for printing Al alloys, for which tool steel-based print heads have demonstrated their efficacy with marginal tool wear and long lifetime. Tool wear becomes more prominent if ceramic particles like SiC or ZrO_2 are placed in the metal matrix. In terms of geometry, minor surface protrusions can be added to the bottom surface of the tool steel-based print head, with the aim to improve vertical material flow and mixing.

For printing high-temperature materials such as Ni-based superalloys, Ti-6Al-4V, and steels, the peak temperature of the print head can be remarkably high, e.g., more than 1000°C. The yield strength of the print head should be sufficient at such high temperatures, making tungsten-based alloys or composites an ideal choice. WC–Co cermet was initially employed for fabricating the print head (Agrawal et al., 2021). However, due to the risk of catastrophic failure, it has been gradually replaced by lanthanated tungsten (Farabi et al., 2022), which has shown efficacy in depositing large amounts of titanium and steel alloys. Another choice is the W–Re alloy, in which the addition of Re decreases the ductile to brittle transition temperature, preventing embrittlement of W when it is cooled to ambient temperature. The W–Re alloy has shown success in friction stir welding of Ti alloys and steels but has not been widely used for additive friction stir deposition yet. It is important to note that the size of the print head in additive friction stir deposition (diameter ~38–40 mm) is substantially larger than the shoulder in friction stir welding (diameter ~15–16 mm). Consequently, the fabrication challenge and cost are much higher. In addition, the feed material going through the print head creates an extra challenge in additive friction stir deposition. The expansion of the feed-rod upon compression causes internal pressure onto the print head, possibly increasing the tendency for tool failure. An economical solution for printing high-temperature material is critical for widespread usage of additive friction stir deposition, but it has remained elusive.

Fig. 5.3A shows a schematic of a possible print head geometry and its location relative to the feed-rod and deposited material. The X-axis aligns with the width of the deposit and spans from the retreating side to the advancing side. The Y-axis aligns with the print head travel direction. The Z-axis is the build direction where $Z = 0$ mm represents the top surface of the substrate plate. For the tool steel-based print head, the dashed lines represent where sets of protrusions would be located. For the W-based, high-temperature print head, no protrusions have been added due to manufacturing challenges. Fig. 5.3B–D show drawings and pictures of three tool designs that have been used for the commercialized MELD machine:

- A flat tool with no features on the bottom surface except for the central feed-rod channel for material feeding (Fig. 5.3B).
- A tool with two small protrusions (height ~ 1.5 mm) on the bottom surface near the edge of the feed-rod channel (Fig. 5.3C).

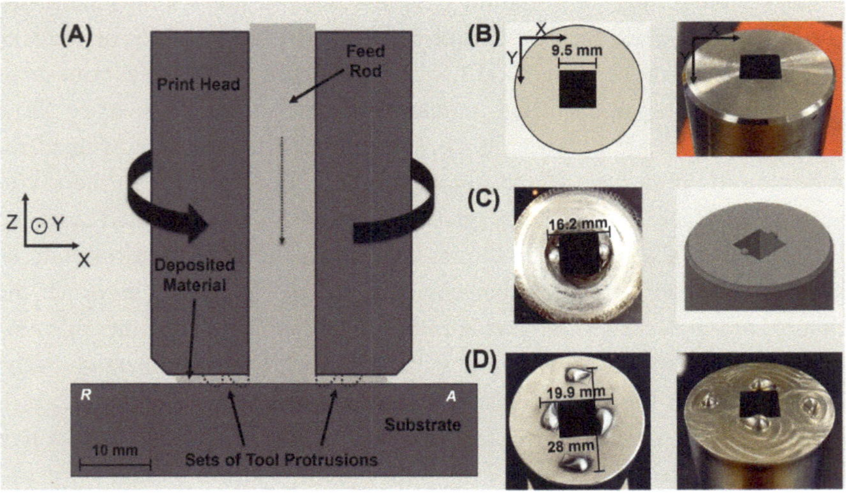

Figure 5.3 Tool geometry in additive friction stir deposition. (A) A schematic showing the print head and its relative location to the deposit and feed-rod, wherein the dashed line refers to possible surface protrusions. R and A refer to the retreating and advancing sides, respectively. Actual print heads that have been used in additive friction stir deposition are shown for (B) tool with 0 protrusions, (C) tool with 2 protrusions, and (D) tool with 4 protrusions. *Adapted from Perry, M.E.J., Yoder, Jake K., Garcia David, Sietins J. M., Yu, et al., 2022. Effects of print head geometry on additive friction stir deposition of dissimiar aluminum, Under Review.*

- A tool with four teardrop-shaped protrusions. Two protrusions are near the edge of the feed-rod channel, while the other two are located at a radius further away towards the outside surface. The orientation of the first two protrusions is perpendicular to the other two. The protrusions have a maximum height of 2.3 mm (Fig. 5.3D).

For all the tools, the opening for the feed-rod has a square cross-section of 9.525 mm × 9.525 mm; the overall diameter of the print head is 38 mm. Flat tools allow for thinner layer deposition, whereas tools with protrusions can promote deep material mixing. The consequential interfacial morphology is examined and compared next.

5.3 Effects of tool geometry on interface morphology

In this section, we present a comparative study of the interface morphology resulting from three different print head geometries: a flat bottom surface, a bottom surface with two small protrusions, and a bottom surface with four teardrop-shaped protrusions. The geometrical details of each design have been shown in Fig. 5.3B–D. This comparative study involves the deposition of an Al-Cu alloy (AA2024) onto an Al-Mg-Si alloy (AA6061) using a MELD R2 machine. Regardless of the print head geometry, all depositions are carried out at the rotation rate of 300 RPM and travel velocity of 2 mm/s. The feed rate varies from 0.9 mm/s to 1.7 mm/s to facilitate good print quality in each case. X-ray computed tomography is used to examine the surface and interface morphology (Perry et al., 2022).

Fig. 5.4 compares the surface morphology of the resulting deposition tracks. In all cases, we see a rough, circular area at the end of deposition (the top of each subfigure) as a result of the feed-rod breaking off from the deposition track. In Fig. 5.4A, the rough circle has a diameter close to the diagonal distance of the square cross-section of the feed-rod. In Fig. 5.4B, the center has a similar rough circle shape, but it is surrounded by a circular trough that represents the path of two protrusions. This is because, during deposition, the protrusions spin through the depositing material and are partially inserted into the substrate; when the print head is lifted, indents are left behind. With four teardrop-shaped protrusions, Fig. 5.4C shows separate imprints for each set of two protrusions located at different distances from the center, resulting in a defined ring of material raised above the troughs. The different morphologies at the end of deposition are a signature of specific tool geometries through tool-material interactions.

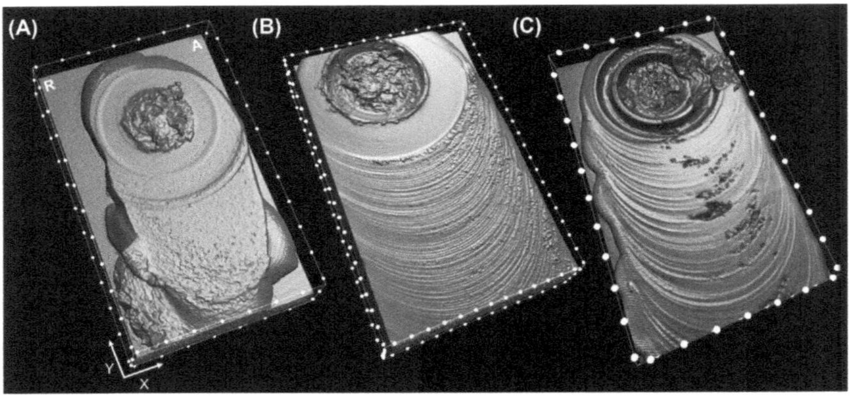

Figure 5.4 A top-down view of the surface morphology based on X-ray computed tomography for deposited AA2024 on AA6061 substrates made via additive friction stir deposition using a tool with (A) 0 protrusions, (B) 2 protrusions, and (C) 4 protrusions. The width of the deposition track is approximately 31, 32, and 47 mm, respectively. *Adapted from Perry, M.E.J., Yoder, Jake K., Garcia David, Sietins J. M., Yu, et al., 2022. Effects of print head geometry on additive friction stir deposition of dissimiar aluminum, Under Review.*

For the depositions using featured print heads (i.e., with protrusions on the bottom surface), the surface exhibits the typical "onion skin" feature seen in other friction stir-based techniques. This semicircular shape is caused by the repeated circular motion of the material as well as the interaction between the tool edge and vertical material flow, which has been proposed and explained in Chapter 3. In theory, the spacing in the onion skin should be equal to the advance per revolution of the print head. Here, regardless of the tool geometry, the main processing parameters are all the same: the rotation rate is 300 RPM; the travel velocity is 2 mm/s. Given these numbers, the calculated advance per revolution is 0.4 mm. With two small protrusions in Fig. 5.4B, the average spacing of the onion skin markings is measured to be 0.41 mm, which is very close to the theoretical value. While this spacing is consistent, it is worth noting that there are more noticeable ridges visible every 1.5–2 mm. With four teardrop-shaped protrusions in Fig. 5.4C, the fine spacing markings are invisible; instead, only 1.5–2 mm apart ridges are observed. This is possibly due to the convolution of the tool edge-material interaction with the complex material flow caused by large protrusions. With no protrusions in Fig. 5.4A, neither fine markings nor ridges are observed. This is because, in the absence of surface protrusions on the print head, the vertical material flow is minimal, resulting in a weak, if any, imprint effect by the tool edge on the trailing side.

The interface morphology is revealed in X-ray imaging by making the substrate transparent and examining the underside of the deposited material. Fig. 5.5 shows the general shape of the interface as well as important features produced by additive friction stir deposition using the three tool geometries. It is worth noting that by flipping the images, the end of deposition is now near the bottom of each sub-figure. As we change from the flat tool (Fig. 5.5A) to the tool with two small protrusions (Fig. 5.5B) and the tool with four teardrop-shaped protrusions (Fig. 5.5C), the interface morphology changes dramatically from a mostly flat shape to a bowl shape with fins and serrations, and finally to a more chaotic shape with multiple bulges and lobes.

Fig. 5.6 highlights two portions of each interface from a tilted angle, corresponding to the regions boxed by the yellow and blue dashed lines in Fig. 5.5. The interface by the flat tool is shown in Fig. 5.6A, in which a slight impression of a swirl is observed from the center out around the retreating side and to the advancing side at the end of the deposit. Even when there are no features on the bottom of the print head, a region in the center is more deformed due to compression from feeding, although the majority of the interface is flat.

The interface by the tool with two small protrusions is shown in Fig. 5.6B, in which a smooth surface is observed on the retreating side; from the center to the advancing side, semicircular features are repeated. The shape of the feature is similar to overlapping or nested pleats, with

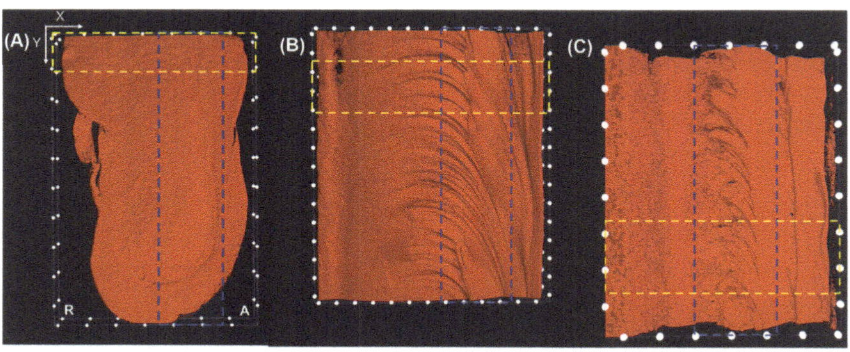

Figure 5.5 Interface morphology arising from additive friction stir deposition using (A) the flat print head, (B) print head with two small protrusions, and (C) print head with four teardrop-shaped protrusions. R and A refer to the retreating side and advancing side, respectively. *Adapted from Perry, M.E.J., Yoder, Jake K., Garcia David, Sietins J. M., Yu, et al., 2022. Effects of print head geometry on additive friction stir deposition of dissimiar aluminum, Under Review.*

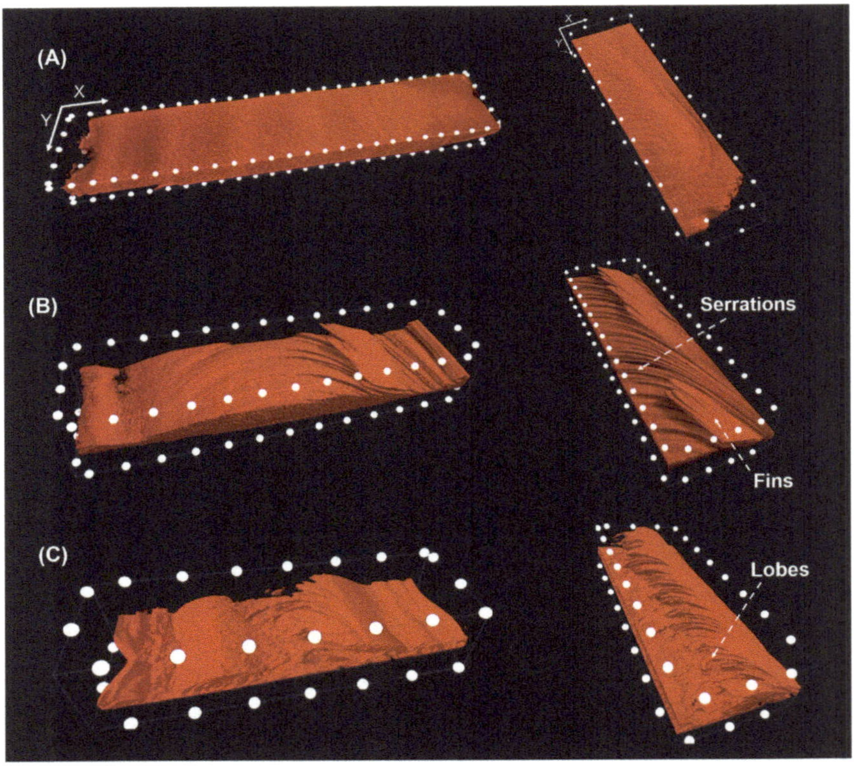

Figure 5.6 Tilted angle view of the two highlighted regions of each interface. (A) Results from the flat print head; (B) results from the print head with two small protrusions; (C) results from the print head with four teardrop-shaped protrusions. The left column corresponds to the regions boxed by the yellow dashed lines in Fig. 5.5, whereas the right column corresponds to the regions boxed by the blue dashed lines. *Adapted from Perry, M.E.J., Yoder, Jake K., Garcia David, Sietins J. M., Yu, et al., 2022. Effects of print head geometry on additive friction stir deposition of dissimiar aluminum, Under Review.*

the deposited material resembling a jagged or serrated edge from the side. As a result, these features are known as serrations. These serrations have an average spacing of about 1.2 mm. Further away from the center, the advancing side is also dominated by raised shelf-like features known as fins. Each fin blade is 0.25 mm thick, 1.5 mm wide, and 11 mm long on average. These fin shapes are relatively flat and straight, but they are oriented approximately 20 degrees off horizontal. At the root, each fin is connected to the rest of the deposited material. These fins have a 2.6 mm average spacing and are less consistent than the serrations.

The interface by the tool with four teardrop-shaped protrusions is shown in Fig. 5.6C, in which a smooth and sloping surface is also observed on the retreating side. Compared to Fig. 5.6B, instead of developing distinct serrations and fins, the resulting periodic feature appears to be a combination of both; the shapes have the same curvature and orientation as the serrations, but the depth of penetration and the maximum angle from horizontal on the advancing edge is similar to the fins. Such features are referred to as lobes because they have a more rounded appearance than the fins. From the X-ray imaging, each lobe is measured to be 0.79 mm thick, 3.5 mm wide, and 13 mm long on average. Most lobes are spaced about 1.4 mm apart on average, but the tips of the deepest lobes are 3–4 mm apart.

Fig. 5.7 shows the representative transverse cross-sections ($X-Z$ plane) of the depositions created by the three tool geometries. For these dissimilar Al alloy prints, the deposited AA2024 is in light gray and the AA6061 substrate is in dark gray. Using the flat tool, Fig. 5.7A again confirms a flat interface with almost no change to the substrate surface profile. Defects can be seen at the advancing and retreating edges, showing small air pockets between the deposit and substrate. Such flash would be removed for a structural application. Using the featured tools, Fig. 5.7B and C show a bowl-like interface where the center of the deposited material goes deeper into the substrate than the edges. Other than that, complex 3D features are prominent, especially on the advancing side. Due to large protrusion sizes, the vertical mixing is more significant in Fig. 5.7C than in Fig. 5.7B. Furthermore, when printing using tools with zero and two small protrusions, the depositions exhibit a very clear distinction at the interface between the two materials. Printing using the tool with four protrusions reveals strong mesoscale mixing and banding between AA2024 and AA6061, both of which are apparent inside the lobes.

The transverse cross-section can be used to calculate the average total width of the deposition track and the average width without flash. Using tools with 0 protrusions, 2 protrusions, and 4 protrusions, the measured average total widths are 31.06, 31.65, and 46.66 mm, respectively. The average widths without flash are 20.78, 26.58, and 39.97 mm, respectively. The flash widths are 10.28, 5.07, and 6.69 mm, respectively. Apparently, by adopting surface features on the print head, the flash width is reduced, the vertical material mixing is enhanced, and more complex 3D features are formed across the interface, such as the serrations, fins, and lobes.

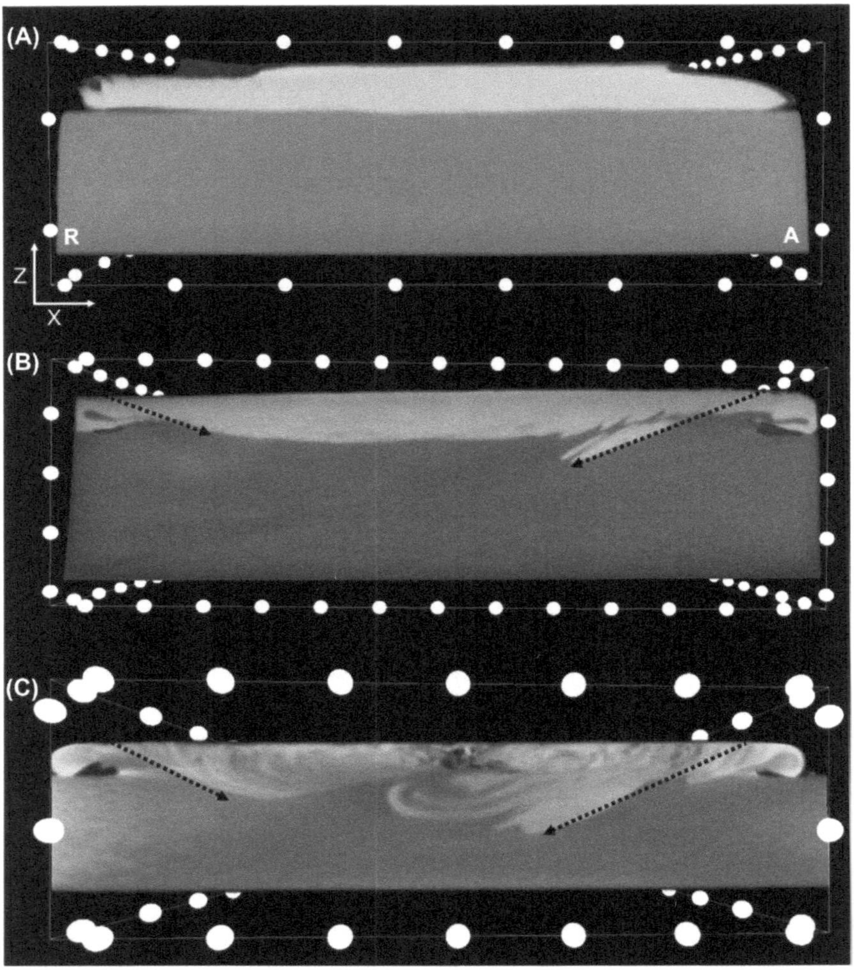

Figure 5.7 The transverse cross-section of AA2024 deposited onto AA6061 using different print head geometries, including (A) the flat print head, (B) the print head with two small protrusions, and (C) the print head with four teardrop-shaped protrusions. R and A refer to the retreating side and advancing side, respectively. *Adapted from Perry, M.E.J., Yoder, Jake K., Garcia David, Sietins J. M., Yu, et al., 2022. Effects of print head geometry on additive friction stir deposition of dissimiar aluminum, Under Review.*

For each deposition, Fig. 5.8 shows various X−Y plane slices at different heights. For the sample created using the flat tool (the first row), the deposition thickness is 2.09 mm. The X−Y plane slices are shown for $Z = 0.13$ mm (Fig. 5.8A), $Z = 0$ mm (Fig. 5.8B), $Z = -0.06$ mm (Fig. 5.8C), and $Z = -1.02$ mm (Fig. 5.8D). The contrast in these X−Y plane slices is typically

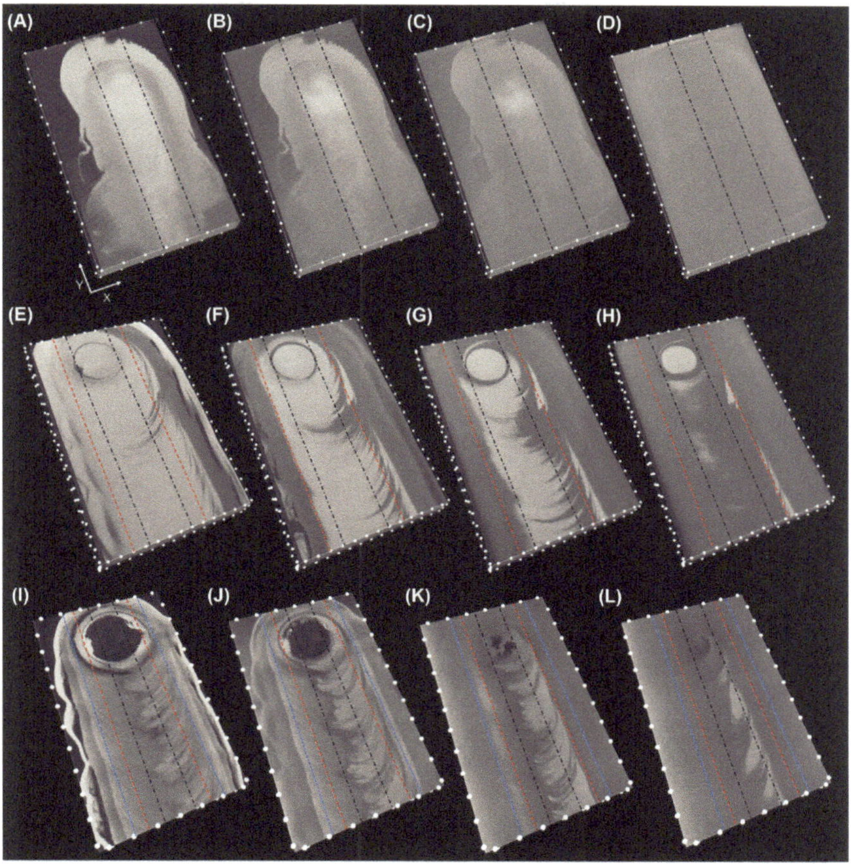

Figure 5.8 Horizontal cross-section (X–Y plane) images for deposited AA2024 (light gray) on AA6061 (darker gray) substrates by additive friction stir deposition using a tool with (A–D) 0 protrusions, (E–H) 2 protrusions, and (I–L) 4 protrusions. For each deposition, the horizontal cross-section images are shown for various *Z* heights. *Adapted from Perry, M.E.J., Yoder, Jake K., Garcia David, Sietins J. M., Yu, et al., 2022. Effects of print head geometry on additive friction stir deposition of dissimiar aluminum, Under Review.*

consistent, indicating no substantial vertical mixing for the flat tool. Having said that, there is a central band of AA2024 at $Z = 0$ mm that matches nicely with the feed material diameter. The region covered by the feed-rod cross-section during deposition is termed the feed-rod zone and is depicted by the black dashed lines. Because of the larger compression in the center, the depth reached by the deposited material is slightly larger in the center than in the perimeter.

For the sample created using the tool with two small protrusions (the second row), the deposition thickness is 1.76 mm. The X–Y plane slices are shown for $Z = 0.37$ mm (Fig. 5.8E), $Z = 0$ mm (Fig. 5.8F), $Z = -0.34$ mm (Fig. 5.8G), and $Z = -0.75$ mm (Fig. 5.8H). There is a significant central band of AA2024 at $Z = 0$ mm that matches well with the protrusion-affected zone, as indicated by the red dashed lines. As the Z height decreases, less AA2024 is visible; in the end, only the fins are visible along the advancing edge of the protrusion-affected zone.

For the sample created using the tool with four teardrop-shaped protrusions (the third row), the deposition thickness is 2.05 mm. The X–Y plane slices are shown for $Z = 0.34$ mm (Fig. 5.8I), $Z = 0$ mm (Fig. 5.8J), $Z = -1.20$ mm (Fig. 5.8K), and $Z = -2.66$ mm (Fig. 5.8L). The black, red, and blue lines correspond to the boundaries of the feed-rod zone, inner protrusion-affected zone, and outer protrusion-affected zone, respectively. There is a central band of AA2024 at $Z = 0$ mm, which can be divided into two parts: the uniform morphology of the retreating half and the layered lobes on the advancing half. Overall, the central band most closely matches the outer protrusion-affected zone, as shown by the blue dashed lines. The contrast between the two materials is not as strong as the second row due to substantial mesoscopic material mixing, which is driven by the large surface protrusions. The lobes exist deep below the substrate surface; even at $Z = -1.20$ mm, the lobes can be seen from the retreating edge of the feed-rod zone to the advancing edge of the inner protrusion-affected zone. On the retreating side, the deepest AA2024 is located between the inner and outer protrusion-affected zones.

5.4 Effects of tool geometry on microstructure

The protrusion–material interactions can generate additional frictional heat and induce significant shear. Therefore, the tool geometry governs the thermal and deformation history of the deposited material as well as the substrate, and therefore the resulting microstructure evolution processes. Fig. 5.9 shows the microstructures of the AA2024 feed-rod and AA6061 substrate before deposition, both of which consist of coarse grains with the grain size on the order of 10^2–10^3 microns. The AA2024 feed-rod is made by drawing, consisting of high aspect-ratio grains elongated along the drawing direction without a strong texture. The AA6061 substrate is made by rolling, consisting of equiaxed grains with a strong (001) texture (Perry et al., 2020).

Fig. 5.10 shows the inverse pole figure maps of the interfacial regions after printing, including the retreating side (left column) and advancing

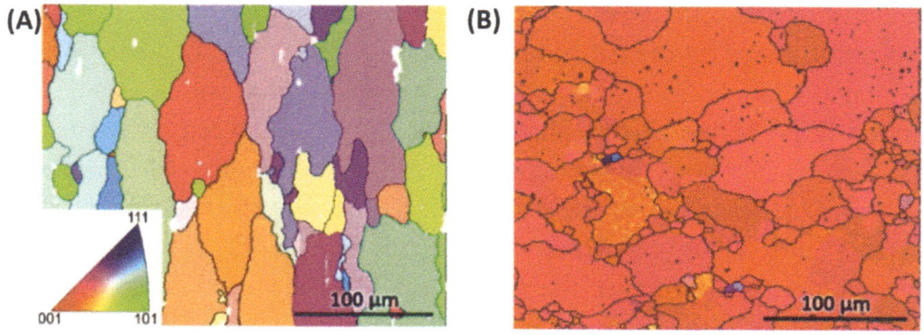

Figure 5.9 Inverse pole figure maps of the (A) AA2024 feed-rod and (B) AA6061 substrate before additive friction stir deposition. *Reprinted with permission from Perry, M.E.J., Griffiths, R.J., Garcia, D., Sietins, J.M., Zhu, Y., Yu, H.Z., 2020. Morphological and microstructural investigation of the non-planar interface formed in solid-state metal additive manufacturing by additive friction stir deposition. Additive Manufacturing 35, 101293.*

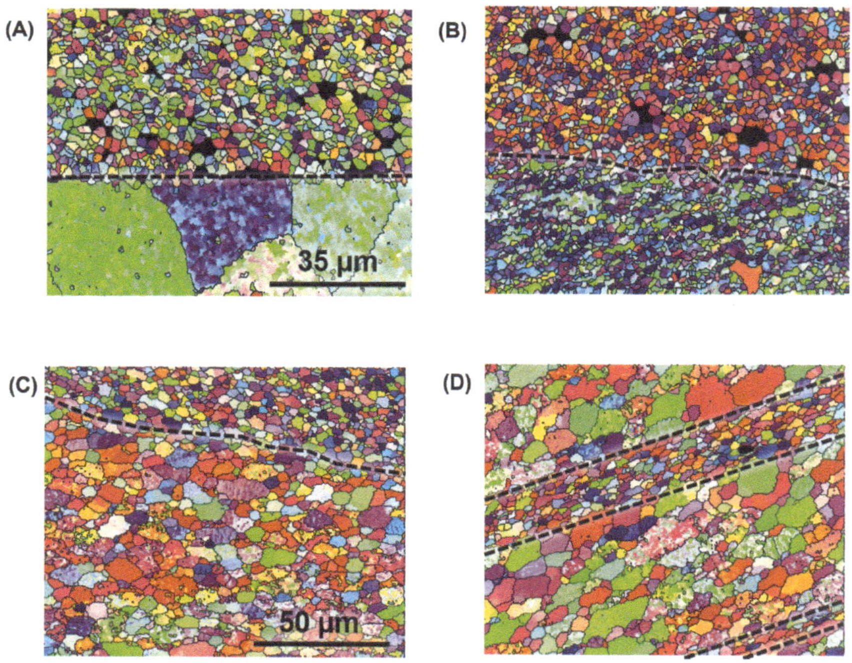

Figure 5.10 Inverse pole figure maps for the interfacial regions after printing AA2024 onto AA6061: (A) using the flat tool, the retreating side; (B) using the flat tool, the advancing side; (C) using the tool with two small protrusions, the retreating side; (D) using the tool with two small protrusions, the advancing side. *Adapted from Perry, M.E.J., Yoder, Jake K., Garcia David, Sietins J. M., Yu, et al., 2022. Effects of print head geometry on additive friction stir deposition of dissimiar aluminum, Under Review.*

side (right column). Fig. 5.10A and B correspond to the dissimilar Al alloy deposition made by the flat tool, whereas Fig. 5.10C and D correspond to the sample made by the tool with two small protrusions. The black dashed lines are used to highlight the boundaries between the deposited AA2024 and AA6061 substrate. In all cases, the deposited AA2024 exhibits a fine, equiaxed grain structure with a grain size on the order of a few microns and without a strong texture, showing evidence of significant dynamic recrystallization of the printed AA2024. When the flat print head is used, we also observe some degree of recrystallization on the advancing side of the substrate surface (Fig. 5.10B), showing reduced grain size but with signs of elongation and intragranular misorientation. However, the grain size is still large on the retreating side (Fig. 5.10A), although the strong (001) texture has disappeared. When the print head with two small protrusions is used, incomplete dynamic recrystallization and substantial grain size reduction are observed on both the advancing and retreating sides of the substrate surface.

Fig. 5.11 depicts the grain average misorientation (GAM) maps for the same four areas as in Fig. 5.10, revealing low GAM values in the AA2024 deposition in all cases. This again confirms significant dynamic recrystallization occurring in the printed material regardless of the tool geometry. Using the flat print head, the advancing side of the substrate surface (Fig. 5.11B) shows fine grains with a mixture of low and high GAM values, suggesting incomplete recrystallization. This phenomenon is observed on both the advancing and retreating sides of the substrate surface using the print head with two small protrusions (Fig. 5.11C and D). However, large grains with consistently high GAM values are detected on the retreating side of the substrate surface resulting from the flat print head (Fig. 5.11A). Therefore, the thermomechanical condition for the feed material of AA2024 is sufficient for complete dynamic recrystallization regardless of whether a flat tool or a tool with two small protrusions is used. For the latter, the presence of rotating protrusions causes substantial, but incomplete recrystallization in the surface layers of the substrate.

In the absence of protrusions, only the advancing side of the substrate surface shows the sign of dynamic recrystallization, whereas the retreating side only shows a decrease of the rolling texture but no significant grain size reduction. This implies that the deformation on the retreating side of the substrate is too low to cause dynamic microstructure evolution. This results in asymmetric microstructures across the deposition width, which can be attributed to the different mechanical constraints between the two sides. When the print

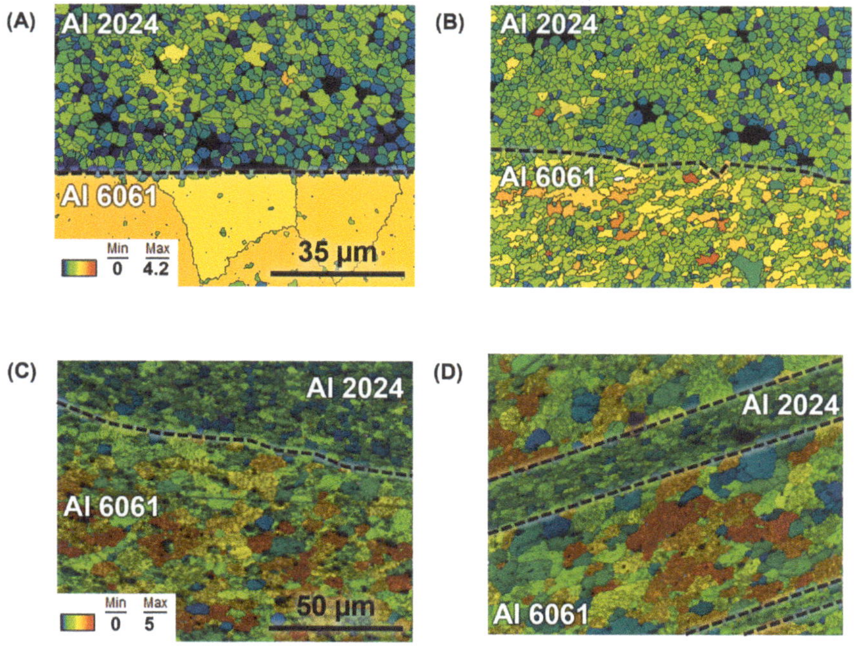

Figure 5.11 Grain average misorientation maps for the interfacial regions after printing AA2024 onto AA6061: (A) using the flat tool, the retreating side; (B) using the flat tool, the advancing side; (C) using the tool with two small protrusions, the retreating side; (D) using the tool with two small protrusions, the advancing side. *Adapted from Perry, M.E.J., Yoder, Jake K., Garcia David, Sietins J. M., Yu, et al., 2022. Effects of print head geometry on additive friction stir deposition of dissimiar aluminum, Under Review.*

head travels, the in-plane expansion of the feed-rod favors the trailing edge over the leading edge. As a result, the influx of material from the trailing edge causes significant material accumulation on the advancing side. To compare, the influx of material flowing from the advancing side to the retreating side has low in-plane restrictions due to the space beneath the print head.

When the tool with four teardrop-shaped protrusions is used, the large dimensions of the protrusions give rise to significant material flow in the substrate surface region to allow for complete dynamic recrystallization. As shown in Fig. 5.12A, both the deposited AA2024 and AA6061 substrate exhibit a fine, equiaxed grain structure. Fig. 5.12B shows that the grain boundary misorientation distributions in AA2024 and AA6061 are close to the Mackenzie distribution that would be expected from polycrystals with a random distribution of the crystallographic orientation. This indicates that the dynamic recrystallization in both materials is nearly complete, as a result of large shear deformation at elevated temperatures.

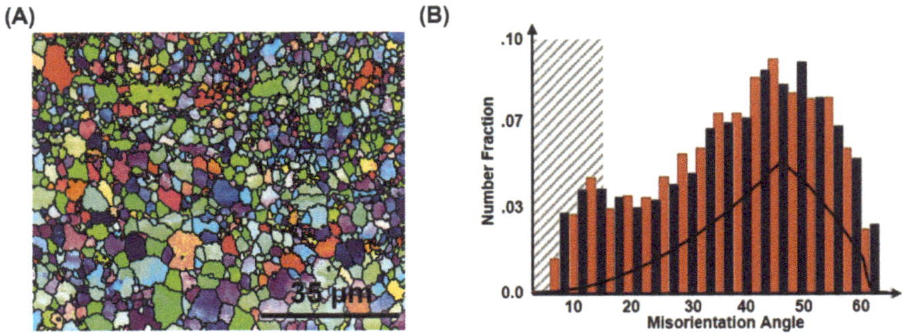

Figure 5.12 Additive friction stir deposition using the tool with four teardrop-shaped protrusions. (A) The inverse pole figure map of the interfacial region. (B) Grain boundary misorientation distribution of the two materials in the interfacial region. *Adapted from Perry, M.E.J., Yoder, Jake K., Garcia David, Sietins J. M., Yu, et al., 2022. Effects of print head geometry on additive friction stir deposition of dissimiar aluminum, Under Review.*

5.5 Summary

Similar to friction stir welding, the tool type and geometry play a critical role in additive friction stir deposition. For printing low-temperature materials like Al, Mg, and Cu alloys, the print head is generally made of tool steel, with the bottom surface being featureless or featured by several protrusions. For printing high-temperature materials like Ni-based superalloys, Ti-6Al-4V, and steels, the W-based print head with a flat bottom surface is typically used. A comparative study has been conducted for printing AA2024 onto AA6061, including three tool designs with the bottom surface containing 0, 2, and 4 protrusions. As the number and size of the protrusions increase, more vertical material flow and mixing are observed, resulting in more complex 3D interfacial features, like fins, serrations, and lobes. In the presence of protrusions, the flash width is seen to be reduced. Regardless of the tool geometry, the deposited AA2024 is seen to undergo significant dynamic recrystallization, resulting in a fine, equiaxed grain structure. The deformation level of the substrate surface and therefore the degree of dynamic recrystallization are found to be highly influenced by the protrusions.

References

Agrawal, P., Haridas, R.S., Yadav, S., Thapliyal, S., Gaddam, S., Verma, R., et al., 2021. Processing-structure-property correlation in additive friction stir deposited Ti-6Al-4V alloy from recycled metal chips. Additive Manufacturing 47, 102259.

Farabi, E., Babaniaris, S., Barnett, M.R., Fabijanic, D.M., 2022. Microstructure and mechanical properties of Ti6Al4V alloys fabricated by additive friction stir deposition. Additive Manufacturing Letters 2, 100034.

Griffiths, R.J., Garcia, D., Song, J., Vasudevan, V.K., Steiner, M.A., Cai, W., et al., 2021. Solid-state additive manufacturing of aluminum and copper using additive friction stir deposition: process-microstructure linkages. Materialia 15, 100967.

Ilangovan, M., Rajendra Boopathy, S., Balasubramanian, V., 2015. Effect of tool pin profile on microstructure and tensile properties of friction stir welded dissimilar AA 6061–AA 5086 aluminium alloy joints. Defence Technology 11, 174–184.

Khan, N.Z., Khan, Z.A., Siddiquee, A.N., 2015. Effect of shoulder diameter to pin diameter (D/d) ratio on tensile strength of friction stir welded 6063 aluminium alloy. Materials Today: Proceedings 2, 1450–1457.

Memon, S., Fydrych, D., Fernandez, A.C., Derazkola, H.A., Derazkola, H.A., 2021. Effects of FSW tool plunge depth on properties of an Al-Mg-Si Alloy T-joint: thermomechanical modeling and experimental evaluation. Materials (Basel) 14, 4754.

Mishra, R.S., De, P.S., Kumar, N., 2014. Friction Stir Welding Configurations and Tool Selection, Friction Stir Welding and Processing: Science and Engineering. Springer International Publishing, Cham, pp. 95–108.

Perry, M.E.J., Griffiths, R.J., Garcia, D., Sietins, J.M., Zhu, Y., Yu, H.Z., 2020. Morphological and microstructural investigation of the non-planar interface formed in solid-state metal additive manufacturing by additive friction stir deposition. Additive Manufacturing 35, 101293.

Perry, M.E.J., Gotawala N., Garcia D., Sietins J.M., Yu, H.Z., 2022. Effects of print head geometry on additive friction stir deposition of dissimilar aluminum alloys, Under Review.

Rai, R., De, A., Bhadeshia, H.K.D.H., DebRoy, T., 2011. Review: friction stir welding tools. Science and Technology of Welding and Joining 16, 325–342.

Singh, K., Singh, G., Singh, H., 2018. Review on friction stir welding of magnesium alloys. Journal of Magnesium and Alloys 6, 399–416.

Yu, H.Z., Jones, M.E., Brady, G.W., Griffiths, R.J., Garcia, D., Rauch, H.A., et al., 2018. Non-beam-based metal additive manufacturing enabled by additive friction stir deposition. Scripta Materialia 153, 122–130.

CHAPTER 6

Beyond metals and alloys: additive friction stir deposition of metal matrix composites

The whole is greater than the sum of its parts.

—Aristotle

The friction stir principle can be leveraged for processing materials beyond metals. For example, friction stir processing is known to be an efficient method for producing *surface* metal matrix composites. Similarly, additive friction stir deposition can be employed to produce *bulk-scale* metal matrix composites. During printing, the soft metal matrix undergoes severe plastic deformation at elevated temperatures, featuring extensive material flow. The metal matrix has a lower melting temperature than the reinforcement phase, which is usually made of refractory metal or ceramic particles. The compression and shear stress states, along with the extensive metal flow around the reinforcement, typically render good quality in the as-printed composites. Not just from an additive standpoint but also from a bulk processing perspective, additive friction stir deposition holds promise for the high-quality manufacturing of metal matrix composites in the future. Its major limitation lies in the maximum attainable volume fraction of the reinforcement.

In this chapter, we present an overview of research on additive friction stir deposition of metal matrix composites, with comparisons to conventional bulk processing and additive manufacturing approaches. We start with an introduction to metal matrix composites, including their types, applications, and advantages over metals and polymer matrix composites. Next, we discuss the bulk processing (both liquid and solid routes) and additive manufacturing approaches (e.g., powder bed fusion, directed energy deposition, and sheet lamination). We then introduce the potential of additive friction stir deposition and present the state-of-the-art experimental work on printing Al−SiC, Al−ZrO_2, Cu−ZrO_2, and Cu−SiC composites. The discussion focuses on the feeding mechanism, print principle, and print quality.

Additive Friction Stir Deposition
DOI: https://doi.org/10.1016/B978-0-12-824374-9.00007-1

6.1 Introduction to metal matrix composites

Modern engineering applications place strict multifunctional criteria on the materials being used. A supporting structural material, for example, should be able to bear significant compressive stress, while isolating thermal flow to maintain temperature stability. The shielding material should be able to absorb a lot of energy while remaining lightweight to allow for mobility. By integrating the benefits of multiple constituent materials, a composite offers promise to address these challenges, especially for the automotive, aerospace, defense, and infrastructure sectors.

By definition, composites are materials made up of two (or more) distinct phases, one of which serves as a matrix and the other as reinforcement (Chawla and Chawla, 2013a). Metal matrix composites consist of a metallic matrix, which is usually a lightweight alloy, and a reinforcement phase, which can be in the form of hard particles, whiskers, chopped fibers, or continuous fibers. Fig. 6.1 illustrates several geometrical configurations. Owing to their superior strength-to-weight and modulus-to-weight ratios, Al, Mg, and Ti-based alloys are the most used matrix materials (Valente et al., 2019). The reinforcement phase can be ceramics, refractory metals, or intermetallics. For example, silicon carbide (SiC) and alumina (Al_2O_3) ceramics are widely utilized in particle-reinforced metal matrix composites (Campbell, 2003). The inclusion of reinforcement improves the mechanical and thermal performance of the base metal while preserving many of its physical properties, such as low density and good

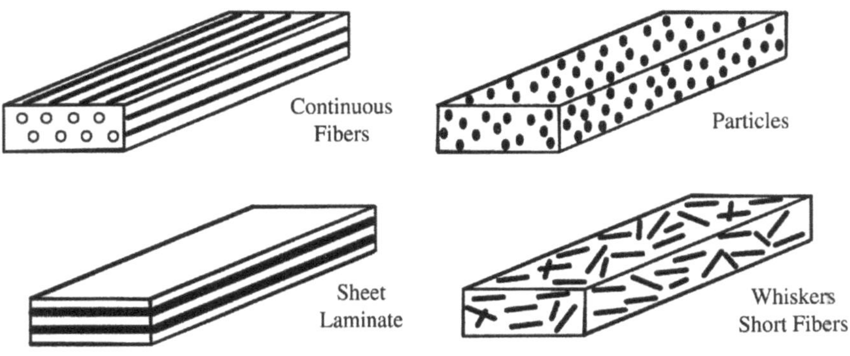

Figure 6.1 Geometrical configurations of metal matrix composites. Examples are shown for various reinforcement forms, including continuous fibers, laminate sheets, particles, and whiskers or short fibers. *Reprinted with permission from Chawla, N., Chawla, K.K., 2013a. Introduction. In: Chawla, N., Chawla, K.K. (Eds.), Metal Matrix Composites. Springer New York, New York, pp. 1–4.*

electrical conductivity. A range of metal matrix composite applications has been demonstrated, such as ballistic armors, tools, structural parts, and thermal regulators (Rajak et al., 2019). Typically, the reinforcement volume fraction is limited to 15%–25% because greater volumes result in poor ductility and low fracture toughness.

Due to their low density, environmental resistance, and superior mechanical qualities, Al-matrix composites have been the most commercially developed metal matrix composites to date (Torralba et al., 2003). The melting temperature of Al is high enough to suit the needs of many critical applications while being sufficiently low for reliable processing. Al also supports a wide variety of reinforcing phases. Although much of the early study was focused on continuous fibers, the majority of current research is focused on discontinuous fibers or particles since they are easier and cheaper to manufacture with isotropic properties. For particle-reinforced Al matrix composites, the reinforcement phase can be made of SiC, Al_2O_3, Si_3N_4, TiB_2, and graphite. The latter has been developed for tribological applications given their superior antifriction, wear resistance, and antiseizure capabilities (Baradeswaran and Perumal, 2014).

The advantages of metal matrix composites are evident through comparisons to alternative materials. Concerning metals, metal matrix composites offer major weight savings, which is due to the higher specific mechanical properties, as well as more resistance to high-temperature exposure. Concerning polymer matrix composites, metal matrix composites offer prominently higher strength and stiffness, and resilience under harsh conditions. Fig. 6.2 compares the material resistance against distortion from thermal or mechanical loads among metals, ceramics, polymer matrix composites, and metal matrix composites (Miracle, 2005). Graphite/epoxy offers the best mechanical distortion resistance but the worst resistance to thermal distortion. Ceramic materials, diamond, and Be alloys all have high resistance to thermal and mechanical loads, but they are unsuitable for most uses: ceramics have poor fracture toughness, diamond is of high cost, and Be alloys are of high health risks. Metal matrix composites outperform all major structural metals in both thermal and mechanical distortion resistance. Metal matrix composites are also characterized by multifunctional capabilities in thermal conduction, electrical conduction, and material joining (Ellis, 1996).

Despite these salient advantages, commercial applications for metal matrix composites are indeed sparse. This is largely owing to the challenges in processing. One challenge is related to the processing quality;

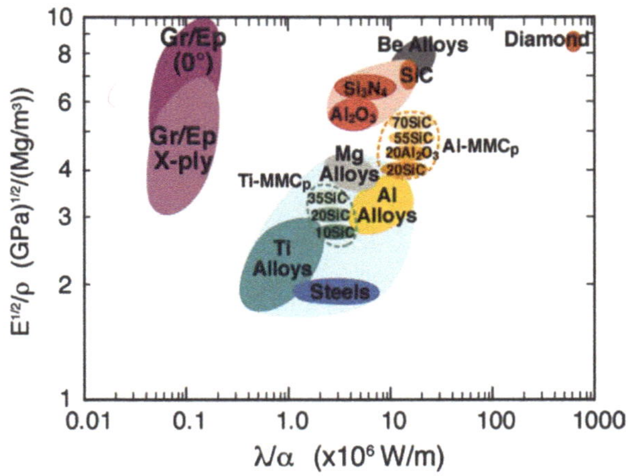

Figure 6.2 Comparison of materials for mechanical (vertical axis) and thermal (horizontal axis) distortion resistance necessary for precision devices. *Reprinted with permission from Miracle, D.B., 2005. Metal matrix composites—from science to technological significance. Composites Science and Technology 65, 2526–2540.*

issues such as porosity and air gap, uneven reinforcement dispersion, and the weak interface between matrix and reinforcement, are prevalent. Another challenge is related to the processing cost, which is considerably higher than the processing of metals or polymer matrix composites.

6.2 Current processing approaches to metal matrix composites

6.2.1 Bulk processing

Bulk processing of metal matrix composites can be implemented by liquid or solid approaches. The former includes stir casting, squeeze casting, and pressure infiltration. The latter is mainly based on powder metallurgy; thermomechanical processing like hot extrusion and forging can be subsequently employed to improve the composite quality. Below we provide a brief introduction of each processing approach.

6.2.1.1 Liquid-state processing: stir casting

Stir casting is the most common commercial method of producing Al-matrix composites, wherein the melt is actively stirred while solidifying and mixing with the reinforcing phase. The mixing is typically enhanced by a motor-driven stirring mechanism or an ultrasonic transducer (Kareem et al., 2021). Fig. 6.3 shows a diagram of the stir casting mixing apparatus.

The processing condition should be carefully controlled to minimize the porosity, enhance the wetting of the metal matrix on the reinforcement, and render homogenous reinforcement dispersion. Consider stir casting of Al–SiC composites as an example. Because SiC has a higher density (3.2 g/cm^3) than Al (2.7 g/cm^3), it has a propensity to sink, resulting in nonuniform reinforcement distribution in the solidified composite. A robust stirring strategy is critical throughout the solidification process. In addition to mechanical vibration, alternating currents in a magnetic field can be leveraged to promote uniform reinforcement dispersion (Ahmad et al., 2014). Since the casting temperature is high, chemical reactions may occur between the Al matrix and SiC reinforcement, giving rise to undesired compounds like Al_4C_3. Therefore, alloys of low reactivity with reinforcement should be used as the matrix material. For this reason, Al–Si alloys are commonly used as the matrix in Al–SiC composites. If the casting temperature is elevated, more silicon should be added to the matrix to avoid chemical reactions (Lloyd, 1997).

6.2.1.2 Liquid-state processing: squeeze casting

The nature of stir casting makes it challenging to fabricate intricate geometries with good quality. This necessitates the use of more expensive

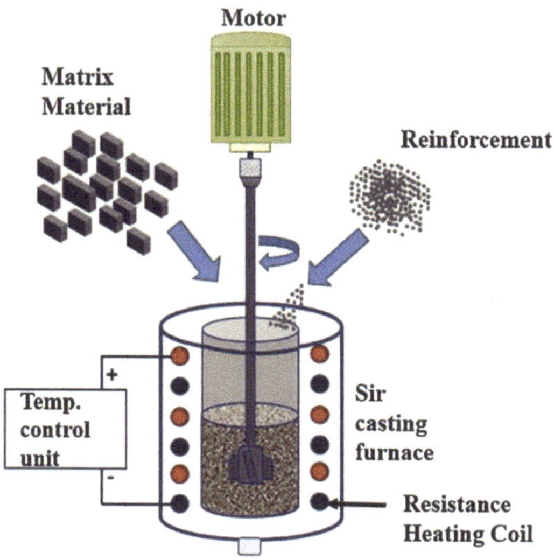

Figure 6.3 A schematic showing the stir casting facility. *Reprinted with permission from Ramanathan, A., Krishnan, P.K., Muraliraja, R., 2019. A review on the production of metal matrix composites through stir casting—Furnace design, properties, challenges, and research opportunities. Journal of Manufacturing Processes 42, 213–245.*

approaches like squeeze casting, which leverages pressure or mechanical force to drive the liquid metal matrix into the reinforcement preform. During solidification of the metal matrix, pressure is typically applied in the range of 70–100 MPa. The reinforcement preform can be manufactured by press forming. This involves pouring an aqueous slurry of particles or short fibers into a mold, pressing the water out, drying, and demolding.

Fig. 6.4A illustrates the squeeze casting setup (Ramanathan et al., 2019). Squeeze casting allows for the fabrication of rather complicated geometries because the applied pressure eliminates the residual porosity or shrinkage voids in difficult-to-reach corners and turnings. One primary advantage of this method over nonpressured casting is the reduced processing time (Chawla and Chawla, 2013b), which is critical for large-scale manufacturing. The reduced processing time also leads to less grain growth and smaller grain sizes. This is beneficial because the large matrix grains could become the source of weak bonding interfaces. Fig. 6.4B shows the microstructure of an AA6061 matrix composite reinforced with 50% SiC particles using the squeeze casting method, demonstrating fine interparticle spacing with no residual porosity. To achieve preform infiltration, the molten metal must have low viscosity and good wettability with the reinforcement. While increasing the casting temperature

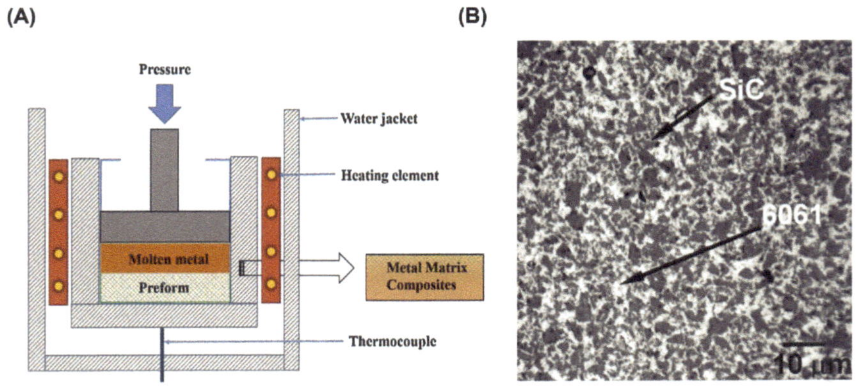

Figure 6.4 (A) Configuration of squeeze casting. (B) A micrograph of AA6061-SiC with 50 vol.% reinforcement made by squeeze casting. *Reprinted with permission from Ramanathan, A., Krishnan, P.K., Muraliraja, R., 2019. A review on the production of metal matrix composites through stir casting—Furnace design, properties, challenges, and research opportunities. Journal of Manufacturing Processes 42, 213–245. Reprinted with permission from Saha, R., Morris, E., Chawla, N., Pickard, S.M., 2002. Hybrid and conventional particle reinforced metal matrix composites by squeeze infiltration casting. Journal of Materials Science Letters 21, 337–339.*

enhances both, it also increases the likelihood of undesired chemical reactions. In this regard, it is critical to keep the preform temperature below the liquid matrix temperature (Saha et al., 2002).

6.2.1.3 Solid-state processing: powder metallurgy

In addition to the liquid phase bonding mechanism, high-quality interfaces can be established between the matrix and reinforcement via diffusion bonding by solid-state processing routes. As an efficient and cost-effective processing choice, powder metallurgy is commonly employed, in which the matrix and reinforcement powders are mixed and blended, consolidated, sintered, and machined to form shaped parts. Before the final machining step, the consolidated billets are often subjected to thermomechanical processing, such as extrusion, rolling, and forging. The thermomechanical processing step helps to eliminate the residual porosity and modify the grain structure in the metal matrix. Using this approach, composites based on Al, Cu, Ti, and other matrices have been successfully manufactured with decent mechanical properties (Guo et al., 2022; Li et al., 2013; Scudino et al., 2009).

Because smaller particle sizes can help to prevent particle breakage and improve interface bonding quality, nanoscale reinforcement, such as ceramic nanoparticles, carbon nanotubes, and graphene, has been incorporated into the metal matrix (Matli et al., 2017). As an example, Fig. 6.5 shows the processing procedure and tensile properties of $Al-Si_3N_4$ nanocomposites created by powder mixing, cold compaction, microwave sintering, and hot extrusion. Si_3N_4 nanoparticles of 0.5, 1, and 1.5 vol.% are added into the pure Al matrix. The hardness of these composites is measured to be 58 $\pm$ 4, 72 $\pm$ 3, and 101 $\pm$ 5 HV, respectively, whereas the hardness of pure Al is only 37 $\pm$ 3 HV. By only adding 1.5 vol.% Si_3N_4 nanoparticles, the Young's modulus increases from 73 to 88 GPa. The yield strength and ultimate tensile strength increase from 105 and 116 MPa to 165 and 191 MPa. At the same time, the tensile elongation drops from 13.6% to 7.2%. In compression, all the composites and pure Al can afford a failure strain larger than 70%. In addition to mixing and blending the matrix and reinforcement ex situ, the nanoscale reinforcement can be directly formed on the metal particles. For example, graphene nanosheets have been grown on Cu particles via chemical vapor deposition, and the resulting Cu-graphene particles are consolidated to form metal matrix composites (Chen et al., 2016). Because of the chemical bonding formed across the interface, the interfacial shear strength is significantly higher than that by other processing means.

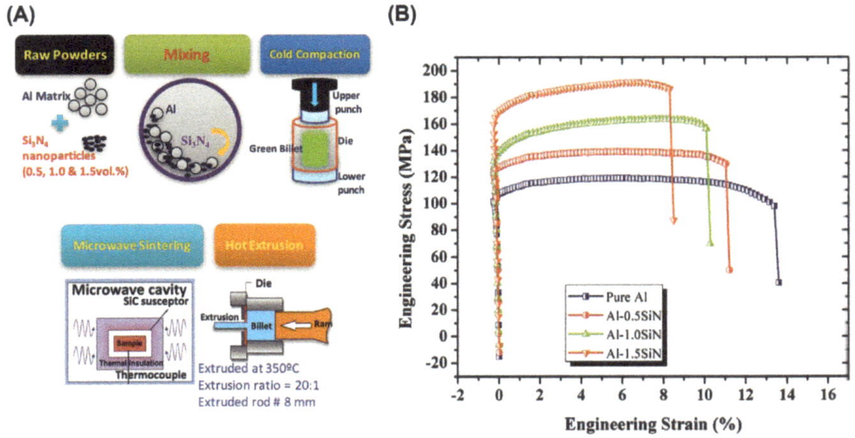

Figure 6.5 Powder processing for fabricating Al-Si_3N_4 nanocomposites. (A) Fabrication procedure involving powder mixing, cold compaction microwave sintering, and hot extrusion. (B) Representative tensile properties of the created composites including 0.5, 1, and 1.5 vol.% nanoparticles, with a comparison to the pure Al. *Reprinted with permission from Matli, P.R., Ubaid, F., Shakoor, R.A., Parande, G., Manakari, V., Yusuf, M., et al., 2017. Improved properties of Al–Si3N4 nanocomposites fabricated through a microwave sintering and hot extrusion process. RSC Advances 7, 34401–34410.*

The key advantages of solid-state processing lie in the absence of melting and solidification, which reduces the element segregation, undesired chemical reactions, and large distortion from solidification shrinkage. Powder processing also enables mass production, good mixing between constituent materials, and material recycling. The main drawbacks include the component shape limits imposed by the die geometry and the high cost of postprocessing.

6.2.2 Additive production

Additive manufacturing brings new paths towards net-shaping or near-net-shaping of metal matrix composites with efficiency. In this section, we briefly review the recent breakthroughs in this realm, including powder bed fusion, directed energy deposition, and sheet lamination.

6.2.2.1 Powder bed fusion

Powder bed fusion refers to technologies that selectively melt the powder in a layer based on a predesign pattern, followed by spreading a new layer of powder and repeating the process until a 3D component is built up

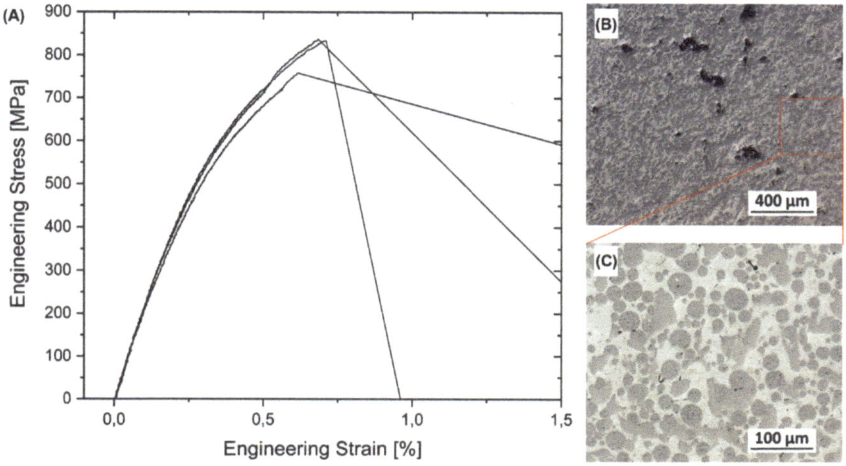

Figure 6.6 Selective melting of Ms1 steel-60 vol.% W composites. (A) Tension test results of three specimens. (B) and (C): Microstructure of the composites at low and high magnifications. *Reprinted with permission from Jäcklein M., Pfaff A., Hoschke K., 2020. Developing tungsten-filled metal matrix composite materials using laser powder bed fusion. Applied Sciences 10, 8869.*

(Gu, 2015b). There are attempts to employ powder bed fusion for additive manufacturing of metal matrix composites (AlMangour et al., 2017; Gu, 2015a; Gu et al., 2009; Wang and Gu, 2015), in which the metal powder is selectively melted and bonded to the reinforcement. For example, selective laser melting has been used to print maraging steel (Ms1)-tungsten composites, with up to 60 vol.% of reinforcement achieved. The printing is carried out using the laser power of 90 W and laser travel velocity of 500 mm/s. Fig. 6.6A shows the tensile testing results from three printed specimens, whose ultimate tensile strengths are measured as 758, 833, and 836 MPa and elongations are between 0.5% and 1%. While some plastic deformation is observed, these numbers are below the ultimate tensile strength of pure Ms1 steel. This is caused by the prevalent presence of porosity in the printed composites, as seen in Fig. 6.6B. The tungsten particle distribution in the steel matrix is shown in Fig. 6.6C. The failure of the composites is found to be driven by cleavage through tungsten particles, indicating decent interface bonding (Jäcklein et al., 2020).

Selective laser melting of metal matrix nanocomposites is also of great interest, as they show greater promise in wear resistance, damping properties, and mechanical strength than their microparticle-reinforced counterparts. For example, aluminum matrix composites reinforced by 3 vol.% Al_2O_3

nanoparticles outperform those reinforced with 10 vol.% Al_2O_3 microparticles or 10 vol.% SiC microparticles (Kang and Chan, 2004; Tjong, 2007). Along with porosity and hot cracking typically seen in systems with microsized reinforcement particles, selective laser melting of nanocomposites faces additional challenges from the small particle size and high specific surface area, which inevitably lead to poor wettability (Yu et al., 2019). In addition, the nanoparticles need to be cost-effective to manufacture, blend well with the matrix powder, and not agglomerate in the molten pool.

6.2.2.2 Directed energy deposition

Directed energy deposition simultaneously applies a high energy beam (e.g., laser) and supplies the feedstock to the same location of interest. The feedstock first melts and then solidifies onto the existing structure. As the print head advances, multiple layers of thin-walled structures or surfaces can be completed. A few metal matrix composites have been printed using this method, such as titanium alloys reinforced with boron carbide particles for improved mechanical properties (Nartu et al., 2020), titanium alloys reinforced with carbon nanotubes for improved tribological behavior (Stenberg et al., 2018), and titanium alloys reinforced with hydroxyapatite for high-performance biomedical implants (Avila et al., 2021). The flexibility of directed energy deposition allows for the inclusion of multicomponent feed materials and multiple feed paths. Premixed or preformed powders can be printed, and the local composition can be easily controlled. This allows for the printing of functionally graded materials and composites (Hofmann et al., 2014; Liu and DuPont, 2003; Oyelola et al., 2018). Fig. 6.7 presents the microstructure of a Ti–TiC functionally graded composite printed by laser engineered net shaping. In the very top layer, there is 95% TiC, which serves as the matrix; Ti appears as islands or particles. The bottom layer mainly consists of Ti (HCP phase) with a very small amount of TiC. Because of the high energy intensity, both Ti and TiC get melted during deposition. Energy dispersive spectroscopy suggests that the dendritic phase is the resolidified TiC. In addition, the microstructure contains a minor number of smaller equiaxed TiC particles from solidification. The un-melted TiC particles are generally large and blocky.

6.2.2.3 Sheet lamination

Additive manufacturing of metal matrix composites may be implemented via ultrasonic additive manufacturing, which belongs to sheet lamination.

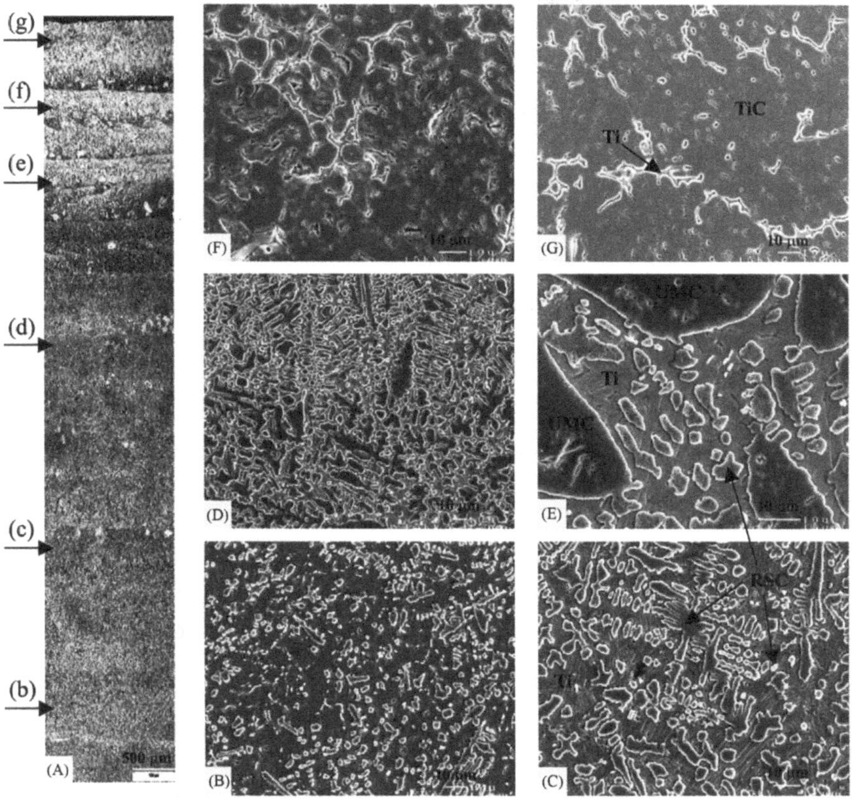

Figure 6.7 Microstructures of the functionally graded Ti–TiC composite printed by laser engineered net shaping. *Reprinted with permission from Liu, W., DuPont, J.N., 2003. Fabrication of functionally graded TiC/Ti composites by laser engineered net shaping. Scripta Materialia 48, 1337–1342.*

This process is based on the ultrasonic welding principle, wherein two ribbons of the same or different materials are placed in intimate contact under a vibrating sonotrode with a normal force. Through plastic deformation and dynamic recrystallization in the interfacial regions, surface oxides are removed, and surface asperities are collapsed and mixed, resulting in metallurgical bond formation across the interface. This process is not a purely additive process as CNC (computer numerical controlled) machining needs to be leveraged for controlling the shape of the bonded layers. With the ultrasonic bonding and CNC machining processes repeated across many layers, large-scale 3D components are readily built with a full density and a high degree of architectural control. Ultrasonic additive manufacturing has been used to create hybrid materials, which are made up of a preformed composite and a pure matrix material. An Al alloy$-Al_2O_3$ fiber composite, for example, has been bonded to alternating layers of an unreinforced Al alloy (Selvaraj et al., 2021). Beyond metallic layers, meshes, fibers, and embedded electronics can be incorporated into the builds, rendering new types of hybrid and composite materials. Fig. 6.8 illustrates how an electronic component with a conductive through-hole via and electrical insulators can be embedded in the metal matrix using the ultrasonic welding process (Bournias-Varotsis et al., 2018).

Despite a few demonstrations in previous studies, the aforementioned additive manufacturing technologies all have fundamental drawbacks that render them unsuitable for printing metal matrix composites reliably. For powder bed fusion and directed energy deposition, the issues arise from the melting and solidification processes, which inevitably compromise the porosity control, columnar and dendritic microstructure development, and residual stress mitigation (Frazier, 2014). It also limits the choice of metals to use as the matrix. For example, Al alloy has a high viscosity in the molten state and therefore low flowability, whereas Cu has high reflectivity that makes energy

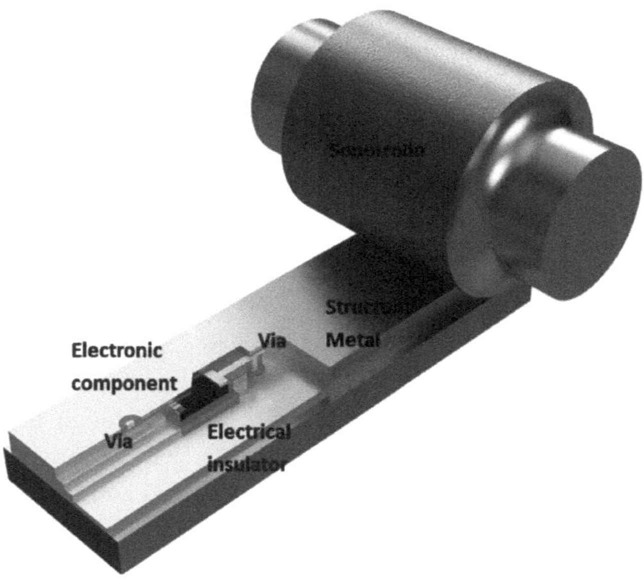

Figure 6.8 Illustration of an electronic component (*black*) with a conductive through-hole via and electrical insulators (*green*) embedded in the metal matrix by the ultrasonic welding. *Reprinted with permission from Bournias-Varotsis, A., Friel, R.J., Harris, R.A., Engstrøm, D.S., 2018. Ultrasonic additive manufacturing as a form-then-bond process for embedding electronic circuitry into a metal matrix. Journal of Manufacturing Processes 32, 664–675.*

absorption inefficient (Horn and Gamzina, 2020). Both Al and Cu have high thermal conductivity, resulting in rapid heat dissipation, high local thermal gradients, and possibly delamination and component failure. Furthermore, a significant number of Al alloys like 2xxx and 7xxx Al are nonweldable and known for hot cracking (Zhou et al., 2020).

More importantly, insufficient wettability of the liquid matrix on the reinforcing surface frequently causes weak interface bonding. Al alloys, for example, have a low wetting capability against the reinforcement materials like TiB_2, TiC, and carbon nanotubes. This likely results in small gaps or microcracks at the matrix-reinforcement interface in the as-printed composite. The reinforcing particles must also resist agglomeration and maintain uniform distribution. However, as the reinforcing particle size decreases in the molten pool, the tendency to agglomerate becomes unavoidable.

Finally, it is generally recommended to avoid chemical reactions on the interface, which can produce undesired compounds that fail the mechanical integrity. For example, Al−SiC composites printed by powder bed fusion would experience a significant increase of crack density when the volume fraction of SiC particles goes beyond 15%. This is possibly due to the chemical reactions between Al and SiC, forming Al_4C_3 between 940K and 1620K and forming Al_4SiC_4 at even higher temperatures. Fig. 6.9A shows the typical X-ray diffraction patterns of the Al-based composite printed by selective laser melting. As the linear laser energy density increases, more compounds like Al_4SiC_4 and Mg_2Si are present. At relatively low linear laser energy density (800 J/m), some residual SiC particles can be observed in Fig. 6.9B. As the linear laser energy density increases to 1000 J/m, in situ formed Al_4SiC_4 starts to be observed in Fig. 6.9C.

To prevent these issues, one may seek solid-state solutions for printing metal matrix composites. While ultrasonic additive manufacturing is a solid-state process that avoids melting, achieving good particle mixing with uniform distribution is extremely challenging because the critical processing step only occurs at the layer interface. In addition, it is a hybrid process in nature that cannot perform site-specific deposition or printing.

6.3 Additive friction stir deposition of metal matrix composites

Additive friction stir deposition is also a solid-state approach, but it is much more intense than ultrasonic additive manufacturing. The entire feed material is stirred during printing, which promotes the uniform

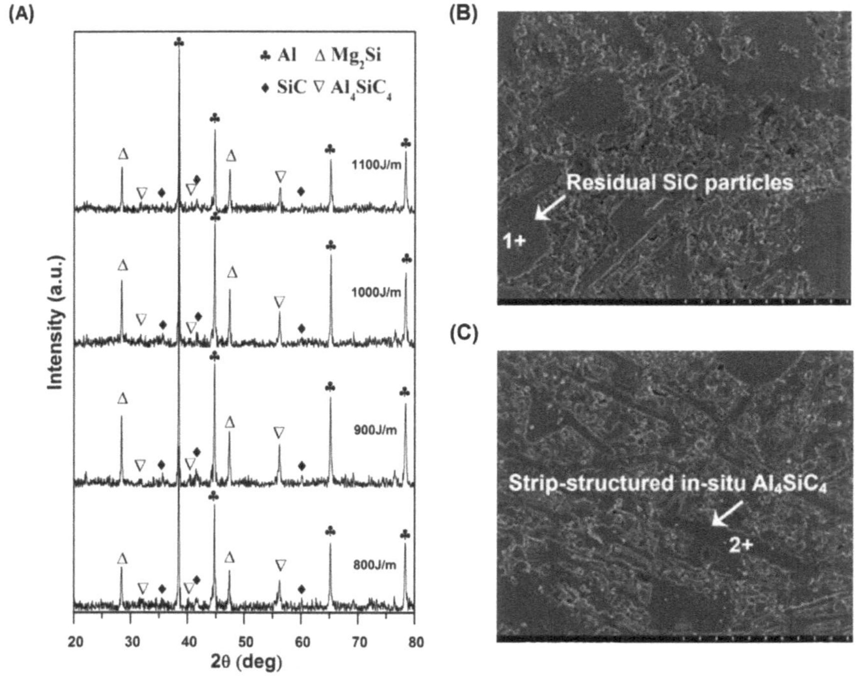

Figure 6.9 Chemical reaction during selective laser melting of Al-based composites. (A) X-ray diffraction patterns as a function of the linear laser energy density. The microstructure of the composite is shown for the laser energy density of (B) 800 J/m and (C) 1000 J/m. *Reprinted with permission from Gu, D., 2015b. Novel Aluminum Based Composites by Selective Laser Melting (SLM) Additive Manufacturing (AM): Tailored Formation of Multiple Reinforcing Phases and its Mechanisms, Laser Additive Manufacturing of High-Performance Materials. Springer Berlin Heidelberg, Berlin, Heidelberg, pp. 201–222.*

distribution of the reinforcement. Under compression and shear, the metal matrix flows extensively at elevated temperatures, eliminating porosity in the matrix or at the reinforcement-matrix interface. Moreover, additive friction stir deposition is a near-net-shaping process that enables site-specific printing. Overall, it has great potential for bulk processing as well as additive manufacturing of metal matrix composites.

6.3.1 Feeding strategy and printing principle

The success of additive friction stir deposition in printing metals can be transferred to the printing of metal matrix composites, for which three feeding strategies can be employed. First, we can use a feed-rod made of preformed metal matrix composites (Fig. 6.10A), which could be purchased or created by conventional composite fabrication techniques. In particular, the metal matrix and reinforcement powders can be premixed, followed by compaction and cold isostatic pressing in a mold that matches the feed-rod geometry. Minor heat treatment is occasionally applied for sintering. The resulting feed-rod is not fully-dense, but the printed metal matrix composites can be free of porosity thanks to the friction stir principle. With the second strategy, a feed-rod made of the matrix material can be hollowed out and packed with loose reinforcement particles (Fig. 6.10B). During deposition, the packed particles are blended into the metal matrix, with the interface adhesion enhanced by the friction stir process. The third option is to use an auger feeding system (Fig. 6.10C). Supplied by two hoppers, the matrix and reinforcement powders are combined inside the hollow print head and fed to the deposition zone in situ. The heat generated by dynamic contact friction creates a thermal gradient in the feed material with a high-temperature zone above the deposition zone, which helps to consolidate the mixed powders prior to deposition. This type of additive friction stir deposition uses a different machine design than earlier approaches, allowing for continuous mixing, consolidation, and printing.

6.3.2 Potential benefits

Additive friction stir deposition of metal matrix composites has several key benefits:

- **Low porosity**. Pores and voids are common issues in composite fabrication causing detrimental effects on the mechanical performance of a part. For instance, they can act as stress risers for crack initiation or as nucleation sites for stress corrosion cracking when containing moisture. In additive friction stir deposition, the significant material flow during printing can heal cracks and fill voids, resulting in low porosity or even fully-dense components as well as good interface bonding. Thanks to the good manufacturing quality, no postprinting heat treatment or high-pressure densification is needed for porosity elimination.
- **Uniform distribution of reinforcement particles**. Clustering of reinforcement particles can lead to local stress concentration or interfacial delamination, whereas a region free of reinforcement particles may be more compliant and susceptible to corrosion damage. If we use the preformed composite feed-rod or the auger mechanism with in situ powder consolidation, the two phases of materials are already mixed

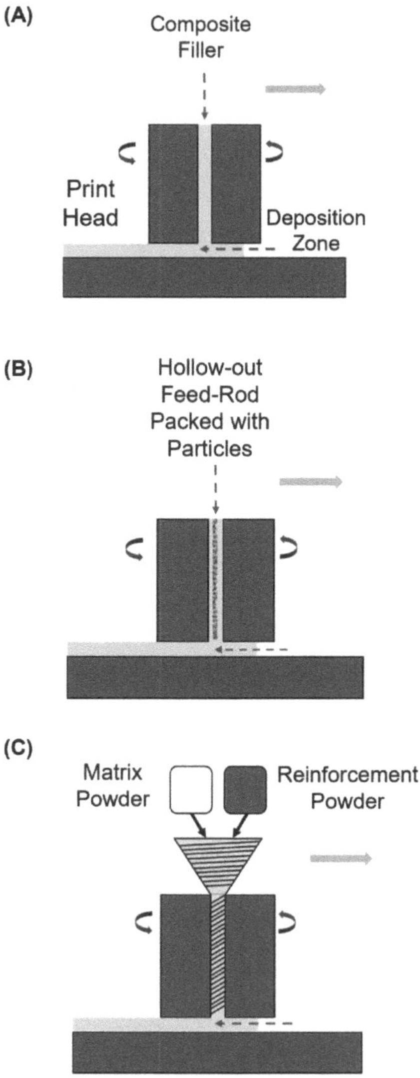

Figure 6.10 Feeding mechanisms for additive friction stir deposition of metal matrix composites. (A) Using composite feed-rod. (B) Packing particles in hollow metal feed-rod. (C) Using an auger feeding mechanism with two hoppers. *Reprinted with permission from Griffiths, R.J., Perry, M.E.J., Sietins, J.M., Zhu, Y., Hardwick, N., Cox, C.D., et al., 2019. A perspective on solid-state additive manufacturing of aluminum matrix composites using MELD. Journal of Materials Engineering and Performance 28, 648–656.*

before the friction stirring takes place. During deposition, the feed material is essentially a self-stirred consumable pin; extensive material flow leads to effective mixing and further promotes uniformity.

- **Scalability and Material Applicability**. As an open-air additive method, additive friction stir deposition is well known to build bulk components without the constraints of a chamber or powder bed. Without melting, a wide variety of metal matrix materials can be used, such as 2xxx, 6xxx, 7xxx Al alloys, Cu, and Mg. The extreme thermomechanical condition during friction stirring leads to a wrought microstructure of the metal matrix, which often has better mechanical qualities than the cast microstructures.

6.4 Examples

6.4.1 Cu-ZrO_2 printed using a composite feed-rod

Zirconia-based ceramics can undergo martensitic transformation upon mechanical or thermal loading, so they have great potential for damping, energy dissipation, and shape memory applications (Lai et al., 2013). Given their intrinsic brittleness, however, they are difficult to use on the bulk scale (Yu et al., 2017). Using ductile metallic "glue" to bind the ZrO_2 particles and forming a metal matrix composite can be a promising route towards bulk-scale implementation. Fig. 6.11 shows the backscattered electron image of a printed Cu–ZrO_2 composite by additive friction stir deposition, showing the homogenous distribution of the ZrO_2 micro-particles within the Cu matrix.

Here, ZrO_2 is doped with CeO_2 to tune the martensitic transformation temperature. Synthesized via coprecipitation, calcination, ball milling, and annealing, each reinforcement particle is a single crystal with a size of ~1–3 μm (Rauch et al., 2019). The synthesized ceramic particles are mixed with Cu in powder form to achieve a 20 vol.% mixture, then homogenized using a ball mill. The mixed powders are placed in a silicone mold and subjected to cold isostatic pressing to achieve the dimensions of the feed-rod, followed by sintering. Starting with such a preformed composite, homogeneous dispersion of the ceramic particles can be easily obtained with friction stirring during deposition.

6.4.2 Al–ZrO_2, Al–SiC, and Cu–SiC composites printed by packing particles in the hollow feed-rod

The creation of composite feed-rods requires multiple powder processing steps. Alternatively, metal matrix composites can be printed by packing particles inside a hollow feed-rod. In this subsection, we present the results based on such a simple and convenient approach, including

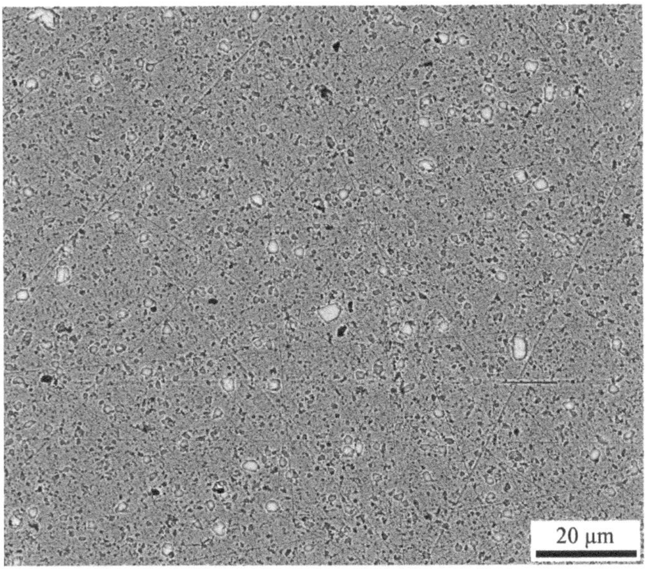

Figure 6.11 A backscatter SEM image of the printed $Cu-ZrO_2$ composite by additive friction stir deposition using a preformed composite feed-rod.

$Al-ZrO_2$, Al–SiC, and Cu–SiC composites. Fig. 6.12 shows the geometry details for feed material design. A metal feed-rod of a square cross-section (9.525 mm × 9.525 mm) is employed; a 6.35 mm diameter circular hole is drilled through the center of the feed-rod and packed with the ceramic powder. In the as-machined feed-rod, the overall fraction of empty space is 34.2 vol.%. With a maximum loose particle packing density of 60%, the final composite typically has a reinforcing volume fraction of less than 20%. Because the metal and ceramic are not mixed prior to friction stirring, the printed composite may be heterogeneous on the mesoscopic level.

Fig. 6.13 presents the microstructure of a printed $AA6061-ZrO_2$-CeO_2 composite using additive friction stir deposition. The ZrO_2-CeO_2 particles are synthesized by the same coprecipitation approach as described in the previous subsection. The feed-rod is made by packing ZrO_2-CeO_2 particles in the hollow AA6061 feed-rod. Fig. 6.13A shows the backscattered electron imaging over a large area of the printed composite, revealing swirl patterns characterized by bright flow lines. Here the matrix phase AA6061 is in the dark, and the bright areas correspond to more content of cerium-doped zirconia. With the reinforcement particle size below 3 μm, the bright flow lines are generally tens of microns wide

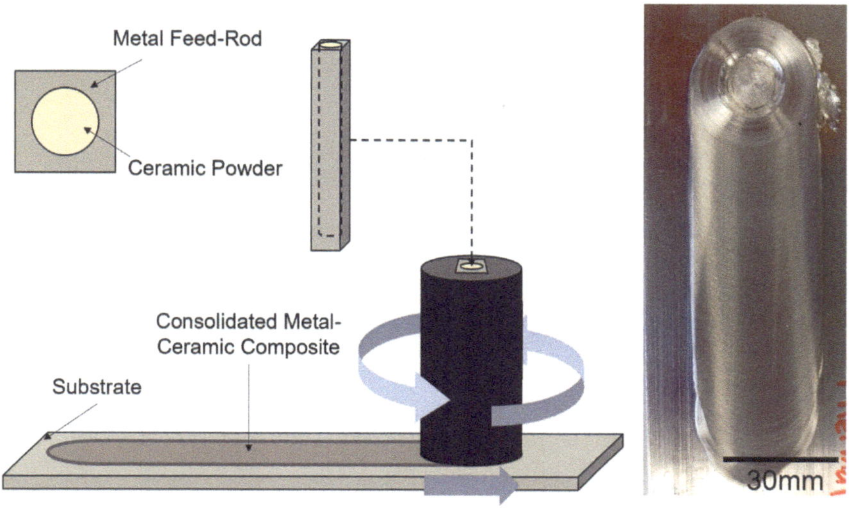

Figure 6.12 Geometrical details of the feed-rod and drilled circular hole for printing metal matrix composites by packing particles in a hollow feed-rod. The right shows the appearance of an as-printed Al matrix composite using this approach.

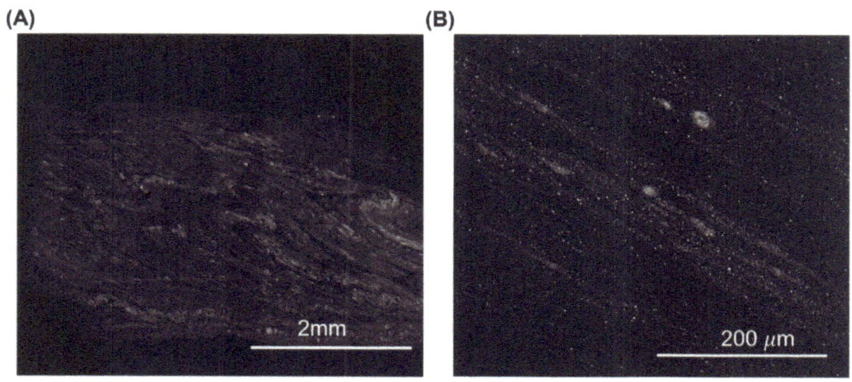

Figure 6.13 Backscattered electron imaging of the printed Al–ZrO_2 composites at (A) low and (B) high magnifications.

and well capture the microscale material flow paths during additive friction stir deposition. The observed swirl material flow pattern is caused by the compression and shear from the rotating print head; similar patterns have been observed in some friction stir welding work as well (Guerra et al., 2002). Fig. 6.13B shows the flow lines at higher resolution, revealing regions of abnormally high particle density or even nebula-like particle clustering.

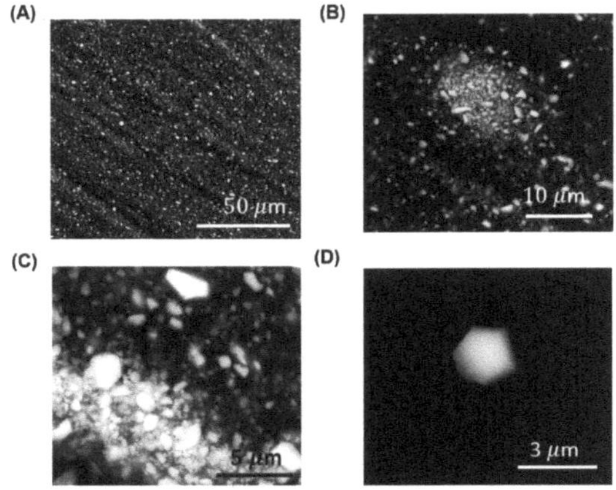

Figure 6.14 Micrographs of the printed Al-ZrO_2 composite with various scales: (A) 130 μm × 130 μm. (B) 35 μm × 35 μm. (C) 17 μm × 17 μm. (D) 9 μm × 9 μm.

Fig. 6.14 presents the microstructure of the printed composite at gradually higher magnifications, all showing no observed porosity. Fig. 6.14A presents the result of an area of 130 μm × 130 μm, in which the spatial distribution of particles and the flow path of the AA6061 matrix can be clearly seen. Interestingly, there are parallel dark lines indicating narrow regions of low ceramic particle concentration. Between neighboring dark lines are belt-like regions with a width ~30 μm, which are characterized by uniform particle distribution. These belt-like regions possibly correspond to the micro-ribbons formed by shear thinning via material-protrusion interactions during printing (Perry et al., 2021), which have a width of 27.5 μm as elaborated in Chapter 3. Fig. 6.14B shows an area of 35 μm × 35 μm, in which agglomeration of zirconia-ceria micro-particles is observed. Despite agglomeration, the particles are intimately surrounded by the AA6061 matrix without interface defects, as seen in Fig. 6.14C. Fig. 6.14D shows an area of 9 μm × 9 μm, highlighting a single crystal zirconia-ceria particle that shows facets formed by high-temperature annealing in powder synthesis. Although there is severe plastic deformation in the metal matrix, this single crystal particle is intact and shows no signs of porosity at the interface.

Fig. 6.15 shows the microstructure of the printed AA6061–SiC composites by packing SiC particles into an AA6061 hollow rod (Garcia et al., 2020). Fig. 6.15A is from a print head with four surface protrusions, labeled the complex geometry tool. Fig. 6.15B is from a print head with no surface protrusions, labeled the simple geometry tool. The overall

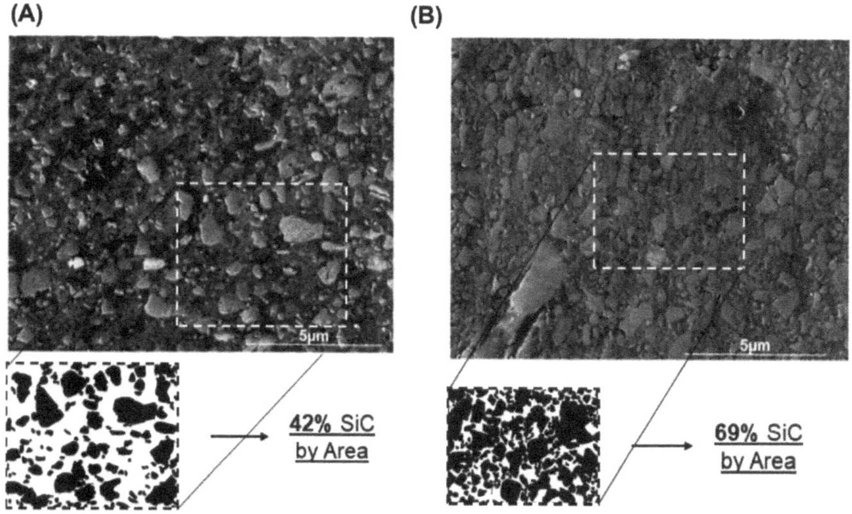

Figure 6.15 Particle loading analysis within the well-mixed regions in the as-printed AA6061–SiC samples for depositions using (A) the complex geometry tool and (B) the simple geometry tool. *Reprinted with permission from Garcia, D., Griffiths, R.J., Yu, H.Z., 2020. Additive friction stir deposition for fabrication of silicon carbide metal matrix composites. In: ASME 2020 15th International Manufacturing Science and Engineering Conference.*

reinforcement volume fraction is ~18% in both composites. However, there are areas of much higher reinforcement concentration. In the composite printed by the complex geometry tool, a local area is found to have 42% SiC; in that printed by the simple geometry tool, a local area consists of 69% SiC. In the latter scenario, the discovery is intuitive because there are no protrusions that would cause SiC dispersion throughout the deposition. In the former scenario, the rotating protrusions should lead to better mixing, but the ultimate dispersion is still uneven. This may be attributed to the highly heterogeneous distribution of the reinforcement in the feed-rod prior to deposition. The heterogeneity becomes more significant on a larger scale. Along the build direction, horizontal band structures with various SiC concentrations are observed. These bands are tight and dense when utilizing the complex geometry tool, but are broad when using the simple geometry tool, indicating higher particle accumulation in the latter.

Following a similar procedure for feed-rod preparation, Cu–SiC composites are printed using additive friction stir deposition. Fig. 6.16A and B show the microstructure of the printed composites using the complex

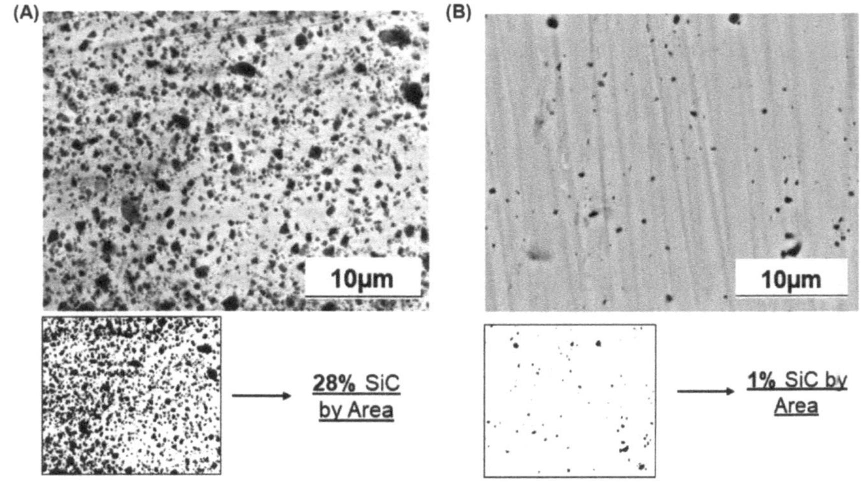

Figure 6.16 Particle loading analysis within the well-mixed regions in the as-printed Cu–SiC composites using (A) the complex geometry tool and (B) the simple geometry tool. *Reprinted with permission from Garcia, D., Griffiths, R.J., Yu, H.Z., 2020. Additive friction stir deposition for fabrication of silicon carbide metal matrix composites. In: ASME 2020 15th International Manufacturing Science and Engineering Conference.*

geometry tool and the simple geometry tool, respectively. By measuring the feed-rod density, the reinforcement volume fraction is measured to be ~15% in both cases. However, when printing using the simple geometry tool, a large amount of SiC powder is left on the surface of the deposit or ejected from the sides of the deposit—rather than mixed into the deposition zone. This can account for the low SiC concentration seen in Fig. 6.16B. There are waves of SiC concentrated regions throughout the width of the deposition track. The depth of mixing is limited in the vertical direction, and kissing bonds are sometimes observed. This characteristic is not commonly observed in additive friction stir deposition of pure copper and can be attributed to the presence of SiC particles. During printing, the hard SiC particles impede the Cu matrix flow, especially between regions of densely packed reinforcement. As the Cu matrix deforms and cools, this can inhibit complete mixing and result in further stress concentrations.

The major difference between printing using complex and simple geometry tools lies in the amount of surface area in contact with the deposition zone. Using the complex geometry tool, the increased surface area generally enhances the mechanical mixing of the material. Similar trends have been observed in friction stir welding, in which modification of the pin geometry leads to varying degrees of material mixing. This explains the better mixing behavior of Al–SiC observed in Fig. 6.15A than Fig. 6.15B. Compared to Al, however, the sticking coefficient between Cu and the print head is lower, so driving Cu flow under the print head is intrinsically more challenging; in additive friction stir deposition of pure Cu, the feed material is simply extruded under the print head without subsequent tool-driven rotation. Moreover, Cu has a lower stacking fault energy, so the dislocations created by plastic deformation during printing can cause significant localized strain hardening around the SiC particles, which in turn reduces material mixing.

6.4.3 Al–SiC printed by auger feeding

Al–SiC composites have also been printed using the auger-based feeding mechanism, with the SiC volume fraction being 10%, 20%, and 30%. The Al and SiC powders are mixed and consolidated in situ before entering the deposition zone. All the samples have a clean surface with a small amount of flash at the edge of the deposits. No surface defect is visually observed. However, each deposition surface is characterized by the onion-skin-like feature, which has been observed in additive friction stir deposition of metals. As discussed in Chapter 3, this feature can be attributed to the interaction of the upward material flow with the trailing edge of the print head. Such observations here suggest that the Al matrix flows well during deposition despite the presence of a significant amount of SiC particles.

Fig. 6.17 characterizes the microstructure of the printed Al–20%SiC composite. Fig. 6.17A shows an optical micrograph, while Fig. 6.17B shows a binary representation map based on energy dispersive spectroscopy, in which Al element is shown in bright and Si element is shown in dark. These images support the conclusion regarding good composite quality and uniform particle distribution. Fig. 6.17C shows a single slice of the printed Al-20 vol.% SiC from X-ray computed tomography, which is representative of the entire scanned volume. Areas of bright contrast correspond to the SiC particles. The slice shows no evidence of porosity or cracking, which would appear as dark spots of contrast. The tiny variation in X-ray absorption between Al and SiC is not effectively contrasted in the original X-ray tomography reconstruction; an adaptive segmentation technique based on the maximization of the likelihood is developed to discriminate the two phases (Griffiths et al., 2019), allowing for identification of any gaps or pores.

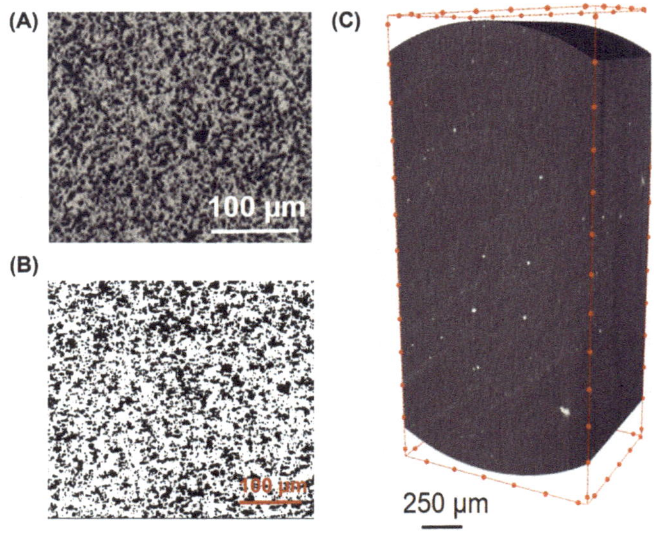

Figure 6.17 Characterization results of Al–20%SiC at higher resolution. (A) An optical micrograph. (B) The binary representation of the energy dispersive spectroscopy map. (C) A slice from the X-ray imaging. *Reprinted with permission from Griffiths, R.J., Perry, M.E.J., Sietins, J.M., Zhu, Y., Hardwick, N., Cox, C.D., et al., 2019. A perspective on solid-state additive manufacturing of aluminum matrix composites using MELD. Journal of Materials Engineering and Performance 28, 648–656.*

Fig. 6.18A shows the 3D distribution of reinforcement particles in the as-deposited Al-20 vol.% SiC composite. The center of each particle is highlighted in the entire imaging volume, showing good dispersion. Fig. 6.18B quantifies the spatial distribution of particles based on histogram and cumulative probability. The measured data is compared with a theoretical composite with significant particle clustering and an ideal composite with uniform particle distribution. The deposited Al–SiC exhibits significantly better performance than composites with particle clustering, while deviating from a completely uniform particle distribution.

From the above examples, we can see a strong correlation between the feeding strategy and reinforcement distribution. While friction stirring promotes dispersion of the reinforcement particles, the final particle distribution greatly depends on the initial state of the feed-rod. If the feed-rod is made of a preformed composite that is already well-mixed, we would expect good uniformity in the as-printed material. Alternatively, the metal matrix and reinforcement may be supplied by two hoppers via the auger feeding mechanism, through which the printed composite also shows

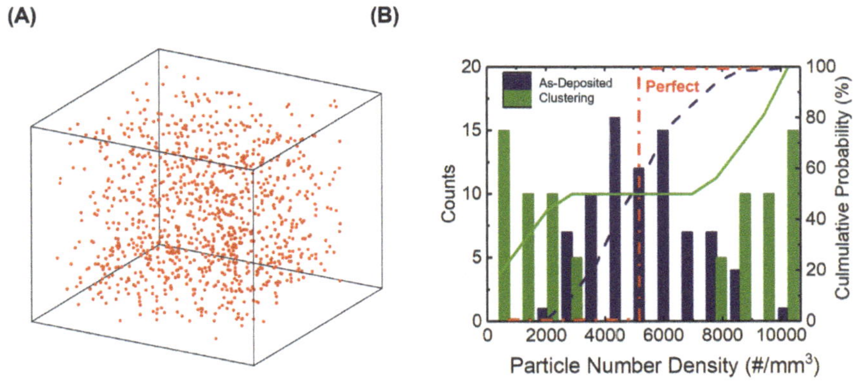

Figure 6.18 (A) In the as-deposited Al−20vol.% SiC composite, the centers of reinforcing particles are shown in red. (B) A statistical examination of the spatial distribution of particle distribution in the as-printed composite, showing the histogram and plot of cumulative probability. Comparisons are made to the composites with perfect particle distribution and particle clustering. *Reprinted with permission from Griffiths, R.J., Perry, M.E.J., Sietins, J.M., Zhu, Y., Hardwick, N., Cox, C.D., et al., 2019. A perspective on solid-state additive manufacturing of aluminum matrix composites using MELD. Journal of Materials Engineering and Performance 28, 648−656.*

good reinforcement distribution. If the feed-rod is made by packing reinforcement particles in the center of the metal matrix, the two phases are not mixed before entering the deposition zone. The resulting composites have some well-mixed areas, but generally show heterogeneous particle distribution across the entire deposition track. Particle agglomeration is also occasionally observed, which poses risks for weak interface bonding.

6.5 Limitations of this printing approach

6.5.1 Maximum volume fraction of reinforcement

The temperature in the deposition zone is generally 60%−90% of the melting temperature of the metal matrix, which is low for the reinforcement. Consider the printing of Al matrix composites. The peak temperature in printing Al-alloys is generally in the range of 75%−90% of the Al melting temperature; this range is equivalent to 23%−28% of the melting temperature of SiC and ZrO_2. Therefore, the reinforcement experiences elastic deformation and may fracture during printing, significantly contributing to the flow stress and elastic modulus of the feed-rod. As the reinforcing volume fraction increases, the flow stress of the composite will dramatically increase, and the ductility will decrease. This imposes a great challenge for successful printing.

From a material flow perspective, the softer metal matrix must flow around the harder reinforcement particles to ensure good mixing, interface adhesion, and continuous plastic deformation in the deposition zone. Increasing the reinforcing loading will obstruct dislocation motion via Orowan stresses, strain gradients, and dislocation build-up, ultimately disqualifying the flow condition for stirring and increasing the likelihood of particle agglomeration and void formation. Efforts to utilize additive friction stir deposition for composite production are still in their early stages. The current printing approach already enables metal matrix composite with a volume fraction of up to 30%, which is comparable to other bulk composite processing approaches. This number is likely to rise considerably through future processes and material optimization.

6.5.2 Tool wear

In contrast to friction stir welding and friction stir processing, there is no nonconsumable pin in additive friction stir deposition. The feed material can be viewed as a consumable pin that spins, creates frictional heat, and coplastically deforms and mixes with the substrate surface. In the work presented in this chapter, the print head is made of tool steel because the matrix materials, like Al or Cu, are of low strength. While the tool wear (i.e., the wear of the print head) is minimal in printing Al alloys, tool wear and fatigue can be substantial in printing metal matrix composites; the reinforcing particles cause significantly faster deterioration than homogenous, low strength materials. Moreover, the material removed from the tool surface is incorporated into the deposited composite, which is usually undesirable and can contribute to contamination. Higher reinforcement loading results in proportionally increased tool wear, which may restrict the widespread use of additive friction stir deposition for composite manufacturing. As shown in Chapter 5, surface protrusions are sometimes employed on the bottom surface of the print head to promote material flow and reinforcement dispersion. Such surface features are particularly susceptible to wear and can be completely eroded away after a number of composite depositions. The degree of tool wear in composite manufacturing has not been systematically investigated and can be an important direction of future research.

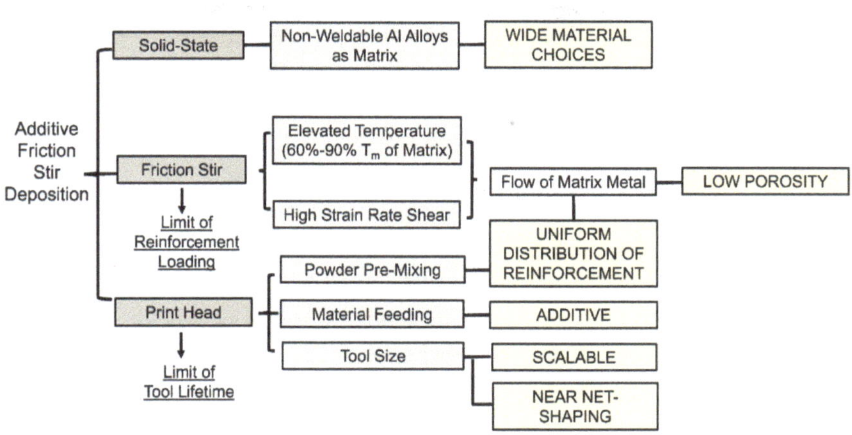

Figure 6.19 Flow chart summarizing the benefits and limitations for additive friction stir deposition of metal matrix composites. *Reprinted with permission from Griffiths, R. J., Perry, M.E.J., Sietins, J.M., Zhu, Y., Hardwick, N., Cox, C.D., et al., 2019. A perspective on solid-state additive manufacturing of aluminum matrix composites using MELD. Journal of Materials Engineering and Performance 28, 648–656.*

Fig. 6.19 summarizes the characteristic features of using additive friction stir deposition for composite fabrication, including both benefits and limitations.

6.6 Summary

Metal matrix composites take advantage of metal properties such as ductility, thermal conductivity, and electrical conductivity, as well as reinforcement properties such as high elastic modulus and resistance to wear and thermal exposure, to implement multifunctions that would be difficult to achieve with any single-phase material. Despite the great potential for several industries, the commercial applications of metal matrix composites have been limited because of the manufacturing constraint, especially the quality and cost.

Current bulk processing approaches include the liquid processing route, exemplified by stir casting and squeeze casting, and the solid processing route, which is mostly based on powder metallurgy. The liquid processing route is limited by porosity formation, element segregation, undesired chemical reactions, and large distortion from solidification shrinkage. The solid processing route relies on postprocessing like sintering and thermomechanical processing (e.g., hot extrusion) to eliminate porosity. Additive manufacturing of metal matrix composite is significantly more difficult. Powder bed fusion and directed energy deposition inherit all the difficulties associated with the traditional liquid processing techniques. Furthermore, the local temperature is greater, which promotes undesirable chemical reactions; the thermal gradient is also larger, which readily causes cracking and severe distortion.

By avoiding the melting and solidification processes, additive friction stir deposition leverages mechanical stirring to enhance particle distribution. Porosity can be eliminated, and good interface bonding can be achieved—thanks to the extensive flow of the metal matrix during printing. One can use a preformed composite feed-rod or employ an auger feeding mechanism with two hoppers for printing. These approaches typically lead to uniform particle distribution in the as-printed composites. Alternatively, one can pack a hollow metal feed-rod with loose reinforcement particles. This convenient and cheap feeding strategy leads to less uniform particle distribution and occasionally particle agglomeration. However, the porosity at the interface between the metal matrix and reinforcement can be still low. Al and Cu have been employed as matrix materials, whereas SiC and ZrO_2 have been used as reinforcement. The main limitations lie in the maximum reinforcement loading, which governs the flowability in printing, and tool wear, which is particularly critical in the case of higher reinforcement volume fractions.

References

Ahmad, A.H., Naher, S., Aqida, S.N., Brabazon, D., 2014. 5.07—Routes to spheroidal starting material for semisolid metal processing. In: Hashmi, S., Batalha, G.F., Van Tyne, C.J., Yilbas, B. (Eds.), Comprehensive Materials Processing. Elsevier, Oxford, pp. 135–148.

AlMangour, B., Grzesiak, D., Yang, J.-M., 2017. Scanning strategies for texture and anisotropy tailoring during selective laser melting of TiC/316L stainless steel nanocomposites. Journal of Alloys and Compounds 728, 424–435.

Avila, J.D., Stenberg, K., Bose, S., Bandyopadhyay, A., 2021. Hydroxyapatite reinforced Ti6Al4V composites for load-bearing implants. Acta Biomaterialia 123, 379–392.

Baradeswaran, A., Perumal, A.E., 2014. Wear and mechanical characteristics of Al 7075/graphite composites. Composites Part B: Engineering 56, 472–476.

Bournias-Varotsis, A., Friel, R.J., Harris, R.A., Engstrøm, D.S., 2018. Ultrasonic additive manufacturing as a form-then-bond process for embedding electronic circuitry into a metal matrix. Journal of Manufacturing Processes 32, 664–675.

Campbell, F.C., 2003. Manufacturing Processes for Advanced Composites. Elsevier Science.

Chawla, N., Chawla, K.K., 2013a. Introduction. In: Chawla, N., Chawla, K.K. (Eds.), Metal Matrix Composites. Springer New York, New York, pp. 1–4.

Chawla, N., Chawla, K.K., 2013b. Processing, Metal Matrix Composites. Springer New York, New York, pp. 55–97.

Chen, Y., Zhang, X., Liu, E., He, C., Shi, C., Li, J., et al., 2016. Fabrication of in-situ grown graphene reinforced Cu matrix composites. Scientific Reports 6, 19363.
Ellis, M.B.D., 1996. Joining of aluminium based metal matrix composites. International Materials Reviews 41, 41–58.
Frazier, W.E., 2014. Metal additive manufacturing: a review. Journal of Materials Engineering and Performance 23, 1917–1928.
Garcia, D., Griffiths, R.J., Yu, H.Z., 2020. Additive friction stir deposition for fabrication of silicon carbide metal matrix composites. In: ASME 2020 15th International Manufacturing Science and Engineering Conference.
Griffiths, R.J., Perry, M.E.J., Sietins, J.M., Zhu, Y., Hardwick, N., Cox, C.D., et al., 2019. A perspective on solid-state additive manufacturing of aluminum matrix composites using MELD. Journal of Materials Engineering and Performance 28, 648–656.
Gu, D., 2015a. Laser Additive Manufacturing of High-Performance Materials. Springer Berlin Heidelberg.
Gu, D., 2015b. Novel Aluminum Based Composites by Selective Laser Melting (SLM) Additive Manufacturing (AM): Tailored Formation of Multiple Reinforcing Phases and its Mechanisms, Laser Additive Manufacturing of High-Performance Materials. Springer Berlin Heidelberg, Berlin, Heidelberg, pp. 201–222.
Gu, D., Shen, Y., Lu, Z., 2009. Preparation of TiN–Ti5Si3 in-situ composites by selective laser melting. Materials Letters 63, 1577–1579.
Guerra, M., Schmidt, C., McClure, J.C., Murr, L.E., Nunes, A.C., 2002. Flow patterns during friction stir welding. Materials Characterization 49, 95–101.
Guo, X.-f, Jia, L., Lu, Z.-l, Xing, Z.-g, Xie, H., Kondoh, K., 2022. Preparation of Cu/CrB2 composites with well-balanced mechanical properties and electrical conductivity by ex-situ powder metallurgy. Journal of Materials Research and Technology 17, 1605–1615.
Hofmann, D.C., Roberts, S., Otis, R., Kolodziejska, J., Dillon, R.P., Suh, J.-o, et al., 2014. Developing gradient metal alloys through radial deposition additive manufacturing. Scientific Reports 4, 5357.
Horn, T.J., Gamzina, D., 2020. Additive manufacturing of copper and copper alloys. In: Bourell, D.L., Frazier, W., Kuhn, H., Seifi, M. (Eds.), Additive Manufacturing Processes. ASM International, p. 0.
Jäcklein, M., Pfaff, A., Hoschke, K., 2020. Developing tungsten-filled metal matrix composite materials using laser powder bed fusion. Applied Sciences 10, 8869.
Kang, Y.-C., Chan, S.L.-I., 2004. Tensile properties of nanometric Al2O3 particulate-reinforced aluminum matrix composites. Materials Chemistry and Physics 85, 438–443.
Kareem, A., Qudeiri, J.A., Abdudeen, A., Ahammed, T., Ziout, A., 2021. A review on AA 6061 metal matrix composites produced by stir casting. Materials 14, 175.
Lai, A., Du, Z., Gan, C.L., Schuh, C.A., 2013. Shape memory and superelastic ceramics at small scales. Science 341, 1505–1508.
Li, S., Sun, B., Imai, H., Kondoh, K., 2013. Powder metallurgy Ti–TiC metal matrix composites prepared by in situ reactive processing of Ti-VGCFs system. Carbon 61, 216–228.
Liu, W., DuPont, J.N., 2003. Fabrication of functionally graded TiC/Ti composites by laser engineered net shaping. Scripta Materialia 48, 1337–1342.
Lloyd, D.J., 1997. Composites Engineering Handbook Marcel Dekker, New York.
Matli, P.R., Ubaid, F., Shakoor, R.A., Parande, G., Manakari, V., Yusuf, M., et al., 2017. Improved properties of Al–Si3N4 nanocomposites fabricated through a microwave sintering and hot extrusion process. RSC Advances 7, 34401–34410.
Miracle, D.B., 2005. Metal matrix composites—from science to technological significance. Composites Science and Technology 65, 2526–2540.

Nartu, M.S.K.K.Y., Mantri, S.A., Pantawane, M.V., Ho, Y.-H., McWilliams, B., Cho, K., et al., 2020. In situ reactions during direct laser deposition of Ti-B4C composites. Scripta Materialia 183, 28–32.
Oyelola, O., Crawforth, P., M'Saoubi, R., Clare, A.T., 2018. Machining of functionally graded Ti6Al4V/ WC produced by directed energy deposition. Additive Manufacturing 24, 20–29.
Perry, M.E.J., Rauch, H.A., Griffiths, R.J., Garcia, D., Sietins, J.M., Zhu, Y., et al., 2021. Tracing plastic deformation path and concurrent grain refinement during additive friction stir deposition. Materialia 18, 101159.
Rajak, D.K., Pagar, D.D., Kumar, R., Pruncu, C.I., 2019. Recent progress of reinforcement materials: a comprehensive overview of composite materials. Journal of Materials Research and Technology 8, 6354–6374.
Ramanathan, A., Krishnan, P.K., Muraliraja, R., 2019. A review on the production of metal matrix composites through stir casting—furnace design, properties, challenges, and research opportunities. Journal of Manufacturing Processes 42, 213–245.
Rauch, H.A., Chen, Y., An, K., Yu, H.Z., 2019. In situ investigation of stress-induced martensitic transformation in granular shape memory ceramic packings. Acta Materialia 168, 362–375.
Saha, R., Morris, E., Chawla, N., Pickard, S.M., 2002. Hybrid and conventional particle reinforced metal matrix composites by squeeze infiltration casting. Journal of Materials Science Letters 21, 337–339.
Scudino, S., Liu, G., Sakaliyska, M., Surreddi, K.B., Eckert, J., 2009. Powder metallurgy of Al-based metal matrix composites reinforced with β-Al3Mg2 intermetallic particles: analysis and modeling of mechanical properties. Acta Materialia 57, 4529–4538.
Selvaraj, S.K., Srinivasan, K., Chadha, U., Mishra, R., Arpit, K., Apurb, K., et al., 2021. Contemporary progresses in ultrasonic welding of aluminum metal matrix composites. Frontiers in Materials 8.
Stenberg, K., Dittrick, S., Bose, S., Bandyopadhyay, A., 2018. Influence of simultaneous addition of carbon nanotubes and calcium phosphate on wear resistance of 3D-printed Ti6Al4V. Journal of Materials Research 33, 2077–2086.
Tjong, S.C., 2007. Novel nanoparticle-reinforced metal matrix composites with enhanced mechanical properties. Advanced Engineering Materials 9, 639–652.
Torralba, J.M., da Costa, C.E., Velasco, F., 2003. P/M aluminum matrix composites: an overview. Journal of Materials Processing Technology 133, 203–206.
Valente, M., Marini, D., Genova, V., Quitadamo, A., Marra, F., Pulci, G., 2019. Lightweight metallic matrix composites: development of new composites material reinforced with carbon structures. Journal of Applied Biomaterials & Functional Materials 17, 2280800019840294.
Wang, H., Gu, D., 2015. Nanometric TiC reinforced AlSi10Mg nanocomposites: powder preparation by high-energy ball milling and consolidation by selective laser melting. Journal of Composite Materials 49, 1639–1651.
Yu, H.Z., Hassani-Gangaraj, M., Du, Z., Gan, C.L., Schuh, C.A., 2017. Granular shape memory ceramic packings. Acta Materialia 132, 455–466.
Yu, W.H., Sing, S.L., Chua, C.K., Kuo, C.N., Tian, X.L., 2019. Particle-reinforced metal matrix nanocomposites fabricated by selective laser melting: a state of the art review. Progress in Materials Science 104, 330–379.
Zhou, S.Y., Su, Y., Wang, H., Enz, J., Ebel, T., Yan, M., 2020. Selective laser melting additive manufacturing of 7xxx series Al-Zn-Mg-Cu alloy: cracking elimination by co-incorporation of Si and TiB2. Additive Manufacturing 36, 101458.

CHAPTER 7

Mechanical properties of the printed materials

Develop flexibility and you will be firm; cultivate yielding and you will be strong.

—Liezi

Additive friction stir deposition leverages compression and shear to enable the printing of metals and metal matrix composites. A combination of these stress states with extensive metal flow typically leads to a fully-dense material in the as-printed state; there is no need for postprocessing treatment for porosity elimination. This is one unique advantage compared to alternative metal additive manufacturing technologies, such as selective laser melting and cold spray. Moreover, additive friction stir deposition has a thermomechanical processing nature such that the resulting properties are close to wrought or forged alloys rather than cast alloys. This is the second advantage that makes additive friction stir deposition particularly promising for applications where the load-bearing capacity is critical.

The capability of additive friction stir deposition to render wrought or forging grade mechanical properties effectively distinguishes it from other metal additive manufacturing technologies. General mechanical properties of interest include:

- Yield and ultimate tensile strength, which is a baseline measurement of the static load-bearing capacity.
- Fracture toughness, which indicates the resistance to crack propagation.
- Fatigue lifetime, which is a measure of the resistance of materials and structures against cyclic loading.

All these mechanical properties are determined by the phase and microstructure of the deposited material, which can be manipulated via process parameters. For example, additive friction stir deposition of precipitation-strengthened alloys will necessitate careful management of the process and postprocess aging conditions to obtain the appropriate precipitate size, spatial distribution, and volume fraction. Additive friction stir deposition of alloys

Additive Friction Stir Deposition
DOI: https://doi.org/10.1016/B978-0-12-824374-9.00009-5

with a strong Hall–Petch effect will necessitate the control of dynamic recrystallization for grain size reduction and optimization.

The goal of this chapter is to provide an overview of the cutting-edge experimental research regarding the mechanical properties of materials printed using additive friction stir deposition. We begin with a review of the fundamentals of the mechanical behavior of materials. Based on the state-of-the-art experimental data, we then introduce the tensile, fracture, and fatigue properties of the alloys produced by additive friction stir deposition. Precipitation-strengthened Al and Mg alloys are the most common examples, for which postprocess aging is used to improve the yield strength. In the as-deposited alloys, the effects of dislocation content, grain size, and phase fraction are also explored. Since additive friction stir deposition can be widely exploited for selective-area reinforcing and cladding, the mechanical properties of bilayer structures are examined too—in the context of uniaxial tension and three-point bending. Finally, the mechanical properties of printed metal matrix composites are discussed.

7.1 Elements of the mechanical behavior of materials

The mechanical behavior of materials describes their response to external loading, which may be in the form of contact forces or body forces. This response depends on the stress state, whether it is tension, compression, shear, or a combination of them. Annealed metals generally exhibit similar behavior under compression and tension. Ceramics can afford high loads in compression but fail easily under tension. Moreover, a triaxial stress state causes catastrophic failure easier than a uniaxial stress state. The mechanical behavior also depends on the external environment, for example, the temperature, and testing condition, for example, the strain rate.

Tension tests are characterized by a uniform uniaxial stress state. Because of its simplicity and efficiency in providing intrinsic mechanical properties, tension testing is widely used to experimentally examine the mechanical behavior of materials. In a tension test, the gauge section is characterized by a uniform tensile stress state. The engineering stress and engineering strain are defined as

$$\sigma = \frac{P}{A_i}, \quad \varepsilon = \frac{\Delta L}{L_i}. \tag{7.1}$$

Here P is the tensile load, A_i is the initial cross-sectional area of the gauge, ΔL is the increase in gauge length, and L_i is the original gauge length. Fig. 7.1 shows a typical engineering stress vs. engineering strain curve in tension testing of a ductile material.

Upon tension loading, the material first deforms elastically with full recoverability. During elastic deformation, the engineering stress scales with the engineering strain; the slope is equal to the Young's modulus of the material. On the atomic scale, elastic deformation corresponds to the stretching or distortion of chemical bonds. Once a critical level of engineering stress is reached, plastic deformation starts, and the stress–strain curve deviates from a straight line. This critical engineering stress defines the yield strength. In most cases, the plastic deformation in metals and alloys is accommodated by multiplication and motion of dislocations, which can be of edge, screw, or mixed type. Dislocations are barriers to dislocation motion. An increase in the dislocation number will necessitate higher applied stress to continue the deformation, so the engineering stress keeps increasing with engineering strain after yielding. This is the strain hardening stage, in which the slope varies as deformation proceeds. In this stage, the relationship between the stress and plastic strain may be empirically described by a power law. The strain hardening exponent *n* is a material property that normally falls between 0.2 and 0.5.

The plastic deformation in the gauge section is uniform until the necking phenomenon occurs, which is a type of instability when the material begins to harden by a lower proportion than the decrease in

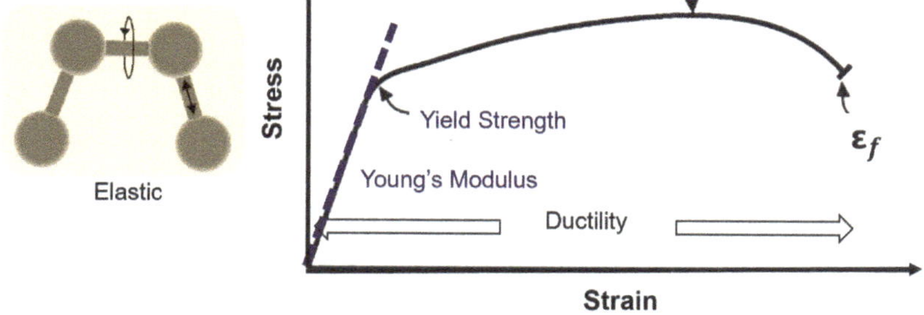

Figure 7.1 A schematic showing the typical engineering stress vs. engineering strain curve in the tension test of a ductile material.

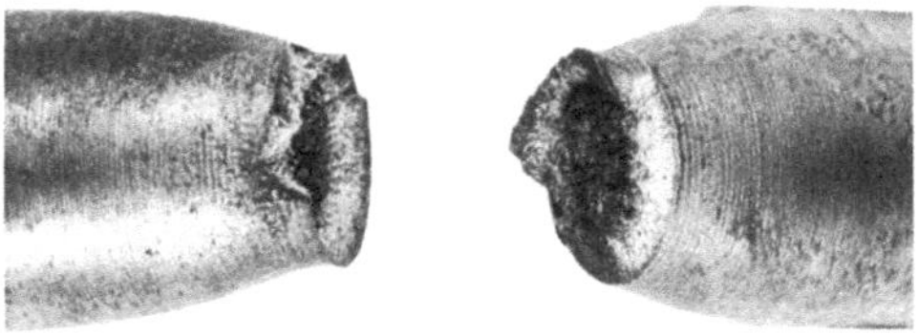

Figure 7.2 Localized deformation of annealed low carbon steel during a tensile test produces a necked region. The micrograph shows necking in a fractured sample. *Reprinted with permission from Askeland, D.R., 1996. Failure—origin, detection, and prevention. In: The Science and Engineering of Materials. Springer US, Boston, MA, pp. 784–819.*

cross-sectional area (Zhao and Lavernia, 2011). The necking phenomenon is characterized by strain localization. The greater the local strain, the smaller the cross-sectional area becomes, producing even more strain concentration. Fig. 7.2 shows an example of necking observed in tension testing of an annealed low carbon steel (Askeland, 1996). Owing to a significant decrease in the cross-sectional area of the necking region, the applied load drops despite an increase of stress in the deformed region. Therefore, necking corresponds to the maximum engineering stress in a tension test, which defines the ultimate tensile strength (UTS) of the material (Kim et al., 2013). Further plastic deformation is constrained to the vicinity of the neck, so the engineering stress-engineering strain curve no longer provides useful information regarding deformation.

For extensive plastic deformation, it is useful to introduce another pair of deformation parameters: true stress $\tilde{\sigma}$ and true strain $\tilde{\varepsilon}$, which are defined as follows:

$$\tilde{\sigma} = \frac{P}{A}, \quad \tilde{\varepsilon} = \ln\left(\frac{L}{L_i}\right). \tag{7.2}$$

Here A and L refer to the actual cross-sectional area and actual gauge length, respectively. In the context of true stress and true strain, the necking instability can be described by the Considère criterion (Yasnikov et al., 2014). Upon the onset of necking, the load reaches the maximum, and we have

$$\frac{d\tilde{\sigma}}{\tilde{\sigma}} = -\frac{dA}{A}. \tag{7.3}$$

A relationship can be established between the engineering stress/strain and true stress/strain before necking (Dowling, 2013):

$$\tilde{\sigma} = \sigma(1+\varepsilon) = \sigma\exp\left(\tilde{\varepsilon}\right), \quad \tilde{\varepsilon} = \ln(1+\varepsilon). \tag{7.4}$$

This relationship cannot be applied after necking, for which the only useful parameter is the cross-sectional area of the necking region: A_{Neck}. With A_{Neck} in situ measured during tension testing, the true stress and true strain can be calculated as:

$$\tilde{\sigma} = \frac{P}{A_{Neck}}, \quad \tilde{\varepsilon} = \ln\left(\frac{A_i}{A_{Neck}}\right). \tag{7.5}$$

After necking, the local stress continues to increase, eventually causing fracture of the tensile specimen. For ductile materials, the fracture is often caused by coalescence of micro-voids.

Intrinsically, the mechanical behavior of materials is highly dependent on the defects. For a perfect crystal (Fig. 7.3A), accommodation of plastic deformation would require simultaneous bond breaking across the slip plane along the shear direction. The required shear stress is on the order of $\tau \sim G/2\pi$, where G is the shear modulus. This value is several orders of magnitude higher than the experimentally measured yield strength. The origin of this discrepancy stems from the existence of dislocations. In actual alloys, the plastic deformation is typically accommodated by the motion of dislocations, which only requires consecutive bond breaking. As demonstrated in Fig. 7.3B, at each step of dislocation motion, bond breaking and bond formation occur only around a single atom (essentially

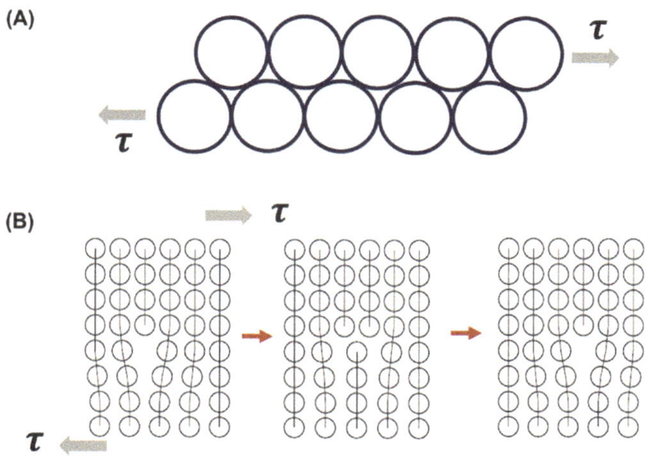

Figure 7.3 A comparison of the required critical shear stress for yielding (A) a perfect crystal and (B) a crystal containing dislocations. For (B), three configurations are illustrated. From left to right: initial state, high-energy configuration state, and final state in a single step of dislocation motion.

a single row of atoms perpendicular to the paper). This takes far less shear stress than deforming a perfect crystal.

The Peierls–Nabarro model (Lu, 2005) suggests that the corresponding critical stress is

$$\tau_{P-N} \propto G\exp\left(-\frac{2\pi w}{b}\right). \tag{7.5}$$

Here, w is the dislocation width and b is the Burgers vector. The Peierls–Nabarro stress can be viewed as a measure of the lattice friction in moving dislocations. In a single crystal, the slip planes and directions are those of the highest packing density. This is because, in such planes and directions, the Burgers vector is of the smallest value, leading to the lowest value of Peierls–Nabarro stress. In polycrystals and multiphase alloys, the compatibility issue and mechanical constraint play an important role, resulting in a much higher yield strength than their single crystal counterparts. Other factors need to be considered as well, such as grain boundaries, solute atoms, and precipitates (Huda, 2022). The resulting yield strength depends on the competition of various strengthening mechanisms, which will be elaborated in the next section—in the context of tensile properties of the printed alloys by additive friction stir deposition.

7.2 Tensile properties of the printed metals and alloys

Materials used in engineering applications are rarely single-crystalline, single-phase, or single-element in nature. Instead, grain boundaries, dislocations, solute atoms, precipitates, or other phases are present. The defects and microstructural heterogeneity can make the alloys strong and tough (Sankaran and Mishra, 2017). In this section, we examine the tensile properties of various types of metals and alloys printed by additive friction stir deposition, including precipitation-strengthened alloys, alloys with considerable Hall–Petch effect, and two-phase alloys. The gradient of mechanical properties across layers is also discussed.

7.2.1 Effects of precipitation strengthening

Under the equilibrium condition, second-phase particles start to form if the concentration of solute atoms exceeds their solubility in the matrix phase. The attributes of these particles can be kinetically controlled. We can heat these alloys above the solvus temperature so that the particles are dissolved in the matrix to form a single-phase alloy. Rapid quenching then will lead to a supersaturated solid solution. This is followed by aging, which refers to heating at an intermediate temperature for a given amount of time. From that, the nucleation and diffusion-mediated growth processes will result in the precipitation of second-phase particles in the matrix phase (Allen et al., 2005). The precipitate size and spacing can be controlled by the aging condition. The interactions between dislocations and finely dispersed precipitates play a major role in determining the yield strength of such alloys. Depending on the nature of the precipitate and the crystallographic relationship between the precipitate and the matrix, dislocations may cut through the particles in the slip plane (see Fig. 4A for an example) or circumnavigate the particles in the slip plane (see Fig. 4B for an example). The precipitates may be termed shearable and nonshearable particles for these scenarios, respectively (Humphreys et al., 2017).

For shearable particles, the matrix dislocations that pass through them are not necessarily the slip dislocations for the precipitates. Cutting a precipitate particle will result in a faulted plane associated with excess interface energy. Based on energy analysis, the shear stress required to cut the precipitate particles can be estimated as

$$\tau_{shear} = \pi r_0 \gamma / 2bl, \tag{7.7}$$

where r_0 is the particle size, l is the particle spacing, and γ is the specific energy of the faulted plane. If the precipitate particles are impenetrable to the dislocations, that is, they are nonshearable particles, shear stress will be required to bend dislocations in a semicircular form between the particles, eventually causing dislocation loops around the particles. This mechanism is illustrated in Fig. 7.4C (Hertzberg et al., 2012). The required shear stress is termed Orowan stress, which can be estimated as

$$\tau_{Orowan} \approx \frac{Gb}{l}. \tag{7.8}$$

In precipitation-strengthened alloys, the yield strength is determined by the competition of the above two mechanisms. If $\tau_{shear} > \tau_{Orowan}$, yielding is associated with dislocation bending and loop formation. This is usually the case of large particles. If $\tau_{shear} < \tau_{Orowan}$, yielding is associated with dislocation cutting through the particles. This is usually the case with small particles. During the aging process, initially, both the precipitate size and volume fraction will increase. After all the excess solute atoms precipitate out from the matrix, the volume fraction will be saturated; continued

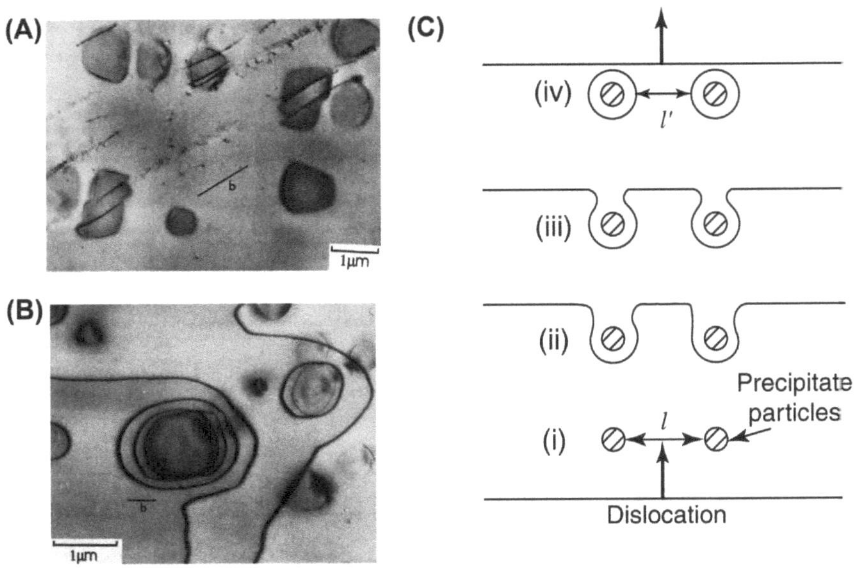

Figure 7.4 Precipitation strengthening. (A) Deformed Ni_3Si particles cut through by dislocations in a Ni-6% Si single crystal. (B) Orowan loops at Ni_3Si particles in a Ni-6% Si single crystal. (C) Illustration of the Orowan mechanism of dislocation bending and loop formation. *(A) and (B) are reprinted with permission from Humphreys, J., Rohrer, G. S., Rollett, A., 2017. Chapter 2—The deformed state. In: Humphreys, J., Rohrer, G.S., Rollett, A. (Eds.), Recrystallization and Related Annealing Phenomena, 3rd ed. Elsevier, Oxford, pp. 13–79. (C) is reprinted with permission from Hertzberg, R.W., Vinci, R.P., Hertzberg, J.L., 2012. Deformation and Fracture Mechanics of Engineering Materials, 5th ed. Wiley.*

aging will cause precipitate coarsening with an increase of the precipitate spacing. For a given volume fraction of precipitates, there is an optimal precipitate size that maximizes the yield strength. Common precipitation-strengthened engineering alloys include 2xxx, 6xxx, 7xxx Al alloys, Mg alloys, Ni-based superalloys, and maraging steels.

The mechanical properties of a number of precipitation-strengthened alloys printed by additive friction stir deposition are further discussed below. Fig. 7.5 shows the hardness measurement results for dissimilar material printing with AA6061 deposited onto an AA6022 substrate. Both are Al–Mg–Si alloys. Because the substrate is very thin, the print head with a flat bottom surface is used in additive friction stir deposition. The Vickers hardness is measured on the longitudinal cross-sections, including the data along the centerline, on the advancing side, and on the retreating side. The feed material is peak-aged, AA6061-T6, with a hardness

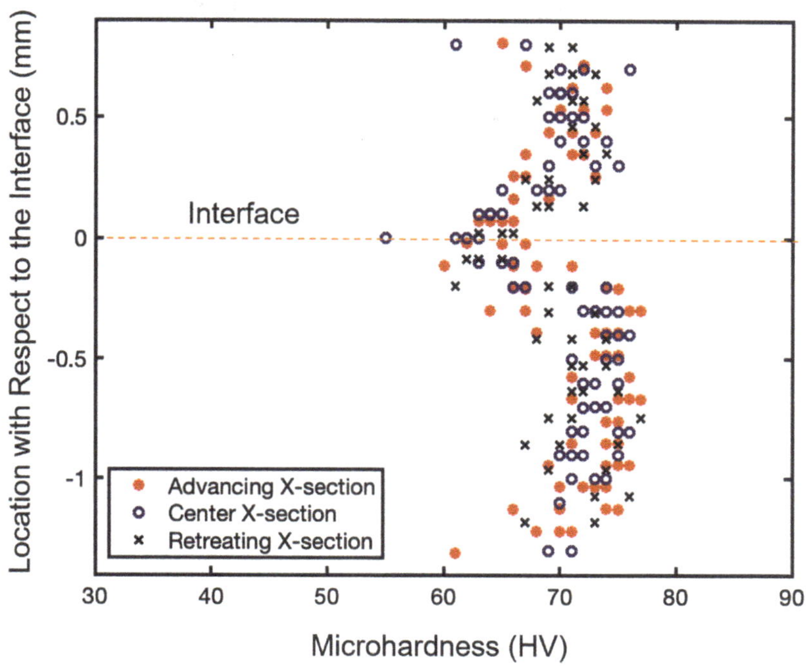

Figure 7.5 Microhardness distribution of AA6061-T6 deposited onto AA6022-T4. Plot of the Vickers hardness along several longitudinal cross-sections, including the data from the centerline, advancing side, and retreating side. *Reprinted with permission from Hartley, W.D., Garcia, D., Yoder, J.K., Poczatek, E., Forsmark, J.H., Luckey, S.G., et al., 2021. Solid-state cladding on thin automotive sheet metals enabled by additive friction stir deposition. Journal of Materials Processing Technology 291, 117045.*

measured to be 115 ± 10 HV. This is consistent with the nominal temper. After deposition, the deposited material exhibits a reduced hardness value of 70 ± 6 HV. This reduction is attributable to the evolution of Mg–Si precipitates during deposition and cooling. During deposition, by combining high temperature and dislocation shearing, the precipitates in the feed material are likely solutionized and dissolved into the aluminum matrix. During cooling, re-precipitation happens, but the size and number density of the resulting precipitates are significantly different from the peak strengthening condition.

In terms of the substrate, the AA6022 is in the T4 condition, which is hardened by solid solution and GP (Guinier–Preston) zones through natural aging (Banhart et al., 2010; Marioara et al., 2001). Deposition results in heat input and deformation in the substrate. However, the precipitation evolution in the substrate only leads to a slightly lower hardness value

(73 ± 3 HV) than the as-received AA 6022-T4 (81 ± 4 HV). This is because the T4 condition is not peak-aged, while natural aging can also happen in the substrate after deposition.

Hardness distribution reflects small differences in local processing conditions, which are defined by the total strain, strain rate, and peak temperature. Microhardness measurements on the advancing side, retreating side, and centerline all exhibit similar values, implying symmetric and homogeneous thermomechanical processing conditions. The majority of the deposition and substrate have similar hardness values, ranging from 70 to 75 HV. However, there is a significant drop near the interface to 62 ± 3 HV. This drop suggests that frictional heating and mixing during metallurgical bond formation lead to greater peak temperatures and more deformation in the interfacial regions.

Fig. 7.6 shows the tensile results of a Mg alloy, WE43, printed by additive friction stir deposition. WE43 is primarily made up of

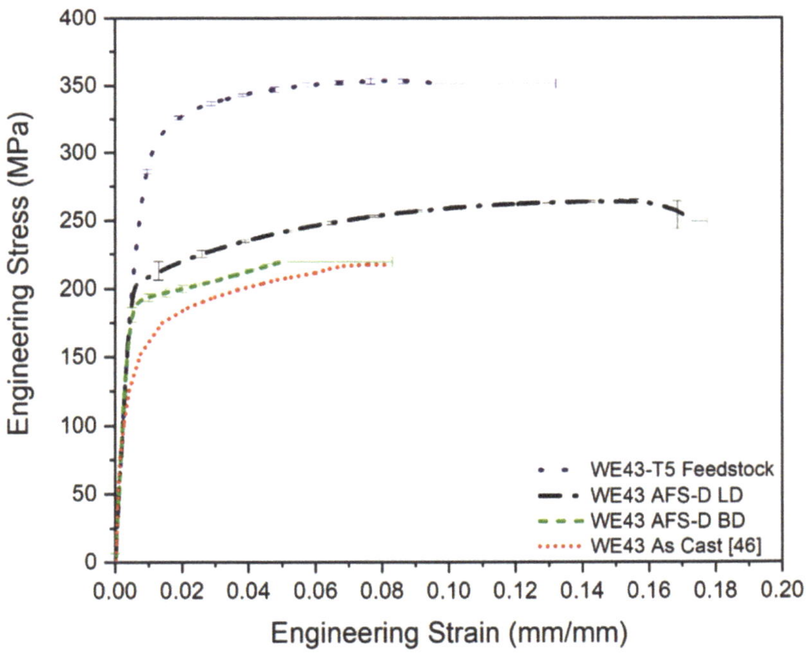

Figure 7.6 Tension testing of as-deposited WE43 in the longitudinal and normal (or build) directions. The engineering stress vs. engineering strain curves for the WE43-T5 feedstock and as-cast WE43 are also presented for comparison. *Adapted from Williams, M.B., Robinson, T.W., Williamson, C.J., Kinser, R.P., Ashmore, N.A., Allison, P.G., et al., 2021. Elucidating the effect of additive friction stir deposition on the resulting microstructure and mechanical properties of magnesium alloy WE43. Metals 11, 1739.*

magnesium, yttrium, and neodymium; it can withstand temperatures of up to 300°C with excellent corrosion resistance. WE series of Mg alloys are strengthened by Mg−Y−Nd precipitates in the form of a disc shape (beta double prime), a disc or globular shape (beta prime), a plate shape (beta 1), and a rod shape (beta) (Kandalam et al., 2015). In this work, a print head with 4 surface protrusions of tear-drop shape is used (Williams et al., 2021). With the deposition carried out at a print head rotation rate of 325 RPM, a feed rate of 1.06 mm/s, and a traverse velocity of 2.54 mm/s, the as-deposited WE43 is subjected to tension testing in the longitudinal and normal directions. The results are compared to the feed material, which is a wrought WE43 alloy under T5 conditions. All depositions show a consistent value of Young's modulus at 44 GPa. With precipitation evolution during deposition and increased thermal cycles in previously deposited layers, the yield strength along the longitudinal direction drops 30% compared to the feed material. However, the elongation to failure, which is a measure of the ductility, also increases by 40%. In other words, the in-plane properties of the as-deposited WE43 are more ductile and less strong than the wrought alloy with T5 age conditions. Regarding the out-of-plane properties in the as-deposited WE43, both strength and ductility are inferior to the feed material but are comparable to or better than the as-cast WE43.

Fig. 7.7 shows the tensile results of a high strength Al alloy, AA7050, printed by additive friction stir deposition in true stress vs. true strain plots (Mason et al., 2021). This includes the tensile response in the top transverse, bottom transverse, and vertical directions of all three sections of the as-built AA7050 structure. When compared to the feed-rod, which is made of wrought AA7050-T7651, the yield strength of the as-built AA7050 is significantly lower in all orientations. AA7050 is an Al−Zn−Mg−Cu alloy mainly consisting of GP zones, η′ ($MgZn_2$), and η ($MgZn_2$) precipitates, in addition to secondary phases and constituent particles (Oliveira et al., 2004). The reduction in mechanical response is caused by the overgrowth of the strengthening phases throughout the microstructure, which stems from repeated thermal cycles experienced by the initial and intermediate layers of the component. Similar to the WE43 work, the out-of-plane (i.e., vertical) direction shows inferior properties. The in-plane direction generally exhibits ductile behavior (up to 20% elongation) with a dimpled fracture surface. In the out-of-plane direction, the material is quite brittle, showing a tensile strain less than 4%. Such low uniform elongation is possibly a result of strain localization caused by

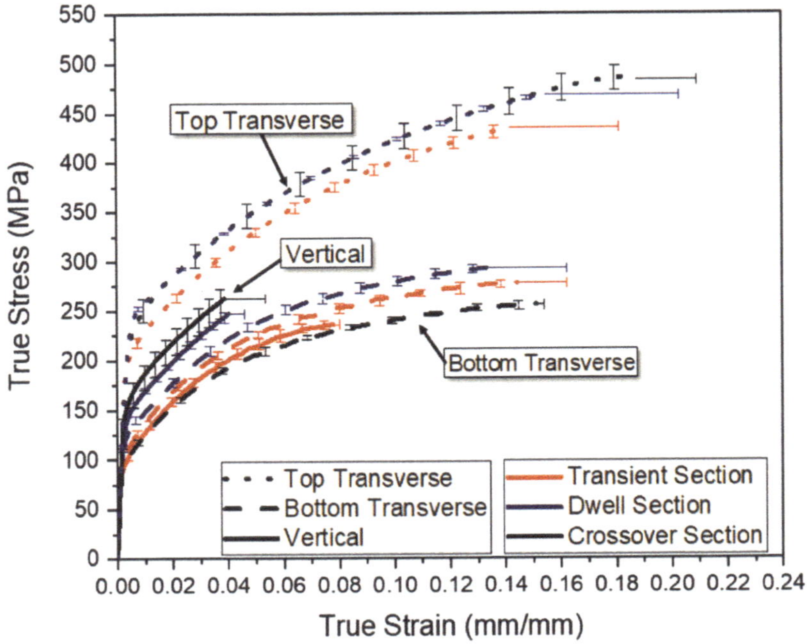

Figure 7.7 Tension testing of the as-deposited AA7050, showing true stress vs. true strain curves. The data is taken from the top transverse, bottom transverse, and vertical orientations. *Reprinted with permission from Mason, C.J.T., Rodriguez, R.I., Avery, D. Z., Phillips, B.J., Bernarding, B.P., Williams, M.B., et al., 2021. Process-structure-property relations for as-deposited solid-state additively manufactured high-strength aluminum alloy. Additive Manufacturing 40, 101879.*

the property gradient among different layers. For all the as-printed AA7050, the strain hardening exponent is consistent, generally in the range of 0.2–0.3.

7.2.2 Effects of postprocess aging

As can be seen from previous examples, additive friction stir deposition of precipitation-strengthened alloys, no matter if Al or Mg, causes significant precipitate evolution involving dissolution, reprecipitation, and coarsening. This typically causes deviation of the precipitate size, spacing, and volume fraction from the optimal attributes, and therefore a reduction in strength. To restore the wrought-like mechanical properties, postprocess aging can be employed.

Fig. 7.8 shows the tensile properties of AA7075 printed by additive friction stir deposition, followed by solutionizing and peak aging (T6)

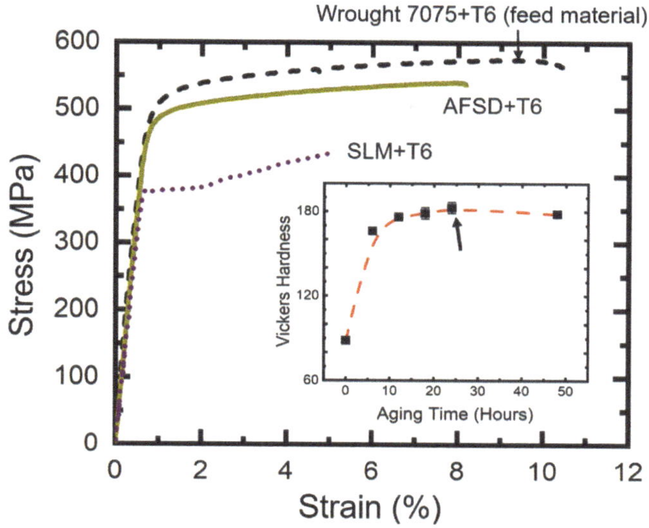

Figure 7.8 Tension testing of AA7075 printed by additive friction stir deposition, with a comparison to the feed material made of wrought AA7075 and AA7075 printed by selective laser melting. T6 aging has been applied to all the cases. Inset: an aging study of the printed AA7075 by additive friction stir deposition. *Reprinted with permission from Yoder, J.K., Griffiths, R.J., Yu, H.Z., 2021. Deformation-based additive manufacturing of 7075 aluminum with wrought-like mechanical properties. Materials & Design 198, 109288.*

(Yoder et al., 2021). For comparison, also plotted are the results for the feed material, which is made of wrought AA7075-T6, and AA7075 printed by selective laser melting followed by T6 aging (Martin et al., 2017). AA7075 is an Al–Zn–Mg–Cu alloy; T6 temper is typically obtained by solution treatment, quenching, and then aging at 120°C for 24 hours. The peak strength is mostly caused by the finely dispersed η and η' precipitates. The inset shows the hardness evolution during the aging of the as-printed AA7075 by additive friction stir deposition. It shows a peak Vickers hardness of ~180 HV near 24 hours, which is similar to the T6 condition for wrought AA7075.

After peak aging, the yield strength, ultimate tensile strength, and elongation of the printed AA7075 are measured as 477 MPa, 541 MPa, and 8.2%, respectively. Because the mechanical behavior of AA7075 is very sensitive to its composition, the mechanical properties of the printed AA7075 must be compared to the wrought AA7075 of the same composition. This requires tension testing of the feed material used for printing, which is characterized by the yield strength of 503 MPa, the UTS of 575 MPa, and elongation of 10.5%. These values are the best that one can obtain in this batch of wrought

AA7075. With peak aging, therefore, the AA7075 printed by additive friction stir deposition retains 95% of the yield strength, 94% of the ultimate tensile strength, and 78% of the elongation.

AA7075 is a nonweldable alloy. To avoid hot cracking in fusion-based additive manufacturing, nanoparticles have been added to the feedstock to promote equiaxed grain structure. Even by altering the composition, AA7075 by selective laser melting and aging only shows a yield strength of 375 MPa and an elongation of 5% (Martin et al., 2017). This reduction is due to vaporization of the solute elements (25% loss of Zn; 32% loss of Mg), which reduces the volume fraction of the primary precipitate for strengthening, $MgZn_2$. To compare, additive friction stir deposition is a solid-state approach without solute loss. Moreover, the as-deposited material is fully-dense, so there is no need for annealing or hot isostatic pressing to reduce porosity after printing.

Fig. 7.9 shows the tensile properties of AA6061 printed by additive friction stir deposition followed by solution treatment and aging, along with the result from the wrought A6061-T651 feedstock (Beck et al., 2021). For comparison, also plotted are the as-printed AA6061 with and without solution treatment (at 565°C) and quenching. Consistent with previous results, the as-printed AA6061 shows a much lower strength but better ductility than the control sample, which is the AA6061-T651 feedstock. This is caused by the dissolution of β'' precipitates (Edwards et al., 1998). Following solution treatment and quenching, the printed AA6061 regains 58% of its strength but the elongation decreases by 10%. Similar to the above work on AA7075, the postaging (at 177°C) tensile properties of the printed AA6061 are significantly improved, comparable to, or even better than the control sample. Remarkably, the UTS increases from 351 MPa in the feed material to 371 MPa in the printed sample. From the fracture surfaces, all the samples are characterized by ductile failure showing distinctive dimples. The most ductile sample is the as-printed AA6061, which shows dimples with fine and shallow characteristics and considerable coalescence. With peak aging, there is enhanced primary void growth and coalescence, causing an increase in the failure rate.

7.2.3 Effects of dislocation content

Dislocations are barriers to the motion of other dislocations. In plastic deformation, the dislocation density increases with the applied strain by dislocation multiplication, for example, via Frank–Read sources or special

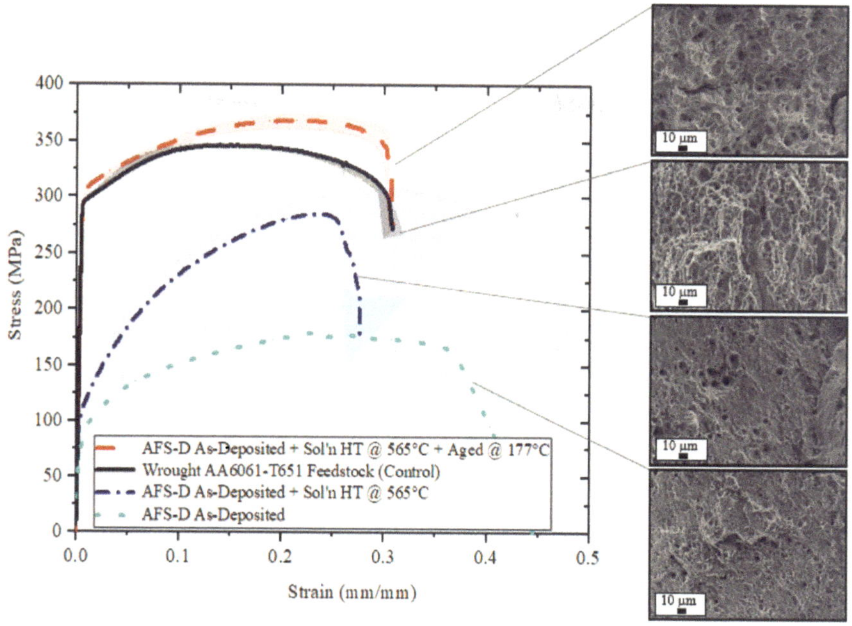

Figure 7.9 Tensile properties of AA6061 by additive friction stir deposition, including as-printed AA6061, printed AA6061 after solution treatment and quenching, printed 6061 after solution treatment, quenching, and aging, and the feedstock used for printing AA6061. *Reprinted with permission from Beck, S.C., Rutherford, B.A., Avery, D.Z., Phillips, B.J., Rao, H., Rekha, M.Y., et al., 2021. The effect of solutionizing and artificial aging on the microstructure and mechanical properties in solid-state additive manufacturing of precipitation hardened Al–Mg–Si alloy. Materials Science and Engineering: A 819, 141351.*

dislocation network topology such as multijunctions (Bulatov et al., 2006). The dislocation density in an annealed metal is typically 10^6 to 10^8 cm^{-2}. After cold working, this number may increase to 10^{12} cm^{-2}. Fig. 7.10A shows the prediction of dislocation multiplication and formation of multijunctions during plastic deformation via dislocation dynamics. The existence of multijunction dislocation topology is validated by transmission electron microscopy experiments in single-crystal molybdenum.

The relationship between the shear strength τ and the dislocation density ρ can be written as

$$\tau = \tau_0 + \alpha G b \sqrt{\rho}. \tag{7.9}$$

τ_0 denotes the stress required to move a dislocation in the absence of other dislocations, whereas α is a constant whose value ranges between

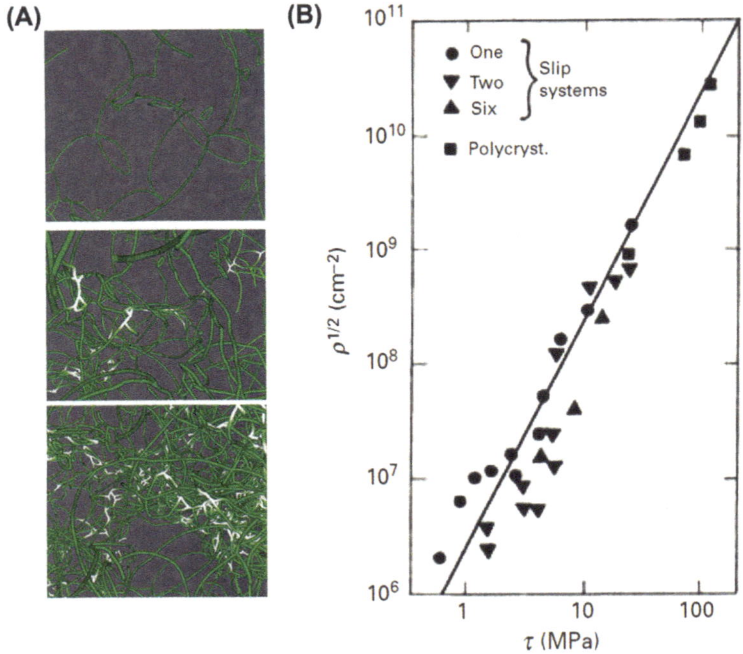

Figure 7.10 (A) Snapshots of dislocation network evolution obtained from a dislocation dynamics simulation, in which continued dislocation multiplication leads to more frequent dislocation interactions and strain hardening effect. Reprinted with permission from Bulatov, V.V., Hsiung, L.L., Tang, M., Arsenlis, A., Bartelt, M.C., Cai, W., et al., 2006. Dislocation multijunctions and strain hardening. Nature 440, 1174–1178. (B) Experimental data on dislocation density vs critical resolved shear stress obtained from monocrystalline and polycrystalline Cu. *Reprinted with permission from Wiedersich, H., 1964. Hardening mechanisms and the theory of deformation. JOM 16, 425–430.*

0.3 and 0.6. This relationship has been found to be valid in the vast majority of situations. Fig. 7.10B plots the square root of dislocation density measured in monocrystalline and polycrystalline Cu as a function of the measured resolved shear stress, in which a linear relationship is found. Although obtained from different microstructures, the experimental data fits Eq. (7.9) quite well (Wiedersich, 1964).

Fig. 7.11 shows the microstructure and mechanical properties of Cu printed by additive friction stir deposition under various process conditions (Griffiths et al., 2021). The Cu feed material is in the H02 or "half-hard" tempered condition. Additive friction stir deposition is performed under three conditions: "Cu-Low/Low" (print head rotation rate of 300 RPM; travel velocity of 1 mm/s), "Cu-High/Low" (600 RPM; 1 mm/s),

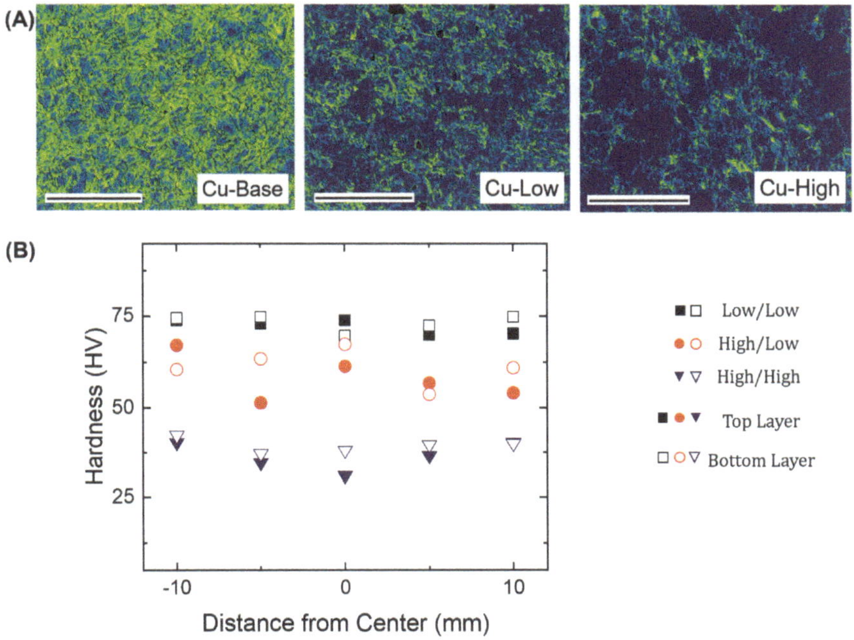

Figure 7.11 Cu printed by additive friction stir deposition. (A) Kernel average misorientation map for the Cu feed material, Cu-Low, and Cu-High. (B) Microhardness testing results for the as-deposited Cu with indentations across the width of the top and bottom layers. *Reprinted with permission from Griffiths, R.J., Garcia, D., Song, J., Vasudevan, V.K., Steiner, M.A., Cai, W., et al., 2021. Solid-state additive manufacturing of aluminum and copper using additive friction stir deposition: process-microstructure linkages. Materialia 15, 100967.*

and "Cu-High/High" (600 RPM; 3 mm/s). The feed rate to travel velocity ratio is kept constant to be 1:3 in all cases. The print head has two minor protrusions on the bottom surface, and a two-layer deposition is performed at each process condition.

The Cu feed-rod (or Cu-base) is hardened via cold-working. From the kernel average misorientation map in Fig. 7.11A, almost all grains in the Cu feed material contain local misorientations, emphasizing the high dislocation content and stored energy due to its cold-worked state. During additive friction stir deposition and postdeposition cooling, discontinuous recrystallization processes happen, effectively removing the dislocation content via high angle boundary migration. In particular, the appearance of large grains with no or little local misorientation in Cu-High/High is evidence of static [or metadynamic (Cho et al., 2001)] recrystallization during cooling. Fig. 7.11B plots the hardness values measured on the transverse cross-section for both the top and bottom layers. With a lower global concentration of dislocations, the as-printed Cu shows a substantial decrease in hardness compared to the feed material. The hardness of the Cu feed material is 112 HV; for Cu-Low/Low, Cu-High/Low, and Cu-High/High, the value is 72.7, 59.6, and 38 HV, respectively. Featuring many "clean" large grains with the least dislocation content, Cu-High/High unsurprisingly has the lowest hardness value.

7.2.4 Effects of grain size

Grain boundary strengthening is a well-known effect, in which the grain size reduction is seen to improve the strength of a material. This was an important motivation that triggered the field of research on nanocrystalline materials. At ambient or moderate temperatures, grain boundaries act as effective barriers to dislocation motion, resulting in pileup of dislocations against boundaries. Moreover, during deformation of polycrystals, the condition of compatibility among surrounding grains requires multiple slips near the areas of grain boundaries for deformation accommodation. Because of the dislocation interactions in these locations, there is a high strain-hardening rate. The grain boundary area per unit volume increases as the grain size decreases, as does the volume percentage of material with a high strain-hardening rate. As a result, the total effect of strain hardening becomes more important.

Empirically, the yield strength and grain size can be described by the Hall–Petch equation:

$$\sigma_y = \sigma_0 + kd^{-1/2}. \tag{7.10}$$

Here σ_y is the yield strength; σ_0 is a material constant; k is the strengthening coefficient depending on the specific material; d is the average grain size. This relationship has been experimentally found widely applicable across a variety of metals and alloys, such as stainless steel, carbon steel, Ni, Mo, Cu, and Al (Meyers and Chawla, 2008). The strengthening coefficient k has a unit of $\mathrm{MPa}\sqrt{m}$; it is a measure of how strong the Hall–Petch effect is. For example, Cu has a relatively weak Hall–Petch effect with $k = 0.12\mathrm{MPa}\sqrt{m}$, whereas Ni_3Al has a strong Hall–Petch effect with $k = 1.70\mathrm{MPa}\sqrt{m}$. Based on experimental data, the Hall-Petch relation is effective for materials with grain sizes ranging from 1 mm to hundreds or even tens of nanometers. However, if the grain size is too small, for example, approximately below 10 nm, the yield strength

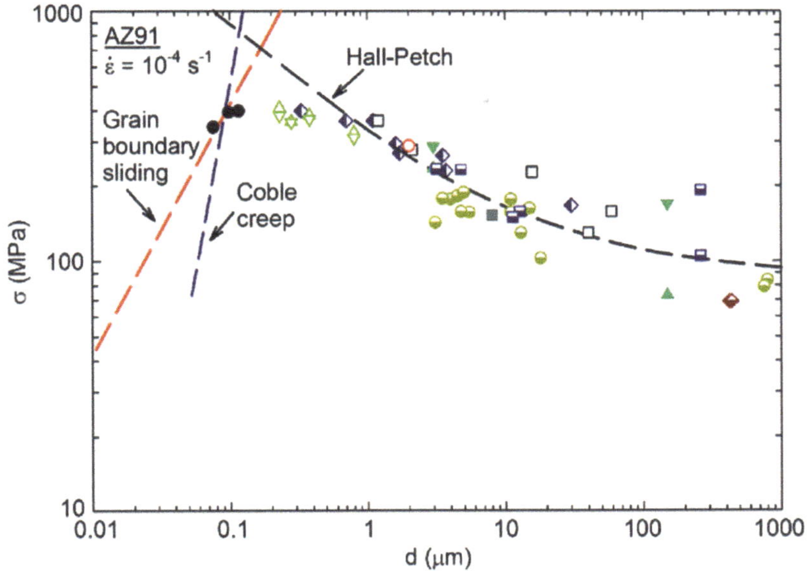

Figure 7.12 The yield strength and grain size relationship of an AZ91 alloy and an AZ91–Al2O3 composite. *Reprinted with permission from Castro, M.M., Pereira, P.H.R., Isaac, A., Langdon, T.G., Figueiredo, R.B., 2020. Inverse Hall–Petch behaviour in an AZ91 alloy and in an AZ91–Al2O3 composite consolidated by high-pressure torsion. Advanced Engineering Materials 22, 1900894.*

will either remain constant or decline as the grains size decreases. Such an inverse Hall–Petch relationship may be caused by dislocation-based, diffusion-based, or grain-boundary shearing-based mechanisms (Carlton and Ferreira, 2007). Fig. 7.12 shows the Hall–Petch and inverse Hall–Petch behavior in an AZ91 alloy and an AZ91–Al2O3 composite consolidated by high-pressure torsion (Castro et al., 2020).

Dynamic microstructure evolution is an important signature of additive friction stir deposition, in which grain size is substantially reduced by continuous or discontinuous dynamic recrystallization. In Al alloys, the stacking fault energy is high, so the microstructure evolution often proceeds via continuous dynamic recrystallization. As a result, the final grain size is governed by the subgrain size. Under the thermomechanical condition typical for additive friction stir deposition, the subgrain size is on the order of several microns. Therefore, even if the feed material has a grain size on the order of hundreds of microns, the final grain size reaches a few microns in the as-printed Al alloys. Despite such drastic grain size reduction, the yield strength always decreases from the feed material. This is because precipitation strengthening, rather than grain boundary strengthening, is the dominant mechanism in such Al alloys. In fact, the strengthening coefficient for Al is quite low: $k = 0.09\text{MPa}\sqrt{m}$ (Cordero et al., 2016).

Inconel 625 is a Ni-based superalloy with high strength at elevated temperatures due to solid solution strengthening and precipitation strengthening. Additive friction stir deposition of Inconel 625 leads to a significant grain size reduction via dynamic recrystallization. As shown in Fig. 7.13A, the grain size is reduced from 30 μm in the feed material to about 1 μm in the printed material. Fig. 7.13B plots the tensile properties of the as-printed Inconel 625 by additive friction stir deposition with a comparison to the feed material. Instead of decreasing its strength like that in precipitation-strengthened Al and Mg alloys, here the yield strength and UTS are both increased. This improvement is likely caused by the grain boundary strengthening effect; it is known that Inconel 625 has a relatively strong Hall–Petch effect, with $k = 1.06\text{MPa}\sqrt{m}$ (Getto et al., 2019). With the grain size reduced from 30 μm to 1 μm, the grain boundary strengthening effect would lead to an increase in the strength by ~860 MPa. Therefore, even with weakening due to precipitate evolution, the overall yield strength and UTS are still increased in the as-printed material (Rivera et al., 2017).

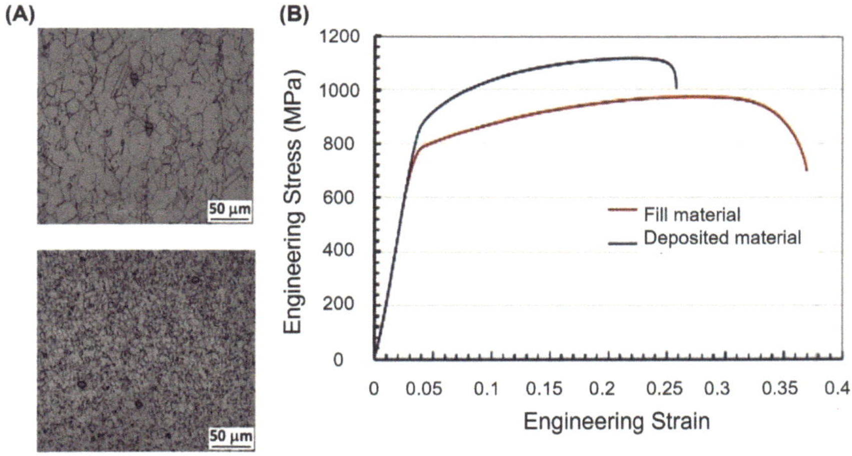

Figure 7.13 Additive friction stir deposition of Inconel 625. (A) A comparison of the microstructure (*top image*) before and (*bottom image*) after deposition. (B) Tensile properties of the as-deposited material and feed material. *Reprinted with permission from Rivera, O.G., Allison, P.G., Jordon, J.B., Rodriguez, O.L., Brewer, L.N., McClelland, Z., et al., 2017. Microstructures and mechanical behavior of Inconel 625 fabricated by solid-state additive manufacturing. Materials Science and Engineering: A 694, 1–9.*

7.2.5 Two-phase alloys

Two-phase alloys are of increasing interest in metallurgy research. As one example, high-strength steel with a ferritic–martensitic microstructure is known as dual-phase steel. With high ultimate tensile strength, low yield strength, high early-stage strain hardening, and macroscopically homogenous plastic flow, dual-phase steel is ideal for sheet forming operations in the automotive industry (Fonstein, 2017). As a second example, Ti-6Al-4V consists of an alpha phase with a hexagonal close-packed crystal structure and a beta phase with a body-centered cubic crystal structure. With high-specific strength and exceptional corrosion resistance, Ti-6Al-4V is utilized in a variety of aerospace and biomedical applications (Inagaki et al., 2014; Murr et al., 2009). Mg alloys (e.g., Mg–8%Li–1%Al) can also exhibit a two-phase structure: the alpha phase with hexagonal close-packed lattice structure and the beta phase with body-centered cubic lattice structure (Liu et al., 2003).

Additive friction stir deposition of Ti-6Al-4V has been conducted with representative tensile properties shown in Fig. 7.14 (Farabi et al., 2022). In this case, the peak temperature during additive friction stir deposition is above the beta transus, resulting in complete alpha to beta transformation. The beta

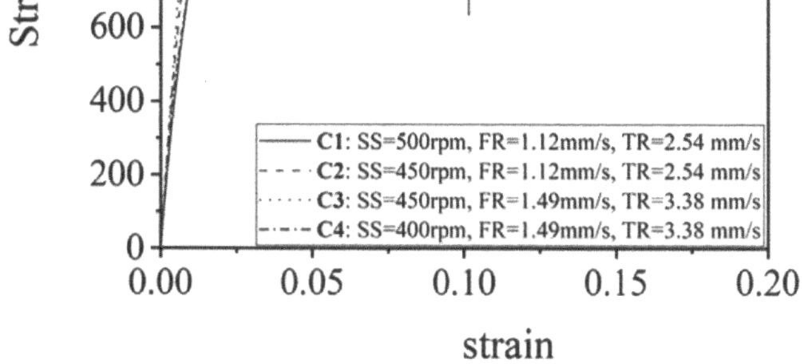

Figure 7.14 In-plane and out-of-plane tensile properties of the as-printed Ti-6Al-4V by additive friction stir deposition. *Reprinted with permission from Farabi, E., Babaniaris, S., Barnett, M.R., Fabijanic, D.M., 2022. Microstructure and mechanical properties of Ti6Al4V alloys fabricated by additive friction stir deposition. Additive Manufacturing Letters 2, 100034.*

phase grains recrystallize with an equiaxed shape under high-temperature deformation in friction stirring. As the beta grains cool, alpha laths form within them, resulting in a two-phase microstructure. If the peak temperature is in the alpha-beta two-phase regime during additive friction stir deposition, dynamic recrystallization results in fine alpha and beta grains. Then, a bimodal structure that consists of equiaxed grains and some lamellar grains may emerge after cooling. Alternatively, given very low cooling rates, the alpha grains may slowly consume beta grains, producing an equiaxed alpha microstructure with beta phase remaining at the grain boundaries.

Four different processing conditions are employed. Significant grain refinement can be achieved at lower rotation rates and related temperatures, reducing lath size and suppressing grain boundary formation. These microstructure characteristics can work together to accommodate deformation in the as-printed state, resulting in a combination of high ductility (20%), high yield strength (1010 MPa), and high ultimate tensile strengths (1233 MPa).

7.2.6 Gradient of the mechanical properties

When building large components involving multiple layers, there are reheating and re-deformation effects on the previously deposited layers. This can make the thermal and thermomechanical conditions in the top layers significantly different from the bottom layers during additive friction stir deposition. As a result, the temperature and strain rate histories of the material are different in each layer. This typically leads to a gradient of the mechanical properties along the build direction. Fig. 7.15 shows the hardness results from the micro-indentation experiments along the build direction of AA2019 printed by additive friction stir deposition (Rivera et al., 2018). AA2219 is an Al–Cu alloy strengthened by the GP zone, the θ'' phase, which is of a thin disk shape fully coherent with the matrix, the θ' ($CuAl_2$) phase, which is of a disk shape semicoherent with the matrix, and the θ ($CuAl_2$) phase, which is of a disk shape incoherent with the matrix (Kamal Babu et al., 2018). Using a solid AA2219-T851 bar as the feed material, a 100 mm long 3D structure made of 6 layers (1 mm thick for each layer) is built up.

The average hardness readings show a steady decline from the top to the bottom of the deposition. According to electron backscatter diffraction, the grain size is relatively homogeneous from the top to the bottom of the as-deposited cross-section. As a result, rather than the Hall–Petch

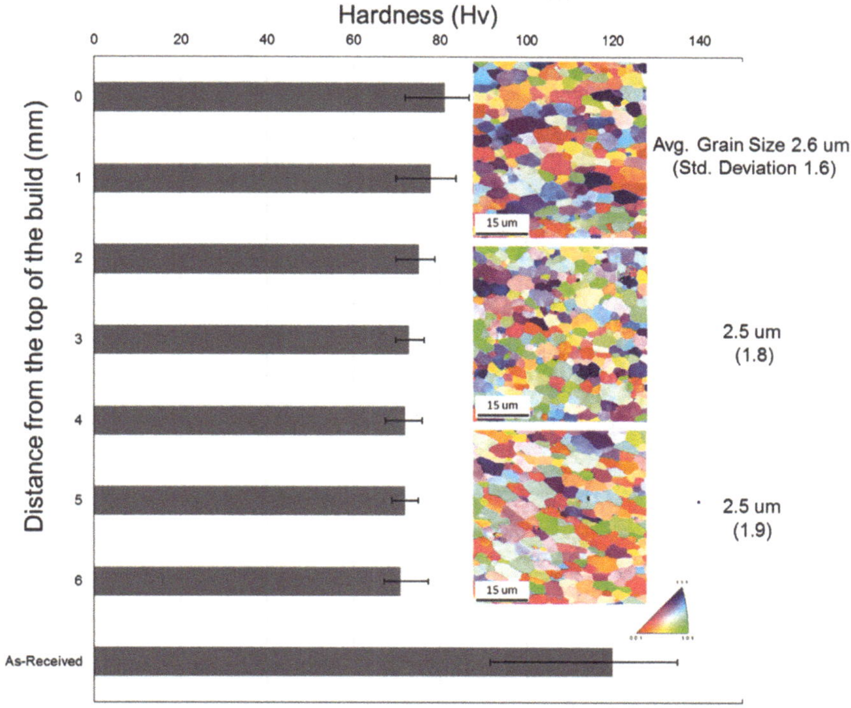

Figure 7.15 On the transverse cross-section of the deposited AA2219, a hardness bar chart is shown across layers with error bars representing minimum and maximum hardness values. *Reprinted with permission from Rivera, O.G., Allison, P.G., Brewer, L.N., Rodriguez, O.L., Jordon, J.B., Liu, T., et al., 2018. Influence of texture and grain refinement on the mechanical behavior of AA2219 fabricated by high shear solid state material deposition. Materials Science and Engineering: A 724, 547–558.*

effect, this property gradient is driven by various precipitate characteristics stemming from changes in the thermomechanical history. The hardness values of the deposited AA2219 are all considerably lower than the hardness of the feed material due to the complex precipitate evolution during deposition and cooling.

Fig. 7.16 shows the tensile test results of Ti-6Al-4V printed by additive friction stir deposition, with a comparison along the build direction. In this work, Ti-6Al-4V metal chips are compacted into the feedstock dimensions for deposition (Agrawal et al., 2021), which is carried out with a rotation rate of 340 RPM, a feed rate of 0.745 mm/s, and a travel velocity of 2.12 mm/s. Each track is 30.5 mm wide and 1 mm thick. The as-printed material is fully dense showing good tensile properties. The hardness of the material is seen to increase from the top towards the

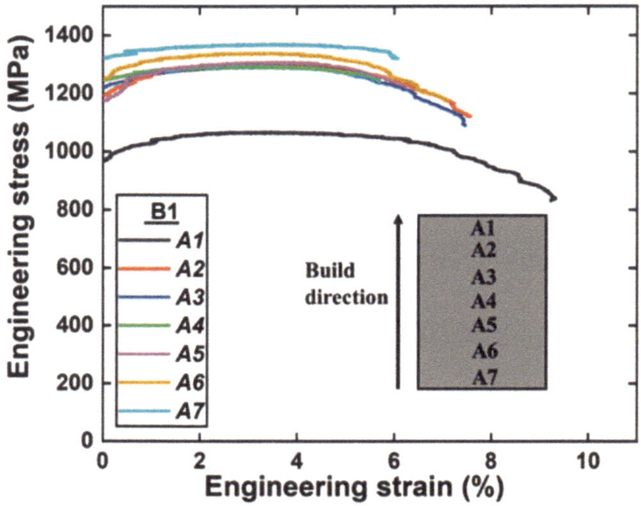

Figure 7.16 Engineering tensile stress—strain curve for the deposited Ti-6Al-4V taken from different heights. *Reprinted with permission from Agrawal, P., Haridas, R.S., Yadav, S., Thapliyal, S., Gaddam, S., Verma, R., et al., 2021. Processing-structure-property correlation in additive friction stir deposited Ti-6Al-4V alloy from recycled metal chips. Additive Manufacturing 47, 102259.*

bottom of the build. The tensile strength follows a similar trend, wherein the very top layer (A1) shows substantially more elongation and lower strength as compared to the rest of the building structure. This can be attributed to a reduction in prior β grain size and lath size from the top to the bottom. A similar trend has been observed in structures printed under other deposition conditions.

7.3 Fracture behavior

Fracture is a material failure phenomenon that depends on the geometry, loading condition and stress state, and intrinsic material properties. In the context of a tension test, a ductile material shows extensive plastic deformation after yielding before it finally fails; a brittle material shows little to no plastic behavior before failure, and the tensile strength is very similar to the yield strength. In a ductile material, the localized stress can be redistributed via local yielding, allowing for continuing deformation to delay the fracture failure. The corresponding failure mechanism involves microvoid coalescence, which proceeds via necking, void nucleation, coalescence of voids and cavities, crack propagation, and fracture. This typically leads to the cup-cone shape with dimpled surface morphology (Fig. 7.17A) (Asgari and Mohammadi, 2018). In a brittle material, once the elastic limit is reached, localized stress can result in a stress concentration that favors crack formation and propagation. The corresponding fracture mechanism is of a cleavage nature, including intergranular (Fig. 7.17B) and transgranular fracture (Fig. 7.17C).

It should be noted that while ductility signifies the capacity of a material to undergo plastic deformation, it is not a fundamental property of materials. This is because external variables, such as the stress state and geometry, affect the fracture type. For example, notches promote a triaxial state and can decrease the apparent ductility. In fact, introducing a triaxial stress state at a sharp notch can cause brittle fracture in a material that displays ductile behavior in a tension test. As a result, fracture behavior must be investigated in the context of the loading condition and crack geometry together. Beyond tension testing, this necessitates the application of more sophisticated approaches for analyzing the fracture phenomena, which is the core of the subject of fracture mechanics.

In fracture mechanics, the material resistance to fracture can be evaluated using two criteria involving the critical strain energy release rate (unit J/m^2) or the critical stress intensity factor (unit $MPa\sqrt{m}$). The first criterion is based on global energy analysis, stating that a crack will propagate if the strain energy release rate reaches a critical value. The analysis is based on the competition between energy cost by creating new surfaces

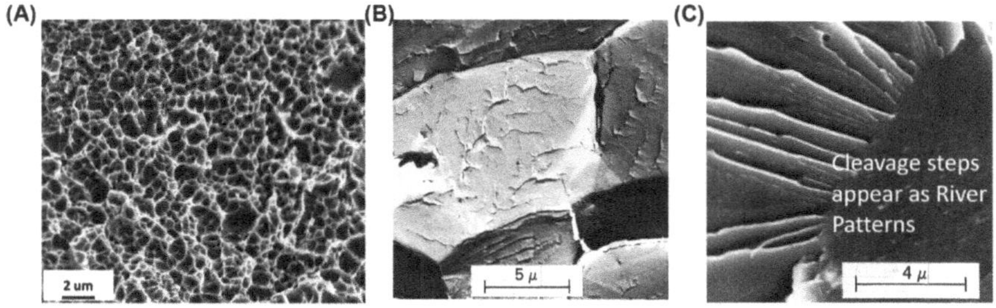

Figure 7.17 Examples of (A) ductile fracture, (B) brittle fracture via intergranular failure, and (C) brittle fracture via transgranular failure. (A) is taken from stainless steel. (B) is from a maraging steel tested in a gaseous hydrogen environment. (C) is taken from a low carbon steel. *(A) is reprinted with permission from Asgari, H., Mohammadi, M., 2018. Microstructure and mechanical properties of stainless steel CX manufactured by Direct Metal Laser Sintering. Materials Science and Engineering: A 709, 82–89. (B) and (C) are reprinted with permission from Hertzberg, R.W., Vinci, R.P., Hertzberg, J.L., 2012. Deformation and Fracture Mechanics of Engineering Materials, 5th ed. Wiley.*

and energy gain via stress relaxation during crack propagation (Anderson, 2005). For a large, thin plate containing a crack of $2a$ length, the strain energy release rate is $G = \pi a \sigma^2 / E$, where σ is the far field tensile stress and E is the Young's modulus. If no plastic deformation and energy dissipation occur during crack propagation, the critical strain energy release rate is equal to $2\gamma_s$, where γ_s is the specific surface energy. Crack propagation occurs when the strain energy release rate is equal to the critical strain energy release rate, that is, $\frac{\pi a \sigma^2}{E} = 2\gamma_s$. Therefore, the critical stress to cause fracture is $\sigma_{critical} = \sqrt{2\gamma_s E / \pi a}$. A large crack length will allow for the crack to propagate under a smaller stress. This clearly shows whether fracture happens depends not only on the applied stress and intrinsic material properties, but also on the geometry of the defects.

The second fracture criterion is based on linear elastic fracture mechanics (Pook, 2000), which describes the crack-tip stress distribution for three major modes of loading (Fig. 7.18A). Mode I is the opening mode, where the crack surfaces move apart directly; Mode II corresponds to the sliding mode; Model III is the tearing mode. In real engineering applications involving cracked components, most of the loading mode is Mode I, for which the distribution of stresses in the vicinity of a crack tip can be written as

$$\sigma_x = \frac{K_I}{\sqrt{2\pi r}} \cos\left(\frac{\theta}{2}\right)\left(1 - \sin\left(\frac{\theta}{2}\right)\sin\left(\frac{3\theta}{2}\right)\right)$$

$$\sigma_y = \frac{K_I}{\sqrt{2\pi r}} \cos\left(\frac{\theta}{2}\right)\left(1 + \sin\left(\frac{\theta}{2}\right)\sin\left(\frac{3\theta}{2}\right)\right)$$

$$\tau_{xy} = \frac{K_I}{\sqrt{2\pi r}} \cos\left(\frac{\theta}{2}\right)\sin\left(\frac{\theta}{2}\right)\cos\left(\frac{3\theta}{2}\right)$$

$$\sigma_z = 0 (\text{plane stress})$$

$$\sigma_z = \nu\left(\sigma_x + \sigma_y\right) (\text{plane strain}) \quad (7.11)$$

The meaning of parameter r, θ, and various stress components is illustrated in Fig. 7.18B. ν is the Poisson's ratio. K_I is defined as the stress intensity factor under Mode I loading. Apparently, the stresses near the crack are all proportional to K_I, which completely defines the crack tip conditions; if

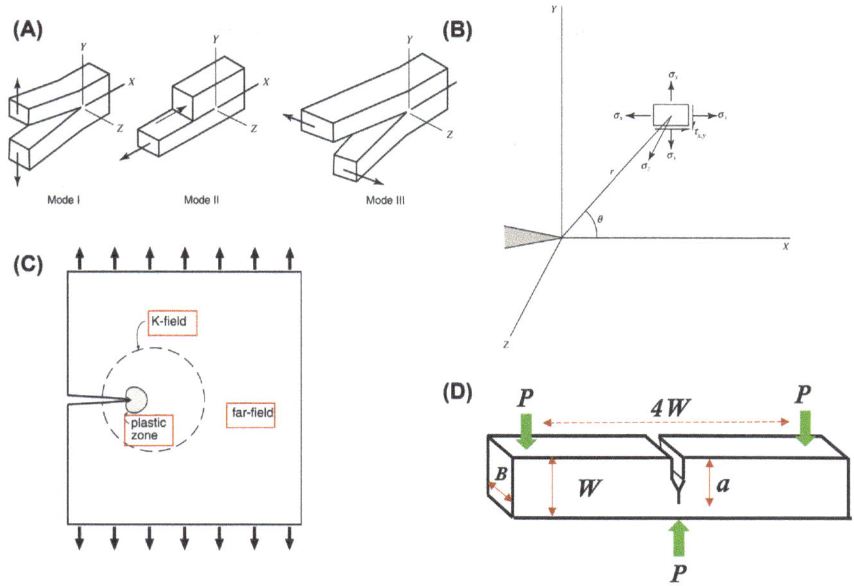

Figure 7.18 Basic concepts in fracture mechanics. (A) Three loading modes. (B) Configuration describing the stress distribution around a crack tip. (C) Plastic zone, K-field, and far-field in a cracked component under loading. (D) The compact tension and single edge notched bending configurations for fracture toughness testing. *(A) and (B) are reprinted with permission from Hertzberg, R.W., Vinci, R.P., Hertzberg, J.L., 2012. Deformation and Fracture Mechanics of Engineering Materials, 5th ed. Wiley.*

K_I is known, it is possible to solve for all components of stress, strain, and displacement as a function of r and θ. Very close to the crack tip, $r \rightarrow 0$, linear elastic fracture mechanics predicts infinitely large stress components. In reality, the high-stress concentration around the crack leads to local yielding in metals, forming a plastic zone (Fig. 7.18C). Outside the plastic zone, there is a region in which the stress state is well predicted by the linear elastic fracture mechanics and Eq. (7.11). This region is termed the "K-field." Outside the K-field, the stress distribution is determined by the far-field applied stress without being affected by the crack geometry.

The magnitude of the stress intensity factor K_I is a measure of the severity of a crack situation. It depends on the applied far-field stress, crack length, and geometry of the component. Generally, it can be written as

$$K_I = FS\sqrt{\pi a}. \tag{7.12}$$

Here S is the far-field stress, a is the crack length (or half length), and F is a geometric factor. For a large, thin plate containing a crack of $2a$

length, the Mode I stress intensity factor is $K_I = \sqrt{\pi a}\sigma$. Comparing this to the expression of strain energy release rate, we can develop a relationship between the strain energy release rate and stress intensity factor: $G=K_I^2/E$. This relationship is valid for thin plates with plane stress conditions. For thick plates with plane strain conditions, the relationship is modified to $G=K_I^2(1-\nu^2)/E$.

The fracture criterion based on linear elastic fracture mechanics simply states that a crack will propagate if the stress intensity factor reaches a critical value, which is often termed fracture toughness. The unit of fracture toughness is MPa$\sqrt{m}$. For Mode I loading, fracture toughness is written as K_{IC}. Fracture toughness can be measured based on ASTM standards, involving specific geometry and configuration requirements. Fig. 7.18D shows one configuration example: single edge notched bending. Another commonly used testing configuration is compact tension. Upon loading, the crack tip opening will become wider without crack propagation until a critical load P_Q or critical far-field stress S_Q is reached. At this critical load, a crack will start to propagate. The critical stress intensity factor then can be calculated based on the geometry and configuration: $K_Q = FS_Q\sqrt{\pi a}$. K_Q can be viewed as the apparent fracture toughness.

It is vital to understand that the measured apparent fracture toughness K_Q is not necessarily a suitable indicator of the fracture resistance of the material. To reveal the intrinsic resistance against crack propagation, the plastic zone size must be minimal in comparison to the specimen characteristic dimensions. The size of the plastic zone is determined by the thickness of the specimen. If the specimen is excessively thin, we will have a plane stress state, which will lead to a large plastic zone. This will require a higher stress intensity factor to trigger the crack propagation. As the specimen becomes thicker, the measured K_Q will decrease. Only after the specimen thickness reaches a threshold value does the measured K_Q becomes stable, reflecting the intrinsic resistance against fracture propagation rather than the geometry effect. Only in this regime does the measured critical stress intensity factor provide a valid measurement of the Mode I plane strain fracture toughness: $K_Q = K_{IC}$. Fig. 7.19A illustrates that a thin specimen exhibits slant fracture dominated by shear due to a large plastic zone, whereas a thick specimen exhibits a flat fracture dominated by tension due to a small plastic zone. Fig. 7.19B schematically shows the variation in fracture toughness with plate thickness.

To characterize the fracture resistance of the interface formed by additive friction stir deposition, the standard single edge notched bending specimen geometry is explored following ASTM E399 (Zhu and Joyce, 2012). As shown in Fig. 7.20A, additive friction stir deposition of many layers (50 + layers) of AA6061 is performed on a wrought substrate made of AA6061-T6. Three beams are cut from the deposition-substrate structure along the build direction with the interface being in the center of the beam. The corresponding fracture testing is used to evaluate the fracture resistance of the interface between the deposition and the bulk substrate. This interface is particularly relevant to additive repair applications, where additive friction stir deposition may be used to selectively deposit material in a variety of damage scenarios including surface damage, through holes, and cracks. The specimen geometry is sketched in Fig. 7.20B.

Specimens are machined and polished before being mechanically notched at the nominal interface between the deposit and the substrate using a diamond-tipped blade. Then, using fatigue loading, a sharp precrack is created so that the fatigue crack extends at least 1 mm beyond the blunt notch. Furthermore, the overall length of notch plus precrack lies within 45%—55% of the size of $(W - a)$. This creates beams qualified for a single edge notched bend test. Under ambient circumstances, fracture tests are performed in a conventional three-point bending configuration at

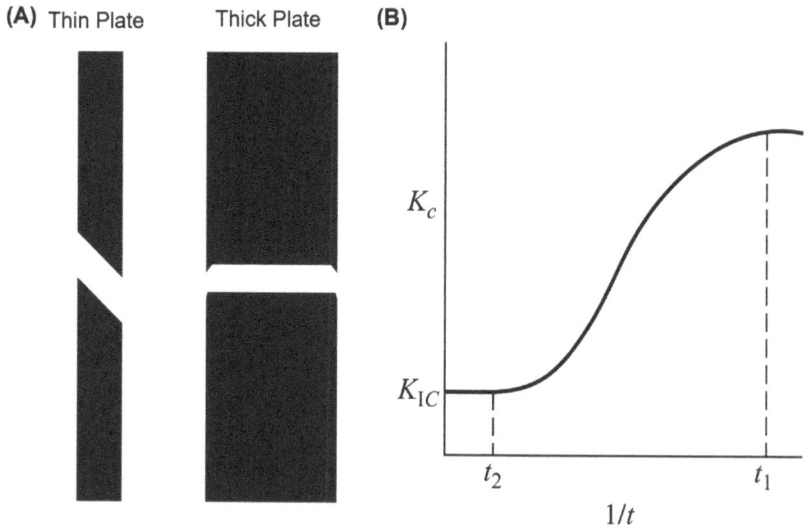

Figure 7.19 Illustration of (A) the failure mode of thin vs thick plates in fracture toughness testing and (B) the measured apparent fracture toughness as a function of the plate thickness. *(B) is reprinted with permission from Hertzberg, R.W., Vinci, R.P., Hertzberg, J.L., 2012. Deformation and Fracture Mechanics of Engineering Materials, 5th ed. Wiley.*

a constant, quasistatic displacement rate of 1 mm/min. Upon loading, the beam specimens are seen to perform in a very ductile manner. In each test, significant crack tip blunting is observed, resulting in a broad plastic zone surrounding the crack tip. Fig. 7.21 shows a time sequence of photographs from the fracture testing of a representative beam specimen, which includes the fracture test from the initial plunger contact to eventual failure. The first few photographs indicate that the crack mouth opens as the loading increases, but the crack tip does not propagate in the early stages of the test. Despite the incredibly sharp fatigue crack, the crack tip gets significantly blunted by plastic yielding, thus generating a plastic hinge.

The resulting load-displacement curve is shown in Fig. 7.22. The onset of crack tip blunting correlates with the divergence from linear elastic behavior. The blunting and energy absorption process continues despite the specimen becoming severely bent. Shortly after the peak load is reached, the material begins to tear from the crack tip, as seen in the first subfigure of the second row in Fig. 7.21. Thanks to the excessive plastic hinging, complete tearing of the specimen is never achieved. The

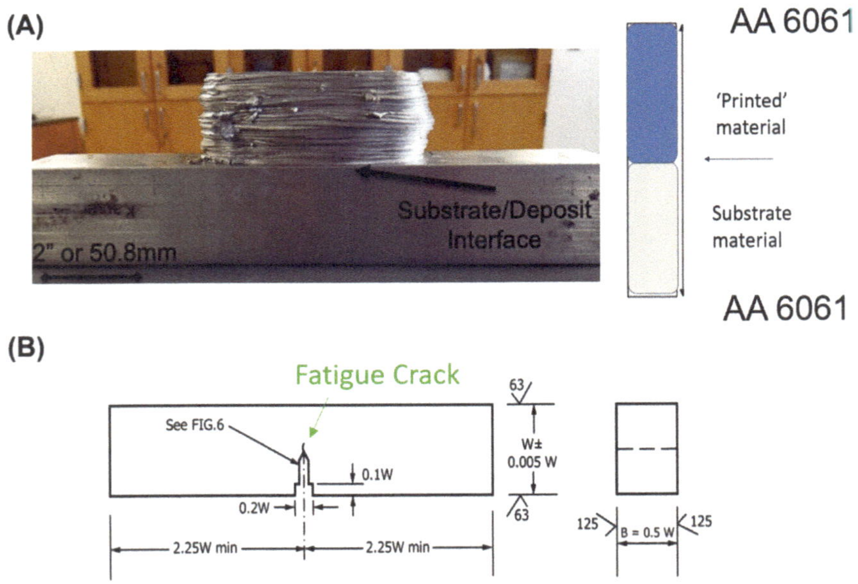

Figure 7.20 (A) A picture of the thick AA6061 deposit on a substrate made of AA6061-T6. (B) Geometry of the beams cut from the deposition-substrate structure. *Adapted from Hartley, W.D., Dillard, David A., Yu, Hang Z., 2022. Estimation of the interface fracture toughness for additive friction stir deposition, In preparation.*

test ends when the specimen eventually falls between the supports rather than splitting into two pieces. Estimation of the apparent fracture toughness K_Q is performed using the 95% offset load P_Q and the procedure defined in the ASTM standard. This leads to a value of $25 - 29MPa\sqrt{m}$ for all the tested specimens (see Table 7.1). As a comparison, the standard K_{IC} value for the base AA6061-T6 material is $29MPa \cdot \sqrt{m}$.

As previously noted, when the plastic zone around the crack is comparable to the specimen dimensions, fracture toughness measurements exhibit a size effect. If the specimen is large enough for the plastic zone to fully develop, the fracture toughness will approach a conservative plateau value. Independent of the specimen size, this value is equal to the plane strain fracture toughness K_{IC}. For a single edge notched bending test, the approximate criterion for plane strain fracture toughness is,

$$a, \quad B, \quad (W - a) \geq 2.5\left(\frac{K_Q}{\sigma_Y}\right)^2, \tag{7.13}$$

where a is the crack length, B is the specimen width, and $W - a$ is the uncracked ligament length. σ_Y is the yield strength. Since K_Q must be measured, the specimen dimensions for achieving plane strain fracture cannot be determined *a priori*. In our situation, the exceptionally ductile behavior of the interface results in a quite large plastic zone. Calculation shows that the measured apparent fracture toughness values of K_Q do not fulfill the criteria in the relationship in Eq. (7.13). Based on the measured K_Q values, a back-of-the-envelope calculation implies that the span of the specimen for single edge notched bend test would need to be hundreds of millimeters long for a reliable plane strain measurement. Methods leveraging the J-contour integral or crack tip opening displacement may be used for quantifying the interface toughness in future work (Rice et al., 1972).

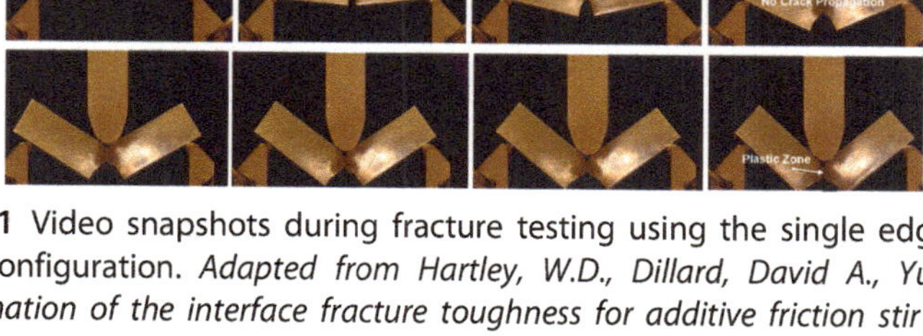

Figure 7.21 Video snapshots during fracture testing using the single edge notched bending configuration. *Adapted from Hartley, W.D., Dillard, David A., Yu, Hang Z., 2022. Estimation of the interface fracture toughness for additive friction stir deposition, In preparation.*

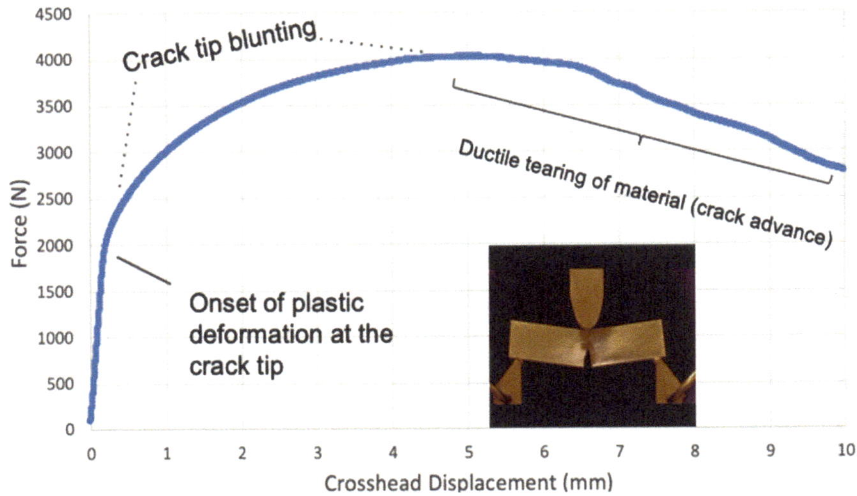

Figure 7.22 Force vs. displacement curve during fracture testing of the printed AA6061-substrate AA6061 beam structure. *Adapted from Hartley, W.D., Dillard, David A., Yu, Hang Z., 2022. Estimation of the interface fracture toughness for additive friction stir deposition, In preparation.*

Although the above K_Q measurements do not meet the most conservative measurement standards, these fracture tests clearly indicate that additive friction stir deposition can lead to excellent interface bonding (Hartley et al., 2022). The fracture mode is seen to be ductile in all cases featuring crack-tip blunting, an intrinsic toughening mechanism. Strength and fracture anisotropy is frequently documented in the additive manufacturing literature, typically with drastically lower mechanical characteristics in the build direction. This is most likely owing to the layer-by-layer manufacturing method, which gives rise to imperfections in each layer interface, like defects, impurities, and inclusions. In additive friction stir deposition, however, strong metallurgical bonding is accomplished by the combined compression and shear as well as the 3D structures formed at the interface elaborated in Chapter 3, Material Flow Phenomena.

7.4 Fatigue behavior

Fatigue, wear, and corrosion are the three main failure mechanisms of metals and alloys in engineering applications. Fatigue failure is defined as the degradation of materials and engineering components as a result of repeated (cyclic) loading. In fatigue loading, the applied stresses can be

Table 7.1 Specimen dimensions, experimental loads, and measured apparent fracture toughness values.

	W (m)	B (m)	a (m)	P_Q (N)	K_Q (MPa $m^{1/2}$)
Specimen 1	0.0128	0.0063	0.0065	804	25.2
Specimen 2	0.014	0.00735	0.0073	1059	28.9
Specimen 3	0.0225	0.016	0.0125	2496	28.5

well below those required to generate fracture on single application of a load. As the cycle number increases, local microscopic damage evolves into microcracks that grow and coalesce to form a large crack. When the crack grows to a critical size, the applied stress will eventually cause catastrophic failure with little or no plastic deformation. Fatigue accounts for at least 90% of all service failures due to mechanical causes (Campbell, 2008). Fig. 7.23 shows the crack growth as a function of cycle number in rotating bending of a precipitation-strengthened Al alloy (Hironobu and Ken-Ichi, 1981).

Fatigue can happen in uncracked components, in which crack initiation is a critical step in determining the fatigue behavior. If the applied stress is below the yield strength, like in wheel and axle engine components, failure normally happens beyond 10^4 cycles, up to 10^7-10^9 cycles. This type of fatigue is referred to as high cycle fatigue. If the applied stress is above the yield strength, like in core components of nuclear reactors, turbine components, and airframes, failure can happen in less than 10^4 cycles. This type of fatigue is referred to as low cycle fatigue. Regarding large structures such as bridges, ships, and pressure vessels, preexisting cracks are inevitable from manufacturing and welding. Fatigue of such cracked structures is governed by crack propagation.

For high cycle fatigue, there are three main steps throughout the process of fatigue loading, (i) crack initiation, (ii) fatigue crack growth, and (iii) ultimate failure. Crack initiation refers to early development of cracks at locations with high stress concentration such as surface scratches, sharp edges, and dents. Although the far-field stress is lower than the yield strength, the micro-heterogeneity of the material can cause local plastic deformation, especially in regions with surface defects. This can eventually result in sharp peaks (or extrusions) and troughs (or intrusions) forming on the surface. Even if these surface offsets are removed by polishing, new surface damage may occur at the same sites upon resumed cyclic loading due to the formation of persistent slip bands (Lukáš et al., 1968).

Once the initial cracks are formed, further cyclic loading will cause stable crack growth. Fatigue crack growth can be divided into two stages. In Stage I, shear stress dominates, and crack growth is along the slip planes. The crack growth rate is low, for example, ~ Å/cycle. The fracture surface is nearly featureless. In Stage II, tensile stress dominates, and crack growth is perpendicular to the applied stress. The crack growth rate is higher, for example, ~ μm/cycle. The fracture surface may develop some features, like clamshell markings on the macroscopic scale and striations on the microscopic scale.

The stage II crack growth rate depends on the cyclic value of the stress intensity factor ΔK:

$$\Delta K = F\Delta S\sqrt{\pi a}, \Delta S = S_{max} - S_{min} \tag{7.14}$$

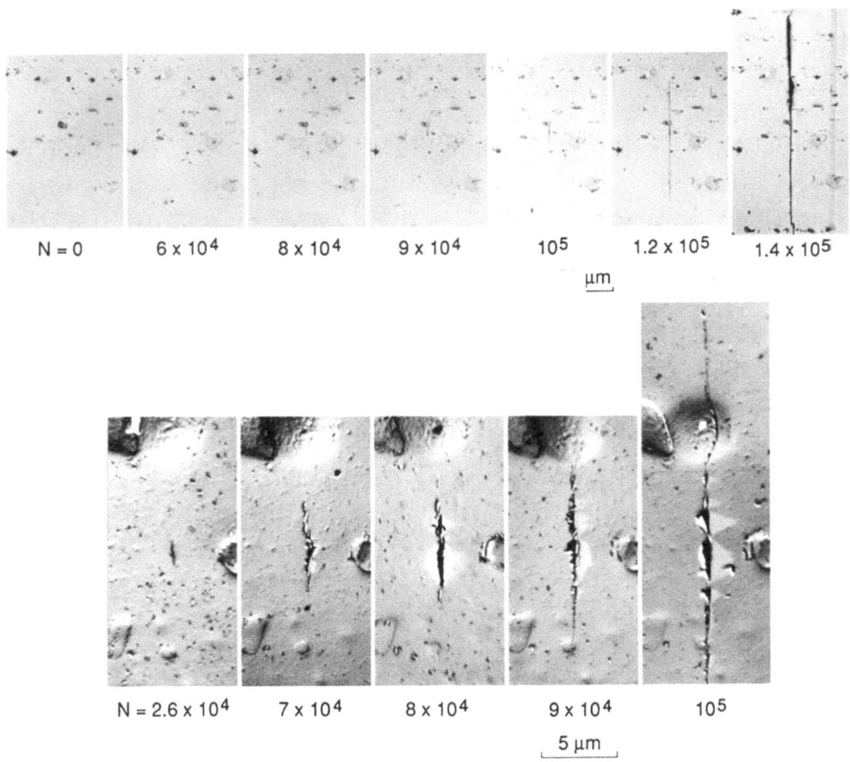

Figure 7.23 Development of a fatigue crack during rotating bending of a precipitation-strengthened Al alloy. *Reprinted with permission from Hironobu, N., Ken-Ichi, T., 1981. Significance of initiation, propagation and closure of microcracks in high cycle fatigue of ductile metals. Engineering Fracture Mechanics 15, 445–456.*

Here S_{max} and S_{min} represent the range of the nominal stress. The R-ratio is defined as S_{min}/S_{max}. ΔK characterizes the severity of a combination of loading, geometry, and crack length for cyclic loading. One can describe the fatigue behavior using the crack growth rate da/dN vs. ΔK curve. This curve is the basis for the damage-tolerant method in engineering applications such as airplane components, where periodic crack examination is used to determine whether repair is required (Fitzpatrick and Edwards, 1998). The Stage II fatigue crack growth generally obeys Paris' law (Paris and Erdogan, 1963):

$$\frac{da}{dN} = C(\Delta K)^m, \tag{7.15}$$

where C and m are constant, and m has a value close to 3 for a lot of metals.

When the crack grows to a critical size a_c, the maximum stress intensity factor in cyclic loading can exceed the fracture toughness of the material K_{IC}. In other words, $S_{max}\sqrt{\pi a_c} > K_{IC}$. Then, brittle fracture will happen, which is the ultimate failure. The fatigue performance can be improved by reducing stress concentrators, reducing mean stress, improving surface quality, and introducing compressive residual stress, for example, via shot peening (Ismail et al., 2017).

The fatigue behavior of the material printed by additive friction stir deposition has been explored in a few precipitation-strengthened Al and Mg alloys, such as WE43 (Williams et al., 2021), AA7075 (Avery et al., 2020), and AA2219 (Anderson-Wedge et al., 2021). As an example, Fig. 7.24 shows the results of strain-controlled fatigue testing for the as-printed AA7075 with a comparison to the AA7075-T651 feedstock. Specimens are tested at $R = -1$ at room temperature with a relative humidity of 50%. For the as-printed AA7075, the fatigue loading is along the build direction (i.e., normal direction) orientation. Clearly, the AA7075-T651 feedstock shows better fatigue performance than the as-printed AA7075; for the same strain amplitude, the cycles to failure in the former is higher than the latter. The overall drop in fatigue resistance may be attributable to a decrease in mechanical strength of the as-deposited AA7075 caused by precipitation evolution as elaborated earlier. It is found that the failure initiation location changes from particles containing Fe and Si in the feed material to coarse grain boundary precipitates in the as-deposited AA7075. The grain boundary precipitates are formed during slow cooling after deposition and are well known to cause brittle failure.

Low cycle fatigue results in a shorter failure life for as-printed WE43 by additive friction stir deposition than the feed material, but the high

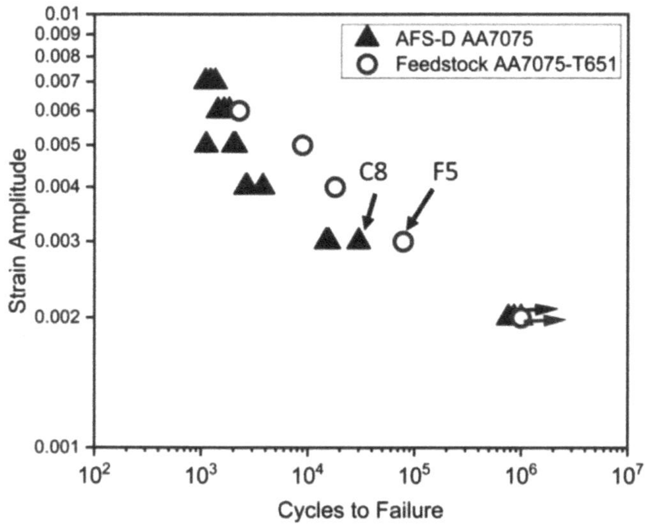

Figure 7.24 Strain-life fatigue results comparing as-printed AA7075 by additive friction stir deposition and the AA7075-T651 feedstock. *Reprinted with permission from Avery, D. Z., Phillips, B.J., Mason, C.J.T., Palermo, M., Williams, M.B., Cleek, C., et al., 2020. Influence of grain refinement and microstructure on fatigue behavior for solid-state additively manufactured Al-Zn-Mg-Cu alloy. Metallurgical and Materials Transactions A 51, 2778–2795.*

cycle fatigue life values are comparable (Williams et al., 2021). Because WE43 is a precipitation-strengthened alloy, coarse precipitates formed during cooling could supply failure initiation sites for the greater stresses associated with low cycle fatigue. The observations in AA7075 and WE43 are not exceptional; other as-printed, precipitate-strengthened alloys also exhibit a reduction in the stress or strain amplitude for a given cycle to failure when compared to literature values of identical materials at full strength. This impact is particularly pronounced in low cycle fatigue, when the nominal stress is larger and beyond the yielding point. In addition to bulk deposition, fatigue testing has been performed on repaired components using additive friction stir deposition. Regardless of precipitate evolution, the debonding issues near the bottom of the repair section can always result in worse fatigue performance.

7.5 Mechanical properties of bilayer structures

In addition to building 3D components, additive friction stir deposition can be widely used for selective area cladding or additive repair. In these applications, not only do the mechanical properties of the deposited

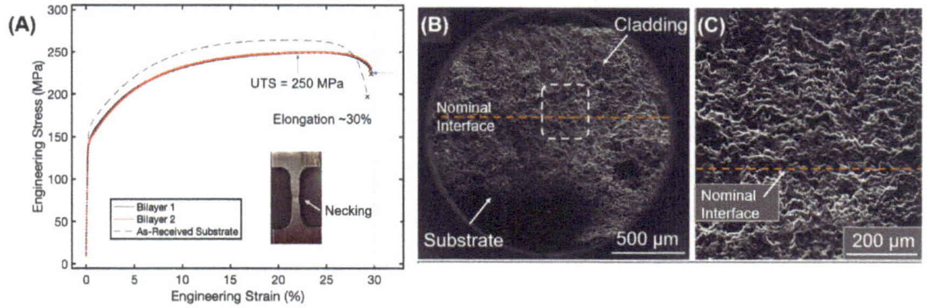

Figure 7.25 (A) The stress–strain curve of tension testing of the reinforced structure with a comparison to the as-received AA6022-T4 substrate. Inset: necking observed in the tensile bar. (B) and (C): Fracture surface showing dimpled morphology at different length scales. *Reprinted with permission from Hartley, W.D., Garcia, D., Yoder, J. K., Poczatek, E., Forsmark, J.H., Luckey, S.G., et al., 2021. Solid-state cladding on thin automotive sheet metals enabled by additive friction stir deposition. Journal of Materials Processing Technology 291, 117045.*

material matter, but it is also important to know how the deposition affects the mechanical properties of the substrate. More importantly, the performance of the resulting deposition-substrate structure as a whole needs to be evaluated, and the interface adhesion should be quantified.

Fig. 7.25A shows the tension testing results of a bilayer structure made of AA6061 (0.9 mm thick)/AA6022 (1.4 mm thick) (Hartley et al., 2021). The as-received thin AA 6022 substrate is characterized by good ductility and formability. It is preferable to retain such mechanical characteristics following printing, so that the resulting bilayer structure can be used for stamping and other forming operations (Bambach et al., 2017). The stress–strain curve of the as-received AA 6022-T4 substrate is also presented for comparison. Both the reinforced bilayer structure and the as-received substrate are ductile, with an elongation to failure of 30%. Tension loading causes necking in the reinforced structure, as illustrated in the inset. The 0.2% offset yield strength and UTS of the reinforced structure are 147 and 250 MPa, respectively. These numbers are only slightly lower (less by 10%) than those measured in the as-received substrate, which are 163 and 265 MPa, respectively. Fig. 7.25B and C depict the fracture surface of the reinforced structure, which has a rough and dimpled morphology typical of ductile failure. Notably, there are no identifying characteristics on the fracture surface of the interface, and cohesive ductile fracture is seen throughout the specimen. This observation proves high-quality interface bonding without porosity enabled by additive friction stir deposition.

A salient advantage of adding a reinforcing layer to the base structure is improved bending resistance. The reinforced structure is subjected to three-point bending to corroborate this characteristic. The mechanical properties of the deposited AA6061 and AA6022 substrate are different, so two distinct specimens are used in the bending test. As illustrated in Fig. 7.26A, specimens are tested with either the AA6061 deposit (Specimen A) or the AA2024 substrate (Specimen B) facing up. Fig. 7.26B shows the force-displacement responses of Specimen A and Specimen B in three-point bending, which exhibit comparable behaviors. The bending stiffness of the reinforced structure is 460 N/mm, which is 4.84 times that of the original substrate, based on the linear elastic component of the curves. This is consistent with elasticity theory, which predicts that bending stiffness scales with the cube of thickness. The thickness of the bilayer structure and substrate is 2.3 and 1.4 mm, respectively; $(2.3/1.4)^3 = 4.434$, which is 92% of the measured value of 4.84. From

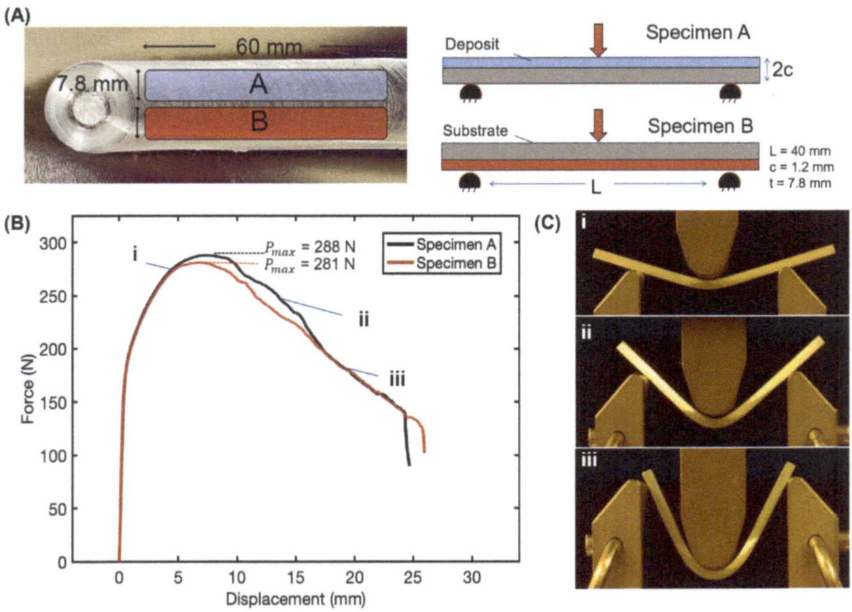

Figure 7.26 (A) Configuration of Specimen A and Specimen B for three-point bending. (B) Load-displacement curves of the two specimens. (C) Representative snapshots during the bending test of the bilayer structure. *Reprinted with permission from Hartley, W.D., Garcia, D., Yoder, J.K., Poczatek, E., Forsmark, J.H., Luckey, S.G., et al., 2021. Solid-state cladding on thin automotive sheet metals enabled by additive friction stir deposition. Journal of Materials Processing Technology 291, 117045.*

the peak load achieved in the bending test (288 N for Specimen A and 281 N for Specimen B), the ultimate bending strength for the two specimens is calculated to be 253 and 247 MPa, respectively, based on the elastic-perfectly plastic material model. These values are close to the measured ultimate strength under tension, 250 MPa.

In three-point bending, the reinforced structure can experience substantial plastic deformation, as demonstrated in Fig. 7.26C. Notably, neither sample fails during the bending tests; the tests are simply terminated when the specimens fall off the supports through the fixture. A posttest visual examination reveals no evidence of cracking or delamination at the interface. This backs up the findings from the tension testing, showing that the ductility of the substrate material is preserved in the bilayer structure.

7.6 Mechanical properties of printed metal matrix composites

In addition to metals, additive friction stir deposition can be used to print metal matrix composites. The inherent friction stir principle promotes uniform particle distribution and enhances interface bonding, resulting in decent mechanical properties in the as-printed state. Tenson tests of the as-printed Al-SiC composites using the auger feeding mechanism have been conducted at room temperature with a strain rate of 10^{-3}/s. Composites having SiC loadings of 10, 20, and 30 vol.% are produced. According to the stress–strain curves, the presence of SiC causes an increase in the flow stress when compared to pure Al. By adding 30 vol.% SiC into Al, the UTS increases from less than 130 to 250 MPa (Su et al., 2017). The volume percentage of SiC also has a considerable effect on the initial work hardening rate, which is greater in composites with a higher volume fraction. Although the ductility of the materials reduces as the volume fraction of SiC increases, Al-30vol.%SiC still has an elongation of more than 12%. This shows that even without postprocessing for porosity reduction, the as-printed Al-SiC composites have a good balance of strength and ductility.

7.7 Summary

In precipitation-strengthened alloys, for example, some Al and Mg alloys, the yield strength is dictated by the precipitate size, shape, and volume fraction, which are all significantly changed during additive friction stir deposition. If the feed material is in the peak aged condition, the as-printed material usually shows lower strength, lower hardness, but better ductility than the feed material. This is associated with precipitate dissolution, coarsening, and re-precipitation during deposition and cooling. If the as-printed alloys are subjected to solution treatment, quenching, and proper aging, peak-aged mechanical properties observed in wrought alloys can be approximately restored. This has been demonstrated in additive friction stir deposition of AA7075 and AA6061. If the feed material is in a worked state with a high dislocation content, additive friction stir deposition will reduce the dislocation density possibly via dynamic and static recrystallization, resulting in lower hardness. If the feed material has a strong Hall—Petch effect, additive friction stir deposition may lead to higher strength because of the grain size reduction via dynamic recrystallization.

The interface between the printed AA6061 and AA6061 substrate is strong, showing extensive plastic yielding around the crack tip in fracture toughness testing based on single edge notched bending. Because the plastic zone size is large compared to the specimen dimensions, however, the apparent fracture toughness cannot be taken as the Mode I plane strain fracture toughness. The fatigue behavior of materials printed by additive friction stir deposition has been explored in a few precipitate-strengthened Al and Mg alloys, which generally show worse fatigue resistance than the feed material due to precipitation evolution. This is especially obvious in low cycle fatigue when the nominal stress may exceed the yield strength. Because of the excellent interface bonding between the printed material and substrate, a bilayer structure can display good tension and bending behavior without delamination. Additive friction stir deposition has been used to print metal matrix composites in a few scenarios. The auger feeding mechanism can result in decent manufacturing quality and good tensile properties in the as-printed state.

References

Agrawal, P., Haridas, R.S., Yadav, S., Thapliyal, S., Gaddam, S., Verma, R., et al., 2021. Processing-structure-property correlation in additive friction stir deposited Ti-6Al-4V alloy from recycled metal chips. Additive Manufacturing 47, 102259.

Allen, S.M., Balluffi, R.W., Carter, W.C., 2005. Kinetics of Materials. Wiley.

Anderson-Wedge, K., Avery, D.Z., Daniewicz, S.R., Sowards, J.W., Allison, P.G., Jordon, J.B., et al., 2021. Characterization of the fatigue behavior of additive friction stir-deposition AA2219. International Journal of Fatigue 142, 105951.

Anderson, T.L., 2005. Fracture Mechanics: Fundamentals and Applications, third ed Taylor & Francis.

Asgari, H., Mohammadi, M., 2018. Microstructure and mechanical properties of stainless steel CX manufactured by direct metal laser sintering. Materials Science and Engineering: A 709, 82–89.

Askeland, D.R., 1996. Failure—origin, detection, and prevention. The Science and Engineering of Materials. Springer US, Boston, MA, pp. 784–819.

Avery, D.Z., Phillips, B.J., Mason, C.J.T., Palermo, M., Williams, M.B., Cleek, C., et al., 2020. Influence of grain refinement and microstructure on fatigue behavior for solid-state additively manufactured Al-Zn-Mg-Cu alloy. Metallurgical and Materials Transactions A 51, 2778–2795.

Bambach, M., Sviridov, A., Weisheit, A., Schleifenbaum, J.H., 2017. Case studies on local reinforcement of sheet metal components by laser additive manufacturing. Metals 7, 113.

Banhart, J., Chang, C.S.T., Liang, Z., Wanderka, N., Lay, M.D.H., Hill, A.J., 2010. Natural aging in Al-Mg-Si alloys—a process of unexpected complexity. Advanced Engineering Materials 12, 559–571.

>Beck, S.C., Rutherford, B.A., Avery, D.Z., Phillips, B.J., Rao, H., Rekha, M.Y., et al., 2021. The effect of solutionizing and artificial aging on the microstructure and mechanical properties in solid-state additive manufacturing of precipitation hardened Al–Mg–Si alloy. Materials Science and Engineering: A 819, 141351.

Bulatov, V.V., Hsiung, L.L., Tang, M., Arsenlis, A., Bartelt, M.C., Cai, W., et al., 2006. Dislocation multi-junctions and strain hardening. Nature 440, 1174–1178.

Campbell, F.C., 2008. Fatigue. In: Elements of Metallurgy and Engineering Alloys. ASM International.

Carlton, C.E., Ferreira, P.J., 2007. What is behind the inverse Hall–Petch effect in nanocrystalline materials? Acta Materialia 55, 3749–3756.

Castro, M.M., Pereira, P.H.R., Isaac, A., Langdon, T.G., Figueiredo, R.B., 2020. Inverse Hall–Petch behaviour in an AZ91 alloy and in an AZ91–Al2O3 composite consolidated by high-pressure torsion. Advanced Engineering Materials 22, 1900894.

Cho, S.-H., Kang, K.-B., Jonas, J.J., 2001. The dynamic, static and metadynamic recrystallization of a Nb-microalloyed steel. ISIJ International 41, 63–69.

Cordero, Z.C., Knight, B.E., Schuh, C.A., 2016. Six decades of the Hall–Petch effect—a survey of grain-size strengthening studies on pure metals. International Materials Reviews 61, 495–512.

Dowling, N.E., 2013. Mechanical behavior of materials: engineering methods for deformation. Fracture, and Fatigue. Pearson.

Edwards, G.A., Stiller, K., Dunlop, G.L., Couper, M.J., 1998. The precipitation sequence in Al–Mg–Si alloys. Acta Materialia 46, 3893–3904.

Farabi, E., Babaniaris, S., Barnett, M.R., Fabijanic, D.M., 2022. Microstructure and mechanical properties of Ti6Al4V alloys fabricated by additive friction stir deposition. Additive Manufacturing Letters 2, 100034.

Fitzpatrick, M.E., Edwards, L., 1998. Fatigue crack/residual stress field interactions and their implications for damage-tolerant design. Journal of Materials Engineering and Performance 7, 190–198.

Fonstein, N., 2017. 7—Dual-phase steels**Every effort has been made to trace copyright holders and to obtain their permission for the use of copyright material. The publisher apologizes for any errors or omissions in the acknowledgements printed in this book and would be grateful if notified of any corrections that should be incorporated in future reprints or editions. In: Rana, R., Singh, S.B. (Eds.), Automotive Steels. Woodhead Publishing, pp. 169–216.

Getto, E., Tobie, B., Bautista, E., Bullens, A.L., Kroll, Z.T., Pavel, M.J., et al., 2019. Thermal aging and the Hall–Petch relationship of PM-HIP and wrought alloy 625. JOM 71, 2837–2845.

Griffiths, R.J., Garcia, D., Song, J., Vasudevan, V.K., Steiner, M.A., Cai, W., et al., 2021. Solid-state additive manufacturing of aluminum and copper using additive friction stir deposition: process-microstructure linkages. Materialia 15, 100967.

Hartley, W.D., Gotawala, Nikhil, Dillard, David A., Yu, Hang Z., 2022. Estimation of the interface fracture toughness for additive friction stir deposition, under review.

Hartley, W.D., Garcia, D., Yoder, J.K., Poczatek, E., Forsmark, J.H., Luckey, S.G., et al., 2021. Solid-state cladding on thin automotive sheet metals enabled by additive friction stir deposition. Journal of Materials Processing Technology 291, 117045.

Hertzberg, R.W., Vinci, R.P., Hertzberg, J.L., 2012. Deformation and Fracture Mechanics of Engineering Materials, 5th ed. Wiley.

Hironobu, N., Ken-Ichi, T., 1981. Significance of initiation, propagation and closure of microcracks in high cycle fatigue of ductile metals. Engineering Fracture Mechanics 15, 445–456.

Huda, Z., 2022. Strengthening mechanisms in metals/alloys. Mechanical Behavior of Materials: Fundamentals, Analysis, and Calculations. Springer International Publishing, Cham, pp. 63–79.

Humphreys, J., Rohrer, G.S., Rollett, A., 2017. Chapter 2—The deformed state. In: Humphreys, J., Rohrer, G.S., Rollett, A. (Eds.), Recrystallization and Related Annealing Phenomena, third ed Elsevier, Oxford, pp. 13–79.

Inagaki, I., Takechi, T., Ariyasu, Y.S.N., 2014. Application and features of titanium for the aerospace industry.

Ismail, S., Ahsan, Q., Haseeb, A.S.M.A., 2017. 2.7 Recent advances in mechanical surface treatment. In: Hashmi, M.S.J. (Ed.), Comprehensive Materials Finishing. Elsevier, Oxford, pp. 171–179.

Kamal Babu, K., Panneerselvam, K., Sathiya, P., Haq, A.N., Sundarrajan, S., Mastanaiah, P., et al., 2018. Influences of metastable θ″, θ′ and stable θ intermetallics formed during cryorolling and friction stir welding process on AA2219. Journal of Alloys and Compounds 732, 624–629.

Kandalam, S., Agrawal, P., Avadhani, G.S., Kumar, S., Suwas, S., 2015. Precipitation response of the magnesium alloy WE43 in strained and unstrained conditions. Journal of Alloys and Compounds 623, 317–323.

Kim, J.H., Serpantié, A., Barlat, F., Pierron, F., Lee, M.G., 2013. Characterization of the post-necking strain hardening behavior using the virtual fields method. International Journal of Solids and Structures 50, 3829–3842.

Liu, T., Zhang, W., Wu, S.D., Jiang, C.B., Li, S.X., Xu, Y.B., 2003. Mechanical properties of a two-phase alloy Mg–8%Li–1%Al processed by equal channel angular pressing. Materials Science and Engineering: A 360, 345–349.

Lukáš, P., Klesnil, M., Krejčí, J., 1968. Dislocations and persistent slip bands in copper single crystals fatigued at low stress amplitude. Physica Status Solidi (B) 27, 545–558.

Lu, G., 2005. The Peierls—Nabarro model of dislocations: a venerable theory and its current development. In: Yip, S. (Ed.), Handbook of Materials Modeling: Methods. Springer Netherlands, Dordrecht, pp. 793–811.

Marioara, C.D., Andersen, S.J., Jansen, J., Zandbergen, H.W., 2001. Atomic model for GP-zones in a 6082 Al–Mg–Si system. Acta Materialia 49, 321–328.

Martin, J.H., Yahata, B.D., Hundley, J.M., Mayer, J.A., Schaedler, T.A., Pollock, T.M., 2017. 3D printing of high-strength aluminium alloys. Nature 549, 365–369.

Mason, C.J.T., Rodriguez, R.I., Avery, D.Z., Phillips, B.J., Bernarding, B.P., Williams, M.B., et al., 2021. Process-structure-property relations for as-deposited solid-state additively manufactured high-strength aluminum alloy. Additive Manufacturing 40, 101879.

Meyers, M.A., Chawla, K.K., 2008. Mechanical Behavior of Materials, second ed. Cambridge University Press, Cambridge.

Murr, L.E., Quinones, S.A., Gaytan, S.M., Lopez, M.I., Rodela, A., Martinez, E.Y., et al., 2009. Microstructure and mechanical behavior of Ti–6Al–4V produced by rapid-layer manufacturing, for biomedical applications. Journal of the Mechanical Behavior of Biomedical Materials 2, 20–32.

Oliveira, A.F., de Barros, M.C., Cardoso, K.R., Travessa, D.N., 2004. The effect of RRA on the strength and SCC resistance on AA7050 and AA7150 aluminium alloys. Materials Science and Engineering: A 379, 321–326.

Paris, P., Erdogan, F., 1963. A critical analysis of crack propagation laws. Journal of Basic Engineering 85, 528–533.

Pook, L.P., 2000. Linear Elastic Fracture Mechanics for Engineers: Theory and Applications. WIT Press.

Rice, J.R., Engineering, B.U.P.R.I.D.O., Commission, U.S.A.E., 1972. Elastic-plastic Fracture Mechanics. Defense Technical Information Center.

Rivera, O.G., Allison, P.G., Brewer, L.N., Rodriguez, O.L., Jordon, J.B., Liu, T., et al., 2018. Influence of texture and grain refinement on the mechanical behavior of AA2219 fabricated by high shear solid state material deposition. Materials Science and Engineering: A 724, 547–558.

Rivera, O.G., Allison, P.G., Jordon, J.B., Rodriguez, O.L., Brewer, L.N., McClelland, Z., et al., 2017. Microstructures and mechanical behavior of Inconel 625 fabricated by solid-state additive manufacturing. Materials Science and Engineering: A 694, 1–9.

Sankaran, K.K., Mishra, R.S., 2017. Metallurgy and Design of Alloys with Hierarchical Microstructures. Elsevier Science.

Su, J., Hubbard, W., Dubrowsky, J., Griffiths, R.J., Yu, H.Z., Hardwick, N., 2017. Additive manufacturing of metal matrix composite parts with high volume fraction reinforcement particles. In: Advances in Powder Metallurgy and Particulate Materials 2017—Proceedings of the 2017 International Conference on Powder Metallurgy and Particulate Materials, pp. 687–692.

Wiedersich, H., 1964. Hardening mechanisms and the theory of deformation. JOM 16, 425–430.

Williams, M.B., Robinson, T.W., Williamson, C.J., Kinser, R.P., Ashmore, N.A., Allison, P.G., et al., 2021. Elucidating the effect of additive friction stir deposition on the resulting microstructure and mechanical properties of magnesium alloy. Metals 11, 1739.

Yasnikov, I.S., Vinogradov, A., Estrin, Y., 2014. Revisiting the Considère criterion from the viewpoint of dislocation theory fundamentals. Scripta Materialia 76, 37–40.

Yoder, J.K., Griffiths, R.J., Yu, H.Z., 2021. Deformation-based additive manufacturing of 7075 aluminum with wrought-like mechanical properties. Materials & Design 198, 109288.

Zhao, Y.H., Lavernia, E.J., 2011. 13—The mechanical properties of multi-scale metallic materials. In: Whang, S.H. (Ed.), Nanostructured Metals and Alloys. Woodhead Publishing, pp. 375–429.

Zhu, X.-K., Joyce, J.A., 2012. Review of fracture toughness (G, K, J, CTOD, CTOA) testing and standardization. Engineering Fracture Mechanics 85, 1–46.

CHAPTER 8

Niche applications

The most obvious characteristic of science is its application: the fact that, as a consequence of science, one has a power to do things.

— Richard P. Feynman

The main advantage of additive friction stir deposition is its capability of rendering fully-dense material in the as-printed state with a uniform recrystallized, equiaxed microstructure. It is by far the only metal additive printing process that can produce wrought or equivalent mechanical properties in an economical manner. Therefore, the most promising applications lie in where excellent mechanical properties are highly desired and minimal postprinting heat treatment is allowed. Regarding materials, additive friction stir deposition has benefits in printing 2xxx and 7xxx high-strength Al alloys, which are prone to hot cracking in fusion-based additive manufacturing. Its major limitation lies in geometric precision owing to its large tooling size and flash formation, so additive friction stir deposition is not the optimal solution for making intricate geometries with small feature sizes. However, also thanks to the large tooling size, additive friction stir deposition is a contender for large-scale additive manufacturing for components on the order of meters. The main industries that will benefit from additive friction stir deposition include aerospace, automotive, defense, and perhaps the recycling and sustainability sectors as well.

This chapter focuses on the niche applications enabled by additive friction stir deposition. We start from the topic of structural repair, which refers to the simultaneous restoration of the original geometry and mechanical performance. This includes examples of hole filling, groove filling, surface and divot repair, and repair of corroded fastener holes. Second, we discuss selective-area cladding on thin automotive sheet metals, in which good quality can be achieved using a flat, featureless print head. The postdeposition curvature development allows for residual stress assessment, from which the thermal stress and thermal mismatch are evaluated to be small. Solid-state material recycling is another unique field where additive friction stir deposition may thrive. Owing to the extensive material flow during deposition, even if the feed material is porous or

Additive Friction Stir Deposition
DOI: https://doi.org/10.1016/B978-0-12-824374-9.00003-4

contains impurities (e.g., made from machine chips), the as-printed material can possess good mechanical properties. Finally, we discuss the applications of large-scale additive manufacturing as well as processing under harsh conditions, like in space or underwater.

8.1 Structural repair

Constantly bearing static and cyclic loads, structural components are vital for a broad range of industrial applications, ranging from aerospace and automotive to infrastructure and defense. In service, they are prone to various types of damages, such as wear, fatigue, erosion, and corrosion. Once a component has been damaged, replacing it with a new one is expensive, both economically and environmentally. This issue may be addressed via repair, which can be further divided into cosmetic repair and structural repair. While conventional cosmetic repair focuses on shape restoration, structural repair aims to simultaneously restore the original geometry and mechanical performance (Gottwald et al., 2021). Compared to replacement, structural repair is a more cost-effective solution with lower carbon intensity for critical components in load-bearing applications.

To implement structural repair, the material in the damaged area must reestablish metallurgical bonds with the newly added material. Based on the additive capabilities and bonding mechanisms, structural repair approaches can be categorized into four types. As shown in Fig. 8.1, these are nonadditive, melting-based (Quadrant 1), nonadditive, nonmelting-based (Quadrant 2), additive, nonmelting-based (Quadrant 3), and additive, melting-based (Quadrant 4). Quadrant 1 refers to fusion welding, which uses thermal energy to melt the materials and form metallurgical bonds during cosolidification (Bhadeshia et al., 2017). Examples include gas metal arc welding, gas tungsten arc welding, laser welding, induction welding, etc. Based on a similar bonding mechanism, fusion-based additive manufacturing in Quadrant 4 enables high geometric precision by continuously supplying the material to the repair spot (Rahito et al., 2019; Saboori et al., 2019). Notable examples include laser engineered net shaping (Keicher et al., 1997) and wire arc additive manufacturing (Cunningham et al., 2018). Fusion welding and fusion additive manufacturing have historically demonstrated a good capability for geometry restoration. However, due to the underlying liquid–solid transformation, the mechanical performance of postrepair components is challenged by solidification porosity, residual stress, hot cracking, stress corrosion

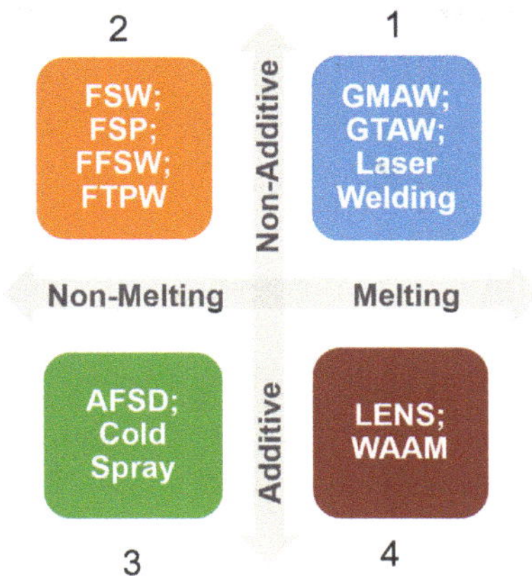

Figure 8.1 Structural repair technologies divided into four categories. Nonadditive, melting-based welding processes such as gas metal arc welding (GMAW), gas tungsten arc welding (GTAW), and laser welding are included in Quadrant 1. Nonadditive, nonmelting-based processes, such as friction stir welding (FSW), friction stir processing (FSP), filling friction stir welding (FFSW), and friction taper plug welding (FTPW) are included in Quadrant 2. Nonmelting additive manufacturing technologies, such as additive friction stir deposition (AFSD) and cold spray are included in Quadrant 3. Melting-based additive manufacturing technologies, such as laser engineered net shaping (LENS) and wire arc additive manufacturing (WAAM) are included in Quadrant 4. *Reprinted with permission from Gottwald, R.B., Griffiths, R.J., Petersen, D.T., Perry, M.E.J., Yu, H.Z., 2021. Solid-state metal additive manufacturing for structural repair. Accounts of Materials Research 2, 780–792.*

cracking, dendritic microstructure formation, and elemental segregation (Jenney et al., 2001).

The problems with the liquid-phase bonding mechanism can be addressed by using nonmelting-based processes in Quadrant 2 and Quadrant 3. Nonadditive, nonmelting-based repair techniques like friction stir welding (FSW) and friction stir processing (FSP) enable strong interface bonding via severe plastic deformation and extensive material flow (Mishra et al., 2014). However, they cannot repair keyholes or volumetric damages since no filler material is used. More recently developed techniques like filling friction stir welding (Han et al., 2013) and friction taper plug welding (Cui et al., 2014) can repair volumetric damages, but they rely on prefabricated pieces that are incapable of autonomous repair. To enable autonomous repair with precise geometric control, continuous feeding capabilities need to be incorporated into the solid-state repair approaches, and we should investigate Quadrant 3.

Quadrant 3 refers to solid-state additive manufacturing technologies that enable continuous feeding and site-specific deposition, including cold spray and additive friction stir deposition. Note that sheet lamination processes, such as ultrasonic additive manufacturing and friction stir additive manufacturing, are challenging to be used for structural repair due to the lack of free-forming capability. Regarding the repair procedure, we will start by removing the damaged or corroded material on the component surface. Then, as shown in Fig. 8.2, solid-state additive manufacturing is applied to deposit layers of material onto the damage region, which may be in the form of divot, groove, or hole. The excess deposited material will be machined away followed by surface polishing. The postrepair component will be subjected to loading again and damage may arise after a certain period of service. The structural repair procedure will be repeated until the postrepair mechanical performance falls well below the performance of the original component.

Additive friction stir deposition is based on continuous plastic deformation and is better suited for volumetric damage restoration. In

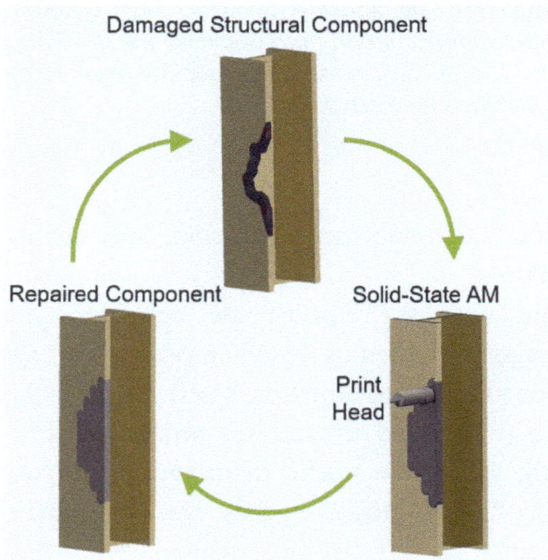

Figure 8.2 Illustration of the structural repair procedure using solid-state additive manufacturing. This involves removing corroded material, applying solid-state additive manufacturing for deposition, machining away the excess material to restore the original component geometry, placing the postrepair component to service, and repeating the repair cycles. *Reprinted with permission from Gottwald, R.B., Griffiths, R. J., Petersen, D.T., Perry, M.E.J., Yu, H.Z., 2021. Solid-state metal additive manufacturing for structural repair. Accounts of Materials Research 2, 780–792.*

comparison, cold spray, which relies on impact interaction between a substrate and accelerated metal particles (Grujicic et al., 2004), is good for surface repair. Also, cold spray typically leads to porosity and an incompletely bonded interface, requiring postdeposition annealing (Cizek et al., 2013). Thanks to the coplastic deformation and mixing between the deposited material and base material, the as-repaired component by additive friction stir deposition can be fully-dense with decent mechanical performance. Additive friction stir deposition is especially useful for the repair of high-strength Al alloys, such as 2xxx and 7xxx, which suffer from hot cracking using fusion welding and fusion additive manufacturing. Additive friction stir deposition is a near-net-shaping technology resulting in flash formation and excess material, so machining is necessary after deposition. Therefore, if postrepair machining is not allowed, other repair technologies should be considered. However, if the postrepair mechanical performance is the top priority, additive friction stir deposition can provide an optimal solution.

Additive friction stir deposition can deal with various types of geometries in repair applications. Fig. 8.3 summarizes the most encountered damage geometries, including shallow divot, volumetric loss, feature damage, and surface crack. As shown in Fig. 8.3A, shallow divots can be repaired by deposition of a thin layer on damaged surfaces after removing the corroded or eroded material. Repair of this type does not require a large volume of filler material. In contrast, for volumetric filling of deep trenches, holes, or grooves depicted in Fig. 8.3B, a large amount of material needs to be deposited and pushed to the bottom of the damage geometry. The challenge lies in the bonding and mixing at the deeper portion of the repair. In Fig. 8.3C, feature reconstruction requires the addition of material with position-level precision; for additive friction stir deposition, postrepair machining is essential. In Fig. 8.3D, surface cracks can be repaired by friction stirring of the crack region using the filler material, which may be supplied at a low feed rate.

8.1.1 Through-hole filling

Additive friction stir deposition has been applied to fill through-holes of various sizes and depths. Fig. 8.4 shows a repair practice involving a small through-hole with a diameter of 1/8 inch (i.e., 3.125 mm) in a 1/4-inch thick AA7075-T651 plate. Note that the size of the hole is much smaller than the size of the AA7075 feed-rod, which has a square cross-section of

Figure 8.3 Repair geometries enabled by additive friction stir deposition: (A) surface layering or shallow divot repair, (B) volumetric fill, (C) feature reconstruction, and (D) crack repair. Schematics are shown for (*left*) original damage, (*center*) immediately following repair, and (*right*) the component after machining and final grinding. *Reprinted with permission from Gottwald, R.B., Griffiths, R.J., Petersen, D.T., Perry, M.E.J., Yu, H.Z., 2021. Solid-state metal additive manufacturing for structural repair. Accounts of Materials Research 2, 780–792.*

9.525 mm × 9.525 mm. In other words, the feature size of the damage is below the spatial resolution of additive friction stir deposition. However, this does not prevent additive friction stir deposition from successfully filling the hole. First, the deposition is initiated on the AA7075 plate surface near the 1/8-inch hole. Once the feed material is fully plasticized, the print head moves in-plane towards the hole with a constant moving speed of 0.32 mm/s, while the deposition continues with a feed rate of 0.1 mm/s. Once the center of the print head aligns with the center of the hole, the movement stops but the print head continues to rotate and supply new material to fill the hole. Once the deposited material is seen to flow out of the hole, implying no more place for deposition inside, the dwell period is brought to an end. The rotating print head will then be lifted, severing the feed-rod from the deposited material. Thanks to the compressive force from the feeding apparatus, the hole is found to be completely filled to the bottom, with no visible interface boundaries.

Figure 8.4 Video snapshots showing the filling of a 1/8 inch through-hole in a 1/4-inch thick AA7075 plate using additive friction stir deposition.

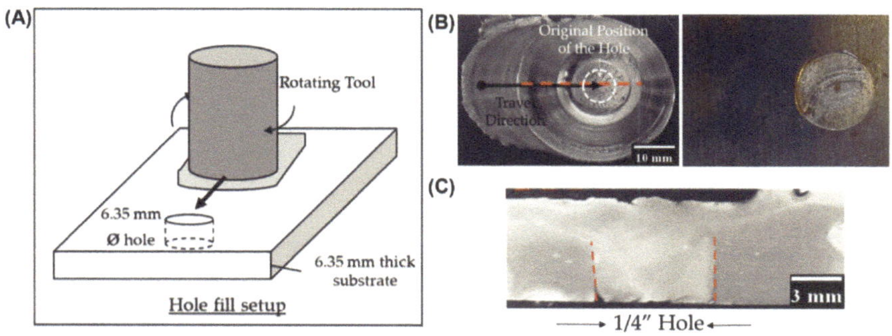

Figure 8.5 Filling of a 1/4 inch through-hole in a 1/4-inch thick AA7075 plate using additive friction stir deposition. (A) A schematic showing the repair procedure. (B) The top and the back side of the repair. (C) Side view of the repair. *Adapted from Griffiths, R.J., Petersen, D.T., Garcia, D., Yu, H.Z., 2019. Additive friction stir-enabled solid-state additive manufacturing for the repair of 7075 aluminum alloy. Applied Sciences 9, 3486.*

In another experiment, repair of a larger through-hole with a diameter of 1/4 inch is performed. As shown in Fig. 8.5A, the repair procedure is similar, involving a moving-and-dwell strategy. Here, the dwelling time is 3 minutes, which is calculated based on the volume of material needed to fill the hole. Fig. 8.5B shows the front and back sides of the repair before the excess material is machined away. The deposition clearly reaches the bottom of the hole. The side view in Fig. 8.5C confirms complete filling, although minor defects are observed at the bottom of the repair. For example, there is a small gap at the left corner of the repair. Defects like these can be avoided by changing the repair path from a simple one-direction translation to multiple-direction traveling involving back and forth deposition.

8.1.2 Groove filling

Additive friction stir deposition can be used to fill large grooves as well. In the repair experiment shown in Fig. 8.6A, a long, wide groove with a rectangular cross-section is created in a 1/4-inch thick AA7075 plate. The groove is 3.175 mm (1/8 inch) deep, 12.7 mm (1/2 inch) wide, and 304 mm (12 inch) long. With the centerline of the print head aligning with the center of the groove, deposition of AA7075 is initiated on the bottom surface of the groove. The print head rotates and supplies the

(A)

(B)

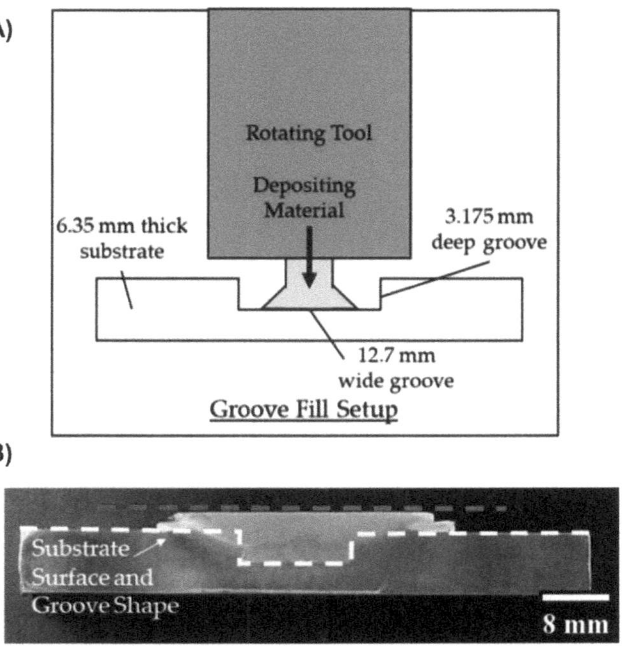

Figure 8.6 Large groove filling using additive friction stir deposition. (A) Illustration of filling the grooves. (B) A cross-sectional image after filling. *Adapted from Griffiths, R.J., Petersen, D.T., Garcia, D., Yu, H.Z., 2019. Additive friction stir-enabled solid-state additive manufacturing for the repair of 7075 aluminum alloy. Applied Sciences 9, 3486.*

AA7075 filler material until it spreads out to reach the sidewalls. Once one end of the groove is filled, the print head will move slowly at 0.42 mm/s along the longitudinal direction of the groove, with a material feed rate of 0.06 mm/s. Multiple layers are deposited inside the groove by going back and forth several times. Each trail deposits a layer with a thickness of 0.76 mm; the total deposition results in a slightly elevated filling surface, as shown in Fig. 8.6B. Significant blending is seen in the upper half of the groove, showing blurred boundary lines between the filler material and the base material. This suggests sufficient mixing and coplastic deformation. The lower half of the groove is seen to be completely filled thanks to the repetitive deposition, which forces the material into the groove, resulting in fully-dense, compact filling.

Evidently, additive friction stir deposition can repair large volumetric damages by completely filling the holes or grooves via continuous plastic deformation. There is no doubt that deep corners of the damage can be reached and good mixing between the base material and deposited material can be obtained in the upper portion of the repair. The most critical question is whether good bonding can be achieved in the deep portion of the repair, which is vital for ensuring good mechanical performance after the repair.

In a more recent experiment, the repair of grooves with different cross-section shapes is investigated. As shown in Fig. 8.7, grooves are drilled to the same depth of 6.35 mm, but with three different shapes, including triangle V, hemi-cylindrical R, and square S. The widths of the S and V grooves are 6.4 mm, while the diameter of the R groove is 12.7 mm. In all cases, the substrate is a 12.7-mm-thick plate made of aluminum alloy 6061-T651. A filler material bar made of AA6061-T6 with a square cross-section of 9.4 mm $\times$ 9.4 mm is used for deposition. To fill these grooves, the print head moves at a speed of 1.3 mm/s, with a feed rate of 2.1 mm/s for the triangle groove, 3.4 mm/s for the hemi-cylindrical groove, and 2.5 mm/s for the square groove, respectively—in order to fit the volume of each groove shape. The layer thickness is 0.5 mm; multiple layers are deposited into each groove until it is completed filled.

The cross-section images obtained by light optical microscopy demonstrate fully-dense filling regardless of cross-section shapes. The contrast patterns in these images show the complex material mixing created by the stirring action. The asymmetric, swirl shape of the mixture is characteristic of the rotating mixing driven by the print head. Interestingly, it seems the

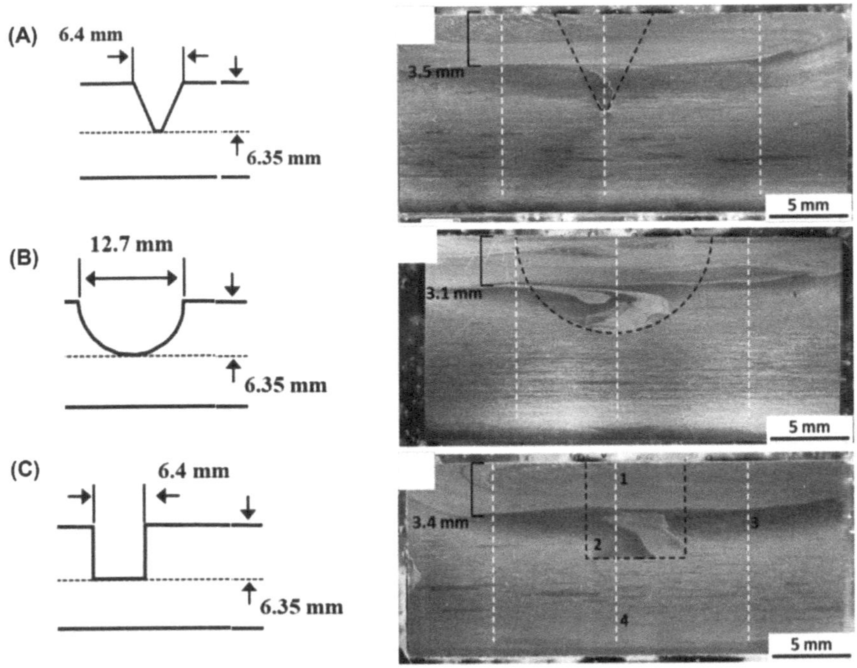

Figure 8.7 Light optical microscopy images of the cross-sections from three filled grooves in the 6061-T651 plate: (A) V groove, (B) R groove, and (C) S groove. *Reprinted with permission from Martin, L.P., Luccitti, A., Walluk, M., 2022. Repair of aluminum 6061 plate by additive friction stir deposition. The International Journal of Advanced Manufacturing Technology 118, 759–773.*

interface between the repair material and base material differs significantly from the original groove boundary. This suggests significant deformation and distortion of the original groove shape caused by friction stirring. Such severe geometry distortion of the base material is not observed in the hole filling and groove filling elaborated above. This is because in the hole filling demonstrated in Figs. 8.4 and 8.5, the deposition is not initiated in the hole; in the large groove filling in Fig. 8.6, the groove bottom is flat, and the groove width is larger than the dimension of the feed-rod. Therefore, the interaction between the stirred feed material and the boundaries of the hole or groove is minimal.

8.1.3 Surface and divot repair

Without substantial material filling, the repair quality of surface damage is evaluated by the interface porosity and adhesion. Fig. 8.8A simulates

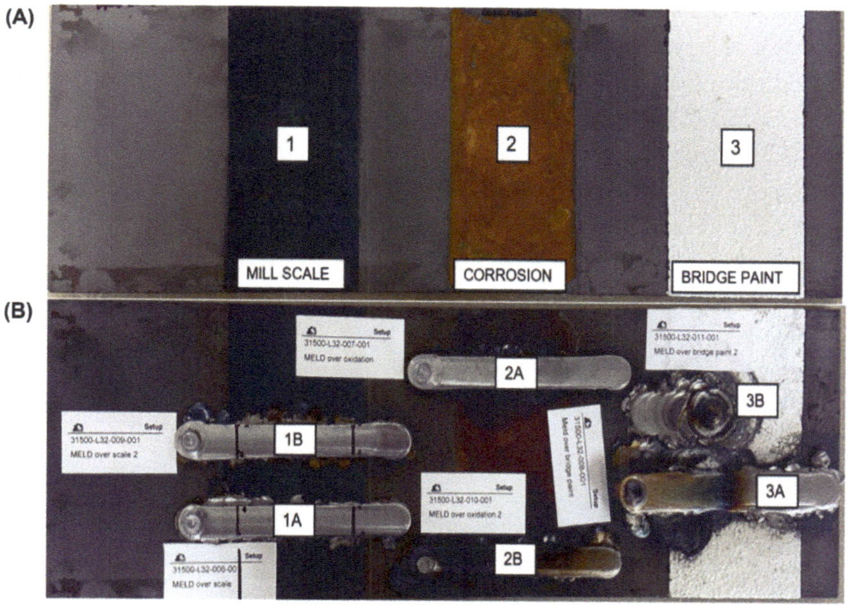

Figure 8.8 Additive friction stir deposition of AISI 1018 steel onto A572 grade 50 steel with three types of surface damage, which were generated by milling, corrosion, and painting: (A) before deposition and (B) after deposition. *Photo courtesy of M.H. Hebdon.*

several realistic surface repair scenarios for an 18-inch-by-6-inch A572 grade 50 steel plate. Three types of surface modification are created by milling, corrosion, or painting. The latter requires the application of the industrial three-coat bridge paint, which includes one layer of zinc primer, one layer of epoxy in the middle, and one layer of polyurethane on top. These scenarios are common in infrastructure and transportation maintenance applications, such as structural bridges, naval vessels, and buildings (Bayliss, 2002). While a blast clean is conventionally implemented to smooth the surface before repair, it is time-consuming and expensive. Here, additive friction stir deposition is directly applied to these damaged surfaces without surface treatment.

Two AISI 1018 steel deposition trials are carried out for each surface, as shown in Fig. 8.8B. During repair, the rough and contaminated surfaces are softened and warped by the rotating print head. This generally leads to intensive mixing between the substrate and deposited material with no apparent interface lines. Although decent deposition is obtained by experimenting with processing parameters, deposition onto corroded and painted surfaces seems more difficult, resulting in excess flash and oxidation. For the painted surface, it is found that the paint material is mostly burned during repair at high temperatures, while the remaining particles are mixed into the deposited material.

Additive friction stir deposition can also repair surface divot. Fig. 8.9A compares the geometry of a beam with a surface divot and a beam after repair. The beams are 3-inch long with a 0.25 inch × 0.25 inch square cross-section. Preliminary experiments have been carried out on these beams, showing no obvious defects on the repair surface. In addition, all

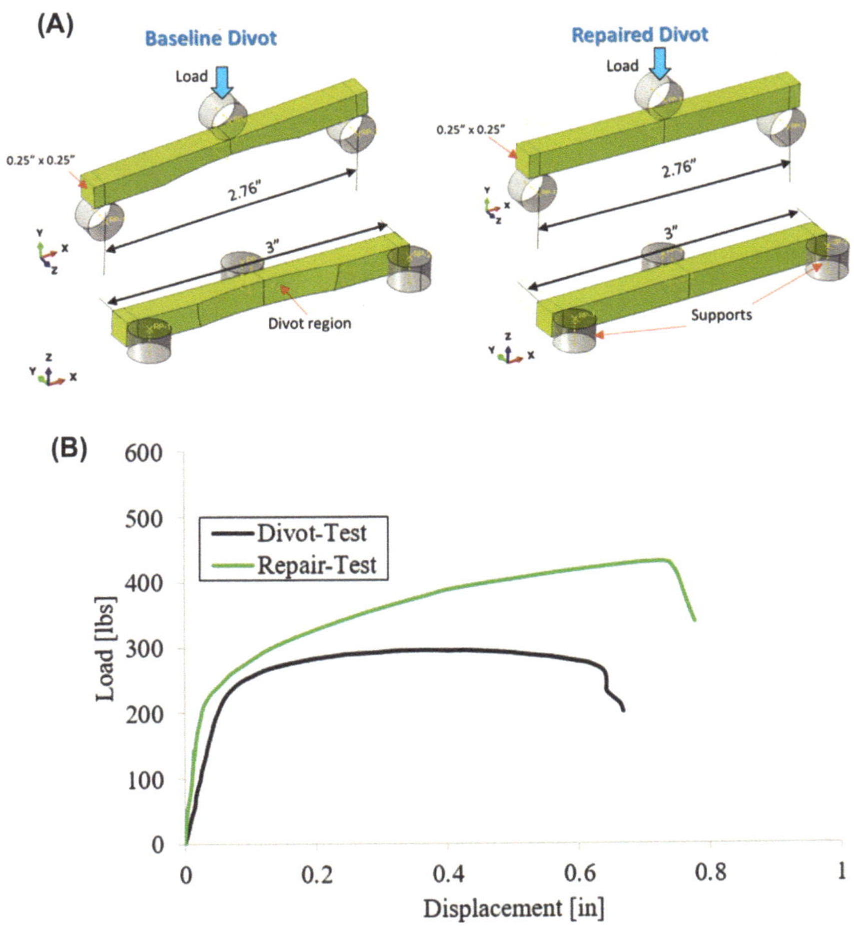

Figure 8.9 Surface divot repaired by additive friction stir deposition. (A) Geometry and configuration of the baseline divot beam and repaired beam. (B) Load-displacement curves in three-point bending.

excess material before machining is at or above the original surface height of the plate. The baseline divot beam and repaired beam are subjected to three-point bending with the divot/repair surface receiving the highest tensile stress during loading. For the repaired beam, the end of testing corresponds to the sample slipping from the grips at a displacement of 20 mm, and no obvious delamination is seen during bending. For the baseline divot, the end of testing is caused by beam fracture. The load versus displacement curves in Fig. 8.9B clearly show that the as-repaired beam can withstand significantly high loads with strain hardening.

8.1.4 Fastener hole repair

Fastener holes have been extensively used in the metallic joints of aerospace structures along with their maintenance (George Nadim, 2018). Corrosion in fastener holes and adhesive-bonded lap joints promotes multisite cracking, which can lead to failure of major aircraft components; in an aggressive environment, corrosion can accelerate the growth of a crack by about a factor of 10. A quick repair option for the detected corrosion damage must be implemented to maintain the structure in safe working conditions up to the next inspection time. As shown in Fig. 8.10, the repair of corroded fastener holes is usually based on the bushing method, involving removal of the corroded material to result in an enlarged hole, manufacturing of a custom bushing, ion-vapor deposition plating, and installation of the bushing. Even though this process restores the original hole geometry, it can take up to two months to complete the entire process and does not save much capital over replacing the entire part. More importantly, the bushing method results in a sharp interface and a smaller fastener hole-edge distance, decreasing the overall airframe static strength and accelerating the risks of fatigue failure.

One application example is the wing structure of a F/A-18 Hornet fighter aircraft. After serving for a few years, practically all the fastener holes will show corrosion around the edge. This is often caused by the galvanic interaction between the Al airframe and carbon fiber composite skin. After removing the corroded material, a stainless-steel bushing is inserted. However, when the aircraft returns to the depot for repair and maintenance next time, the corrosion around the bushing can be significantly worse and fatigue-induced damage may arise. The bushing is then removed, the corrosion is ground out, and a new, bigger bushing is placed. Note that there is no chemical bonding between the bushing and

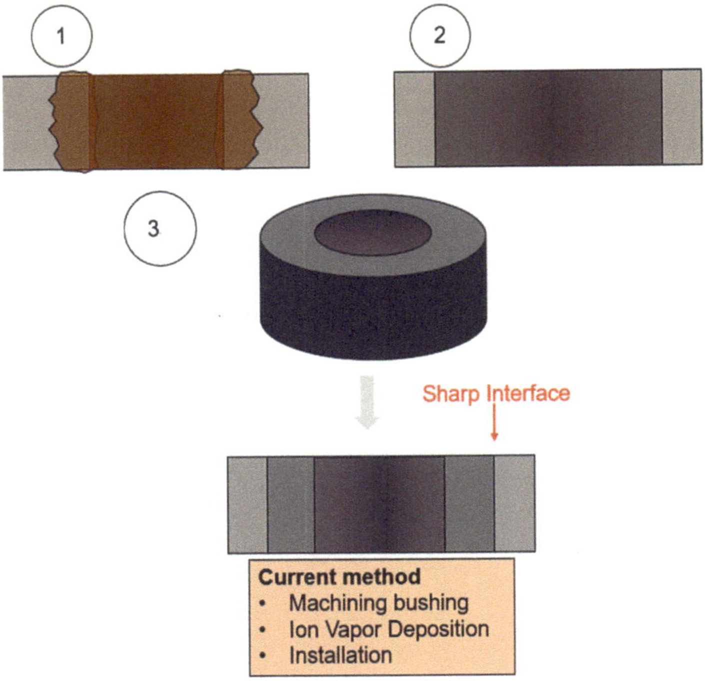

Figure 8.10 Corrosion in fastener holes. (A) Illustration of the current repair method for corroded fastener holes: (1) a corroded hole structure, (2) formation of an enlarged hole after removing the corroded material, and (3) bushing fabrication and insertion to restore the original hole geometry.

sidewall; structural stability relies solely on the mechanical constraint. In addition, there is a distinct change in material properties across the sharp interface between the bushing and the sidewall.

To operate the aircraft within the required safety margin, it is imperative to develop a novel and quick repair method for corroded fastener holes with no sharp interface formation. The ideal repair approach would be able to restore the fatigue life of the component, while maintaining acceptable static strength. Given its promise of hole filling, additive friction stir deposition can be used for repairing corroded fastener holes by adding one more step—to reestablish the original hole geometry via machining and drilling. The salient advantage is that the deposited material and sidewall undergo coplastic deformation and mixing, resulting in indiscernible interfacial regions. This minimizes the stress concentration in the repair regions and potentially improves the fatigue life.

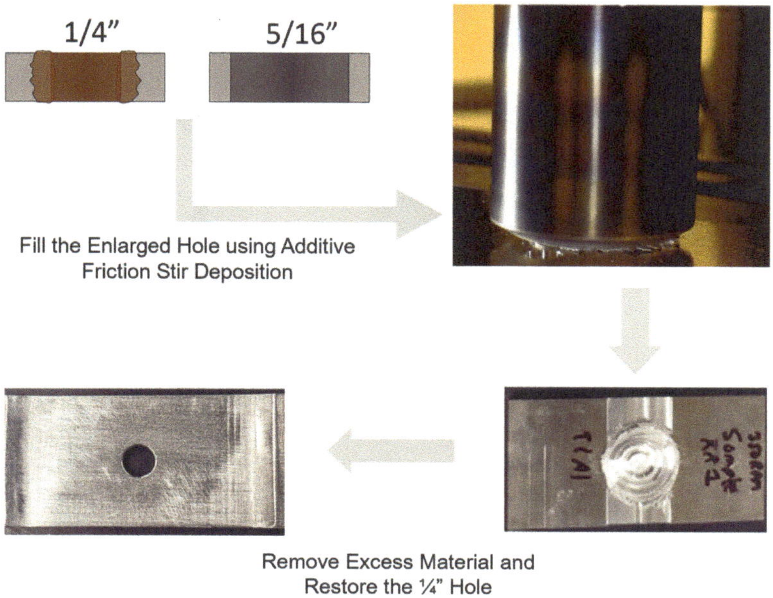

Figure 8.11 Illustration of the repair procedure of corroded fastener holes using additive friction stir deposition.

Fig. 8.11 illustrates the procedure for repairing corroded fastener holes using additive friction stir deposition. We start with a 1/4-inch hole in a 1/4 inch-thick AA7050 plate with corrosion around the hole. After machining the corroded material away, we get an enlarged through-hole with a diameter of 5/16 inch. Additive friction stir deposition is then applied to fully fill the 5/16-inch hole. The deposition is initiated in the center of the enlarged hole, in which the plasticized feed material of AA7050 deforms and spreads under the rotating print head. It takes tens of seconds from the initial plunge phase to the commencement of a pre-written G-code, which involves back and forth in-plane motion to optimize the hole-filling quality. After the hole is filled, the excess material on the top surface will be removed. This results in a seemingly homogeneous AA7050 plate. Then, a new hole of 1/4 inch is drilled at the original location.

Additive friction stir deposition leads to good mixing between the filled material and sidewall, showing a featureless surface; poor mixing would result in a ring structure between the 5/16 inch enlarged hole and the new 1/4-inch hole. Without interfaces between the filled material and sidewall, the fastener hole to edge distance will not decrease. Such repairs have been conducted at different rotation rates of the print head. A high rotation rate will result in excessive thermal exposure and softening of the base material, whereas a low rotation rate may cause insufficient material flow and therefore poor repair quality. At the optimal rotation rate, excellent postrepair fatigue performance is achieved. At an R-ratio of −1, the repaired hole coupons show a substantially improved fatigue life compared with the baseline 5/16-inch hole coupons. The improved performance has been proven to be repeatable. Some of the repaired coupons even exhibit a fatigue lifetime comparable to the pristine 1/4-inch hole coupon before the damage. The fatigue data belongs to an ongoing project by the author and is not presented here.

8.2 Selective-area cladding on thin automotive sheet metals

With a similar configuration to fused deposition modeling, additive friction stir deposition enables selective cladding reinforcement, which has become more and more significant in automotive and aerospace manufacturing. Reinforcement of lightweight components with strong materials in critical regions can substantially improve the overall mechanical performance while remaining lightweight to increase energy efficiency. Since additive friction stir deposition is based on deformation, however, concerns arise regarding cladding on thin substrates, such as thin automotive sheet metals. Will the compression and shear forces from the print head cause severe distortion or damage in the substrate? In this section, we investigate this problem through a case study of selective-area cladding on 6xxx Al automotive sheet metals.

8.2.1 Cladding quality

To reinforce Al−Mg−Si panels used for the automotive industry, an experiment is carried out involving cladding of AA6061 onto a thin substrate cut from a 1.4 mm-thick automotive sheet metal made of AA 6022-T4, which is supplied by the Ford Motor Company (Hartley et al., 2021). Fig. 8.12 presents the cladding results from two different geometries of the print head. In Fig. 8.12A, a featureless flat print head is used with the cladding carried out at rotation rates of 300 RPM, 600 RPM, and 900 RPM, respectively. All these cases involve cladding of a single layer of AA6061 with a thickness of 0.9 mm. While decent cladding quality is obtained for all three different rotation rates, 600 RPM yields the

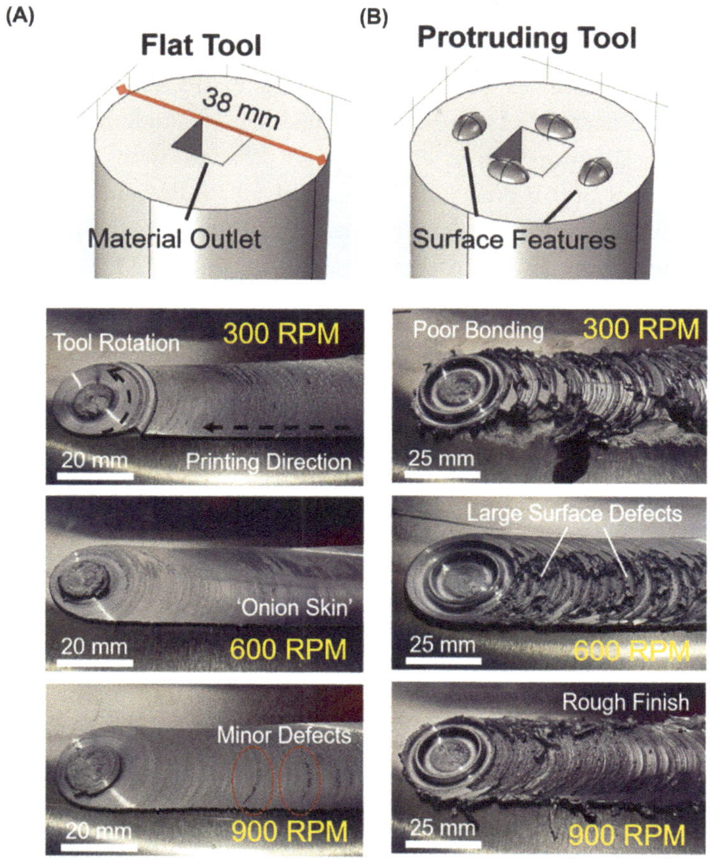

Figure 8.12 Surface quality of the AA6061 cladding on a thin AA6022-T4 substrate, using (A) a flat, featureless print head and (B) a print head with four teardrop-shaped protrusions. In each case, three claddings are performed at rotation rates of 300 RPM, 600 RPM, and 900 RPM. *Reprinted with permission from Hartley, W.D., Garcia, D., Yoder, J.K., Poczatek, E., Forsmark, J.H., Luckey, S.G., et al., 2021. Solid-state cladding on thin automotive sheet metals enabled by additive friction stir deposition. Journal of Materials Processing Technology 291, 117045.*

best appearance with a defect-free surface and a uniform cladding edge. The only notable feature is associated with the "onion skin" pattern on the surface, which is generated by the interaction between the trailing edge of the rotating print head and the top surface of the cladding. This phenomenon has been explained in Chapter 3, Material Flow Phenomena.

In Fig. 8.12B, a print head with four teardrop-shaped protrusions ~2.3 mm tall is used for cladding. In additive friction stir deposition, such a protruding tool is typically employed to improve vertical mixing and generate additional heat. With a thin substrate of 1.4 mm thick, the geometry of the surface protrusions necessitates thicker cladding to prevent substrate damage. As a result, the thickness of single layer cladding is increased to 2.1 mm. Three claddings are created with the print head rotation rate of 300 RPM, 600 RPM, and 900 RPM, respectively. The surface finish and apparent cladding quality are substantially inferior to the claddings using the featureless print head. At 300 RPM, the material is not thoroughly plasticized, resulting in uneven and incomplete bonding of the cladding, whereas, at 600 and 900 RPM, significant surface defects, excess flash, and roughness are evident. Such poor surface quality can be attributed to the reduced heat input from the interfacial friction and volumetric plastic energy dissipation. As the cladding thickness increases, the interface area remains the same, so the interfacial heat generation rate per unit material volume is reduced. The volumetric energy dissipation rate is likewise lower because less material flow is driven by the shear forces imposed by the rotating print head.

From these observations, cladding on thin automotive sheet metals is better to be performed using a flat, featureless print head. Among various processing conditions, a rotation rate of 600 RPM can give rise to good surface cladding quality. Fig. 8.13 presents the cross-sectional and interface characterization results for the cladding made under this condition. Fig. 8.13A shows the optical microscopy images of the cladding-substrate before and after etching. No interface is visible before etching, while a continuous, well-bonded interfacial line is shown after etching. On a closer look at the microstructure, Fig. 8.13B shows a micrograph at the same interfacial location using scanning electron microscopy (SEM). This high-resolution SEM image validates that the cladding and substrate are well mixed to form a sawtooth-shaped interface. With no microscopic defects or porosity, such a nonplanar interface is expected to render strong interface adhesion.

In Fig. 8.13C, the inverse pole figure map from electron backscatter diffraction shows an equiaxed grain structure for both the cladding material and the substrate as well as an indistinguishable interface between them. The nature of this interface is of particular interest, given that bonding takes place without melting and in the presence of a native oxide layer, which can fracture or disintegrate during deposition. The microstructure is consistent and homogeneous throughout the interface, indicating good interfacial quality. Both the cladding and the substrate in this

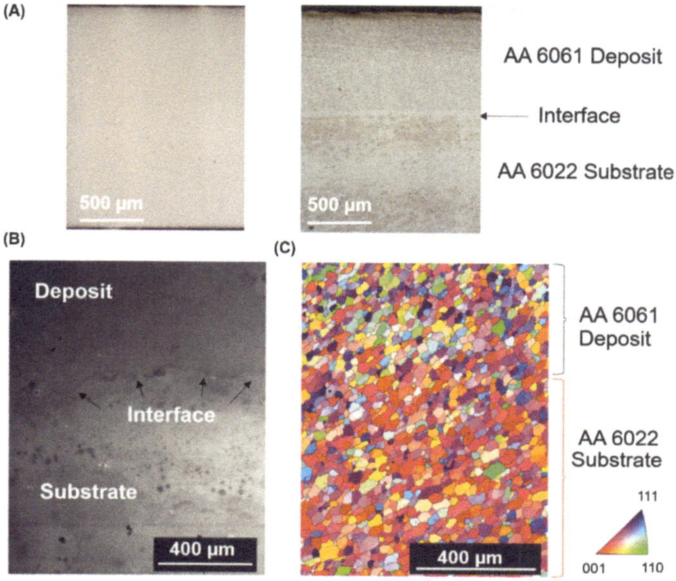

Figure 8.13 Cross-section and interface characterization. (A) Polished, bright-field optical microscopy images (*left*) before and (*right*) after electro-etching. (B) An SEM micrograph of the interface region before etching. (C) The inverse pole figure map showing the grain structure of the same interfacial region. *Reprinted with permission from Hartley, W.D., Garcia, D., Yoder, J.K., Poczatek, E., Forsmark, J.H., Luckey, S.G., et al., 2021. Solid-state cladding on thin automotive sheet metals enabled by additive friction stir deposition. Journal of Materials Processing Technology 291, 117045.*

work are made of Al−Mg−Si alloys. Energy dispersion spectroscopy shows no elemental segregation around the interface region. Instead, the compositions of Mg and Si elements exhibit a progressive transition across the interface.

8.2.2 Thin substrate distortion

Good cladding quality is only one side of the story; minimizing the distortion is also crucial to the performance of the reinforced components. This is especially important for additive friction stir deposition, which imposes compression and shear to enable deposition. In situ observation of the transverse cross-section is used to monitor the distortion of the thin substrate. Fig. 8.14 illustrates the clamping conditions of the substrate plate. From the top view in Fig. 8.14A, the clamping direction is made parallel to the cladding direction (i.e., the travel direction of the print head) using continuous bar clamps. At the same time, the vertical direction is left

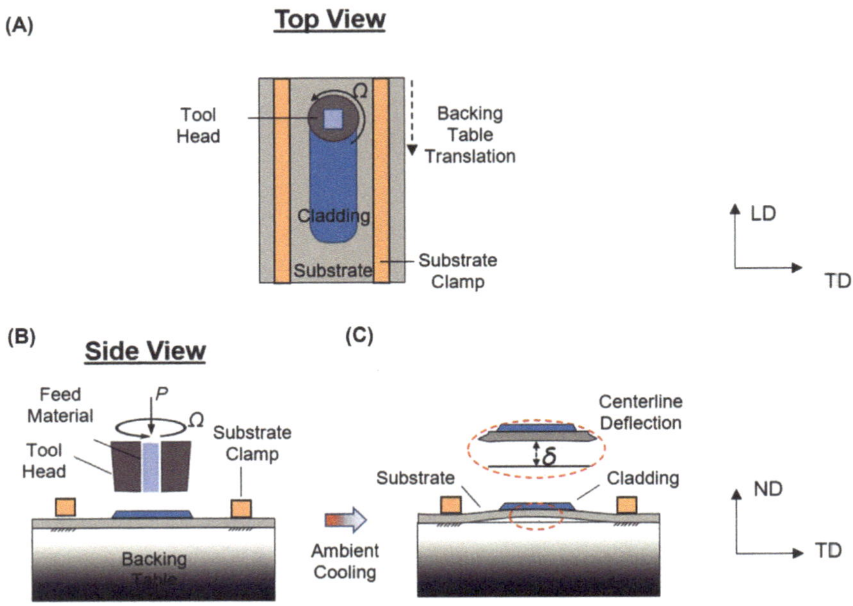

Figure 8.14 Clamping conditions. (A) View from the top, with longitudinal clamping parallel to the cladding direction. (B) A side view of the cladding without distortion. (C) Without vertical constraint, distortion in the transverse cross-section may be observed due to internal stress development. *Reprinted with permission from Hartley, W.D., Garcia, D., Yoder, J.K., Poczatek, E., Forsmark, J.H., Luckey, S.G., et al., 2021. Solid-state cladding on thin automotive sheet metals enabled by additive friction stir deposition. Journal of Materials Processing Technology 291, 117045.*

unconstrained to allow for distortion. As a result, the internal stress developed during or after deposition may lead to an out-of-plane deflection as illustrated in Fig. 8.14B and C.

No local buckling is observed during or immediately after cladding. However, buckling distortion starts to be observed as the system cools down, showing a gradual increase in the out-of-plane displacement. This is a global deformation mode developed in the entire thin substrate. Fig. 8.15 shows the snapshots of the transverse cross-section of the substrate at various periods after deposition. As time passes, the gap between the distorted substrate and backplate appears to widen until it reaches 120 seconds.

Given the minimal differences in composition between the AA6061 cladding and AA6022 substrate, their coefficients of thermal expansion should be close. Therefore, this distortion originates from the temperature difference between the cladding material and the substrate, which

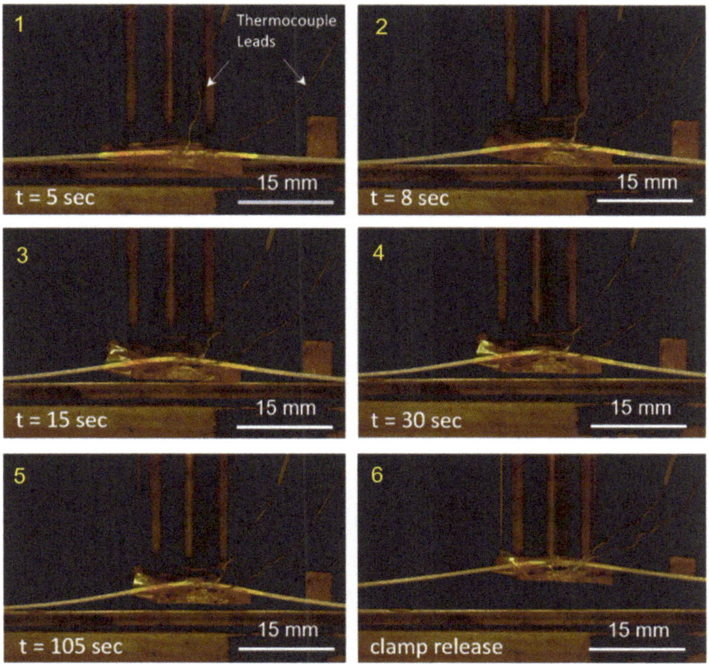

Figure 8.15 In situ monitoring of the transverse cross-section, showing snapshots of the substrate distortion evolution during ambient cooling. *Reprinted with permission from Hartley, W.D., Garcia, D., Yoder, J.K., Poczatek, E., Forsmark, J.H., Luckey, S.G., et al., 2021. Solid-state cladding on thin automotive sheet metals enabled by additive friction stir deposition. Journal of Materials Processing Technology 291, 117045.*

produces different degrees of contraction upon cooling. Because the substrate is restricted longitudinally, the resulting internal stress leads to distortion in the vertical direction. We note that whether local buckling occurs is sensitively dependent on the location and distance of the clamps. When the clamping constraints are too far away from the cladding, we get localized buckling failure in the substrate directly near the print head. When the clamps are too close together, reaction forces might induce deformation around the clamps. Here, we observe no such local buckling or wrinkling in the substrate with an acceptable clamping distance of 100 mm.

Fig. 8.16 plots the out-of-plane displacement of the centerline as a function of cooling time. The number labeled at each data point corresponds to the specific subfigure in Fig. 8.15. Before unclamping, the displacement versus time curve shows an exponential decay shape due to cooling and thermal stress development. The substrate distortion stabilizes after 120 seconds, suggesting that a local thermal equilibrium is reached.

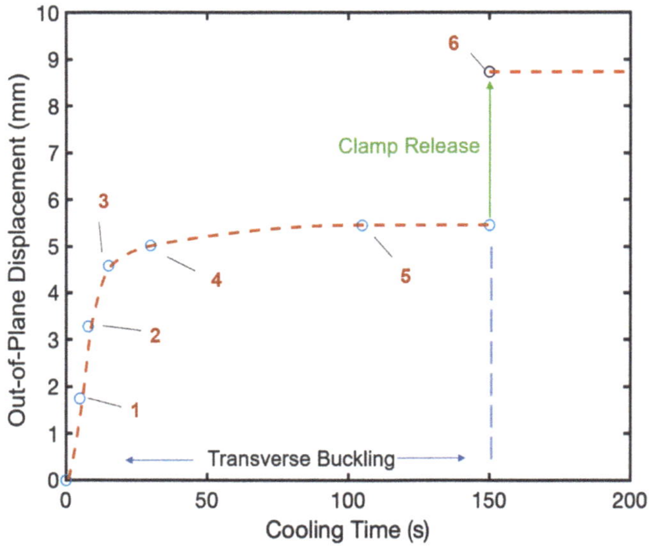

Figure 8.16 Plot of the out-of-plane displacement of the substrate centerline as a function of the cooling time. *Reprinted with permission from Hartley, W.D., Garcia, D., Yoder, J.K., Poczatek, E., Forsmark, J.H., Luckey, S.G., et al., 2021. Solid-state cladding on thin automotive sheet metals enabled by additive friction stir deposition. Journal of Materials Processing Technology 291, 117045.*

On the other hand, a thermocouple mounted at the edge of the substrate plate suggests the thermal equilibrium of the entire system has not been reached even after 500 seconds. This difference may provide insights into the temperature evolution around the cladding area versus the entire substrate. When the clamp is removed, the substrate springs back and leads to an instantaneous increase in the out-of-plane displacement, which is ~3.5 mm. Changes in boundary conditions induce this geometric change, which relieves the stored elastic energy through curvature development. Similar geometric changes are common in sheet metal forming (Carden et al., 2002).

Release of the clamps not only leads to an instantaneous curvature change in the transverse direction as observed in Fig. 8.16, but also an instantaneous curvature development in the longitudinal direction, which is no longer constrained. This is illustrated in Fig. 8.17A. The 3D laser scanning result in Fig. 8.17B shows that the resultant cladding-substrate systems are always characterized by an anticlastic or "saddle-shaped" curvature. During deposition, the temperature is higher in the cladding than the substrate; upon cooling, the substrate ends up being compressed

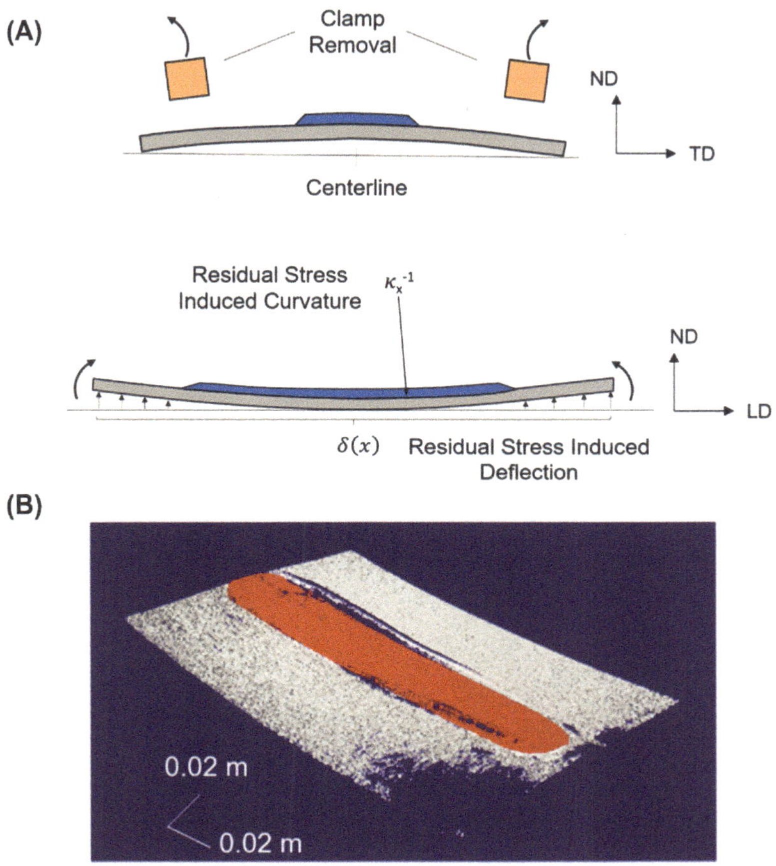

Figure 8.17 Saddle-shaped cladding-substrate system upon release of clamps. (A) Schematic illustration. (B) Curvature measurement using 3D laser scanning. *Reprinted with permission from Hartley, W.D., Garcia, D., Yoder, J.K., Poczatek, E., Forsmark, J.H., Luckey, S.G., et al., 2021. Solid-state cladding on thin automotive sheet metals enabled by additive friction stir deposition. Journal of Materials Processing Technology 291, 117045.*

bi-axially by the cladding, and likewise, the cladding is stretched by the substrate. This results in an unstable flat condition with large internal stresses and considerable potential energy when clamped. When the cladding-substrate system is unclamped, it reconfigures itself by bending to reduce the potential energy, resulting in the observed saddle shape. The convex–concave curvature produces a bending-dominated stress condition in thin geometries, as opposed to shear-dominated angular distortion in thick plates (Wang et al., 2013; Woo et al., 2006).

Starting from the cladding—substrate system, a rectangular piece is cut in the cladding region to result in a bilayer structure, measuring 80 mm × 27 mm in-plane. This is shown in Fig. 8.18A. Although bi-axial stress states frequently induce a spherical bending mode, it is known that departures from a 1:1 in-plane aspect ratio can result in an early transition from spherical to cylindrical bending (Freund and Suresh, 2004). This is corroborated in this study, where we see negligible curvature in the transverse direction along with a little twist. The curvature along the longitudinal direction allows us to estimate the residual stress distribution in this bilayer structure using the classical lamination theory (Finot and Suresh, 1996).

Fig. 8.18B plots the longitudinal stress and transverse stress distribution in the AA6061-AA6022 bilayer structure. For the longitudinal stress, the cladding is mostly tensile, with a maximum stress of 40 MPa at the interface. The substrate develops a compressive stress of -30 MPa near the interface and tensile stress of 43 MPa at the bottom surface. For the transverse stress, the cladding develops a constant tensile stress around 40 MPa, while the substrate develops a constant compressive stress of −30 MPa.

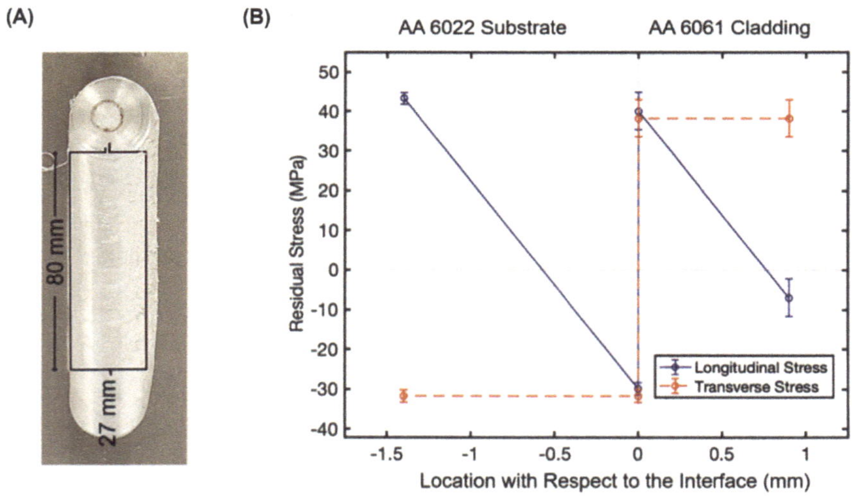

Figure 8.18 Residual stress estimation using the bilayer structure. (A) A picture of the cladding part used for making the bilayer structure. After cutting, the curvature is roughly cylindrical, with a prominent curvature along the longitudinal direction. (B) Results of internal stress distribution based on curvature measurement. *Reprinted with permission from Hartley, W.D., Garcia, D., Yoder, J.K., Poczatek, E., Forsmark, J.H., Luckey, S.G., et al., 2021. Solid-state cladding on thin automotive sheet metals enabled by additive friction stir deposition. Journal of Materials Processing Technology 291, 117045.*

These numbers are generally low compared to those resulting from fusion cladding. We note that the residual stress in the bilayer structure differs considerably from the residual stress in the cladding–substrate system as the boundary conditions have been altered by cutting. However, what matters here is the mismatch strain at the interface, which describes the temperature gradient caused by cladding and is independent of the boundary condition. From the residual stress plots, the mismatch strain is measured to be $\sim 7 \times 10^{-4}$, which is only a third of the elastic limit of the AA6022-T4 substrate. The actual mismatch strain is a gradient with a sigmoidal curve, rather than a discrete number as assumed in the classical lamination theory. Nevertheless, this simple analysis confirms that the thermal exposure by additive friction stir deposition is small, thus resulting in low residual stresses.

8.3 Recycling

8.3.1 Solid-state metal recycling background

It is a top priority to reduce energy usage and carbon footprint in all stages of product manufacturing. Metal manufacturing, as one of the most energy-intensive industries, contributes significantly to excess carbon emissions. For steel production alone, the annual energy consumption accounts for ~7% of the direct greenhouse gas emissions of the world. Although aluminum manufacturing is on a considerably smaller scale, the unit energy cost is several times greater than steel (Energy, 2007; Fruehan et al., 2000).

Metal recycling, which bypasses the energy-intensive raw material production stage, has the potential to significantly lower the energy cost associated with metal product manufacturing and is thus crucial for reaching carbon neutrality (Becker et al., 2020). Metal recycling can be carried out using either melting-based or solid-state processes (Söderholm and Ekvall, 2020). Compared to melting-based recycling, solid-state techniques consume less energy and emit less CO_2 by eliminating the melting step (Shamsudin et al., 2016). Solid-state recycling typically utilizes just 5%–10% of the energy required for melting-based recycling, while it may lead to properties similar to wrought materials. Notably, it is difficult to recycle Al using melting-based methods, which can result in a significant metal loss. Most of this material loss is unavoidable due to the high surface area to volume ratio of the chips, as well as contamination from oxides and machining fluid.

In solid-state metal recycling, the starting material is in the form of scrap chips or swarf, which is cleaned and degreased usually in an ultrasonic bath. They are then dried before being compressed. The quality difference between the final material from the cold and hot-pressed chips may not be significant; cold pressing is favored due to the cost–benefit ratio (Hu et al., 2008). The pressed chips can then be sintered, or subjected to plastic deformation, for example, through extrusion, equal channel angular pressing, forging, rolling, etc., finally resulting in a finished, chip-based product (Fogagnolo et al., 2003; Gronostajski et al., 2000). Among others, hot extrusion is now the most prevalent deformation-based method for processing pressed chips. This method can break down the oxide layer surrounding the chip surface and promote a sufficient weld bond under excessive plastic strain generated during the operation (Haase and Tekkaya, 2014).

The chip geometry has a profound influence on the final product microstructure and quality. Fig. 8.19 shows AZ91D Mg alloy chips with different dimensions, including square chips in (A), measuring (4–6) mm × (3.5–4.5) mm × (1.45–1.55) mm; long chips in (B), measuring

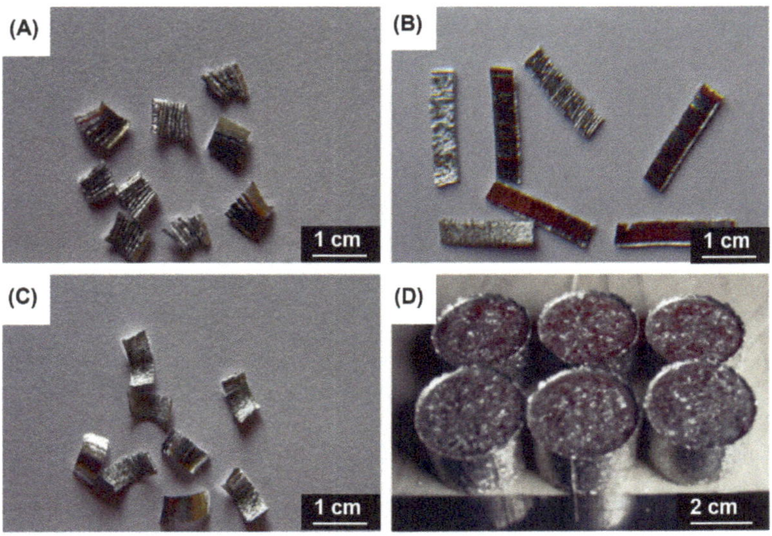

Figure 8.19 AZ91D chips: (A) square chips, (B) long chips, and (C) thin chips. The compressed billets are shown in (D). *Reprinted with permission from Hu, M., Ji, Z., Chen, X., Zhang, Z., 2008. Effect of chip size on mechanical property and microstructure of AZ91D magnesium alloy prepared by solid state recycling. Materials Characterization 59, 385–389.*

(10–16) mm × (1.8–2.2) mm × (0.48–0.52) mm; thin chips in (C), measuring (5–8) mm × (1.8–2.2) mm × (0.18–0.22) mm. After cold pressing, the relative density is 87% using the square chips, 84% using the long chips, and 83% using the thin chips, showing signs of voids and cracks. The pressed chips are then hot extruded to form billets with a diameter of 40 mm as shown in (D). For comparison, hot extrusion is also performed using an as-received AZ91D ingot under identical conditions.

In this specific work, the hot extrusion temperature is 673 K, which is >70% of the melting temperature of AZ91D, so dynamic recovery or recrystallization possibly occurs during processing. For industrial chips, the microstructure of AZ91D magnesium alloy typically consists of an alpha phase of Mg solid solution and a beta phase of $Mg_{17}Al_{12}$. With a grain size of ~ hundreds of microns in the alpha phase, the beta phase is distributed along the grain boundaries in a discontinuous manner. Fig. 8.20 shows the microstructure of the extruded billets, including the ones made from the square chips in (A), long chips in (B), thin chips in (C), and the control sample in (D). All the images are taken on the plane perpendicular to the extrusion direction. Without signs of chip boundaries, these microstructures consist of equiaxed fine grains mixed with elongated large grains. The

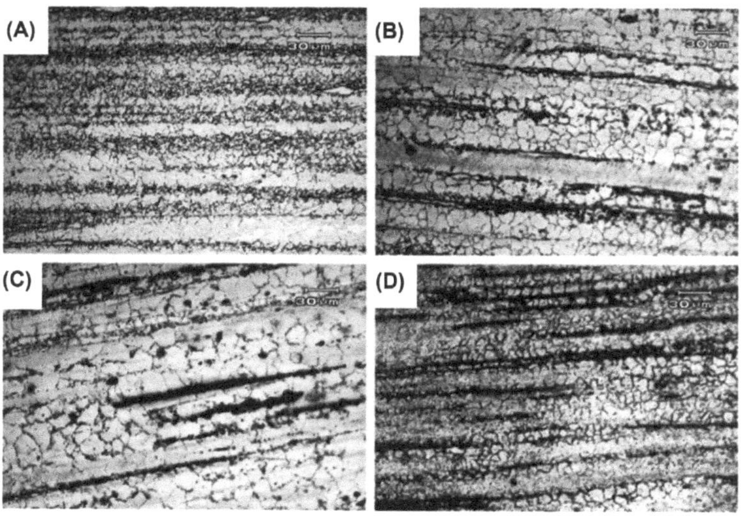

Figure 8.20 Microstructure of the specimens made from (A) square chips, (B) long chips, (C) thin chips, and (D) the control sample. *Reprinted with permission from Hu, M., Ji, Z., Chen, X., Zhang, Z., 2008. Effect of chip size on mechanical property and microstructure of AZ91D magnesium alloy prepared by solid state recycling. Materials Characterization 59, 385–389.*

average grain size resulting from the square chips is 11 μm, which is the same as the control sample, whereas the grain size resulting from the long and thin chips is considerably larger, 22 and 25 μm, respectively. Regarding the tensile properties, the specimen made using the square chips exhibits an ultimate tensile strength of 340 MPa and elongation to failure of 10.5%, both of which are better than the specimens made from the long chips and thin chips. The strength number here is even slightly higher than the control sample because of dispersion strengthening caused by the higher oxygen content in the machine chips. However, as the oxygen content continues to increase, such as in the specimens made by the long and thin chips, the ductility is seen to decrease significantly.

In hot extrusion-based solid-state recycling, the die geometry also plays a key role in controlling the microstructure and properties of the finished products. Common die designs include the flat face, integrated equal channel angular pressing (iECAP) (Haase et al., 2012), and porthole (Güley et al., 2013). When AA6060 chips are extruded via these dies at 550°C, the microstructure and properties exhibit profound differences. Fig. 8.21 (top half)

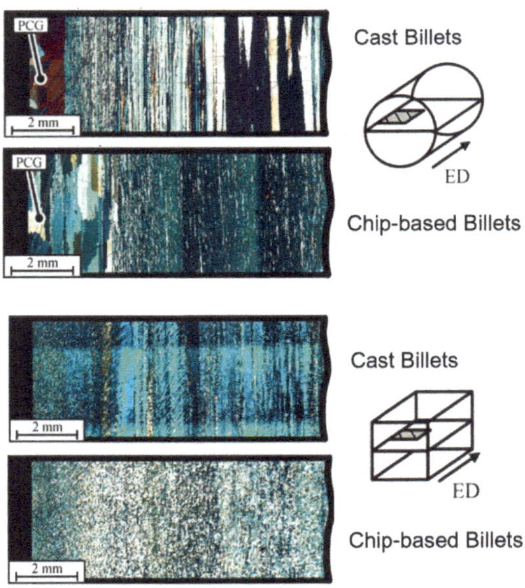

Figure 8.21 Microstructure of the products extruded using the flat-face die and iECAP die. For each die, a comparison is made between two feed materials: cast billets and chip-based billets. *Reprinted with permission from Haase, M., Tekkaya, A.E., 2014. Recycling of aluminum chips by hot extrusion with subsequent cold extrusion. Procedia Engineering 81, 652–657.*

shows the microstructures by extrusion using a flat-face die with a comparison of two feed materials, a cast billet, and a chip-based billet. The former exhibits coarse grains elongated in the extrusion direction along with a zone of peripheral coarse grains (PCG), whereas the latter is characterized by a mixture of fine equiaxed grains and elongated grains. Also shown in the figure (bottom half) are the microstructures formed by extrusion using an iECAP die utilizing these two feed materials, neither of which has peripheral coarse grains. In particular, a homogeneous microstructure consisting of tiny, equiaxed grains are formed when chip-based billets are extruded via the iECAP die.

For chip-based billets, the product extruded using the flat-faced die has a lower tensile strength (~ 130 MPa vs. ~150 MPa) and less ductility (elongation to failure: 25% vs. 37%) than that extruded using the iECAP die. Such differences can be linked to the variances in microstructure: the flat-faced die results in a significant portion of elongated coarse grains, whereas the iECAP die leads to a finer grain structure. Fundamentally, this is probably because the design of the iECAP die allows for more deformation and more complete dynamic recrystallization during hot extrusion.

8.3.2 Friction stirring for solid-state recycling

Hot extrusion-based solid-state recycling relies on high-temperature plastic deformation. Along that line, friction stir-based processes offer a more extreme thermomechanical condition for metal recycling (Baffari et al., 2019; Buffa et al., 2016). The first notable technology in this category is friction stir extrusion (Behnagh et al., 2015), in which a rapidly rotating die is plunged into a cylindrical chamber containing machine chips. The stirring motion of the die generates heat and pressure; the softening machine chips undergo severe plastic deformation. With compression and shear, the gaps between the chips are reduced, eventually resulting in intimate contact that facilitates metallurgical bond formation under pressure. Depending on the die configuration, the subsequent back extrusion may result in a fully-dense rod, wire, or tube.

A relevant technology is simply termed fiction stir consolidation (Baffari et al., 2018), which leverages compression and shear to consolidate a block of material without the need for extrusion. Fig. 8.22A illustrates a typical configuration of this technology. In friction stir consolidation, the quality of the resulting product, which is in the form of a disk, is highly

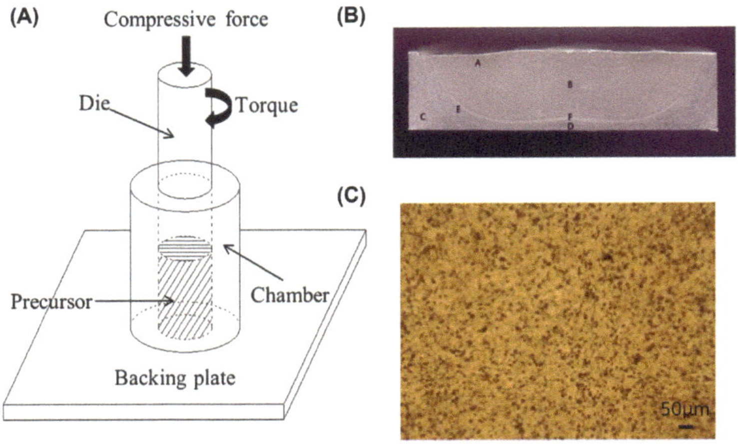

Figure 8.22 (A) Configuration of fiction stir consolidation. (B) A cross-sectional image of the formed billet charge, showing a fully-consolidated region with a bowl shape. (C) Microstructure of the center of the fully consolidated region, showing a fine, recrystallized grain structure. *Reprinted with permission from Li, X., Baffari, D., Reynolds, A.P., 2018. Friction stir consolidation of aluminum machining chips. The International Journal of Advanced Manufacturing Technology 94, 2031–2042.*

dependent on the heat generation rate and pressure (Li et al., 2018). With less processing time or insufficient heat generation, there could still be a significant amount of loose chips at the bottom of the die chamber. Under optimal conditions, however, a billet charge with decent quality can be formed, wherein a fully-consolidated zone with a bowl shape is seen at the top due to the spatial heterogeneities of thermal and deformation conditions. Fig. 8.22B shows this phenomenon for friction stir consolidation of AA6061 machine chips. The volume fraction of the fully-consolidated zone can be controlled by processing parameters, such as the die rotational speed, compressive force, and processing time. As the processing time increases, the deformation zone and heat generation zone expand from the layers adjacent to the contact surface to more distant regions. In the center of the fully-consolidated zone, a fine recrystallized grain structure is observed, as shown in Fig. 8.22C.

As another friction stir-derived technology, additive friction stir deposition can also be leveraged for metal recycling with the potential to skip some steps in material preparation, processing, and shaping. Instead of offering products with constant cross sections by friction stir extrusion and consolidation, additive friction stir deposition can produce variable geometries with its free-forming nature. Since at each time step only a small volume of material is being stirred, high-quality uniform products are feasible through the continuous addition of the consolidated material. As previously discussed in Chapter 6 in the context of printing metal matrix composites, powders or machine chips may be fed through the print head as a compacted feed-rod or via an auger feeding mechanism.

Jordon and coworkers have produced discontinuous machine chips from an AA5083 plate (Jordon et al., 2020), which are then used for additive friction stir deposition by utilizing the auger feeding mechanism (Fig. 8.23A). The printing is finished when the height reaches 6 mm. Consisting of fine equiaxed grains, the as-printed material is found to be ductile with an elongation of 18 percent. The yield strength is lower than the wrought alloy AA5083, but the ultimate tensile strength is higher due to extensive strain hardening (Fig. 8.23B). The observed mechanical properties are remarkable because no postprocess heat treatment is employed. This suggests minimal defects in the as-printed state. By skipping the process of chip compression and preparation, this approach offers an advanced and convenient route for solid-state recycling of metal chips.

Instead of using the auger feeding mechanism, Mishra and coworkers have prepared the feed rod by compacting the metal chips into the feedstock dimensions (Agrawal et al., 2021). Here, the material of interest is Ti-6Al-4V, which is provided by the Boeing Corporation in the form of machine chips. The compacted feed-rod has a dimension of 9.525 mm × 9.525 mm × 152.4 mm and is fed through a hollow rotating tool

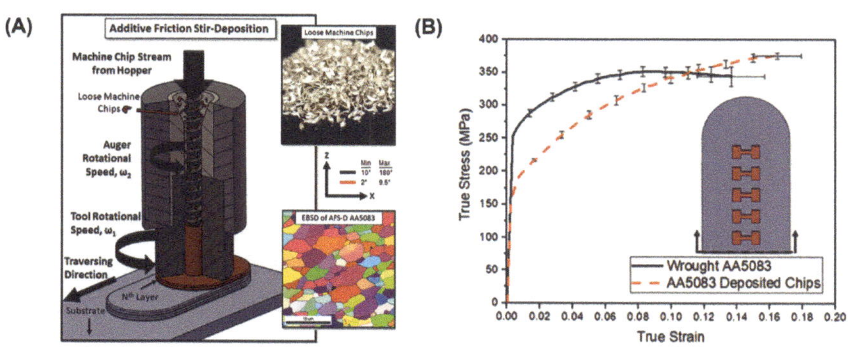

Figure 8.23 (A) Illustration of using additive friction stir deposition to recycle AA5083 chips. (B) A comparison of the tension behavior of the as-deposited chips and typical wrought AA5083. *Reprinted with permission from Jordon, J.B., Allison, P.G., Phillips, B.J., Avery, D.Z., Kinser, R.P., Brewer, L.N., et al., 2020. Direct recycling of machine chips through a novel solid-state additive manufacturing process. Materials & Design 193, 108850.*

made of WC with a diameter of 38.1 mm and a length of 88.9 mm. Additive friction stir deposition results in a final product of 30 mm × 135 mm × 25 mm, as shown in Fig. 8.24. The as-printed Ti-6Al-4V is largely made of the alpha phase (HCP structure) with a small quantity of the beta phase (BCC structure). This suggests that the peak temperature during additive friction stir deposition is above the beta transus temperature; during cooling, phase transformation occurs converting the original beta phase to a mixture of alpha and beta phases. There are in total 12 alpha variants that may form during the phase transformation (Sankaran et al., 2017). Preferred selection of the variant has been observed in selective laser melting (Yang et al., 2017). However, this is not the case in additive friction stir deposition, which shows a relatively random selection of the alpha variants possibly due to the role of dynamic restoration.

Multiple processing conditions are tried by tuning the tool rotation rate Ω, tool travel velocity V, and feed rate F. Regardless of the process condition, the final phase fractions appear to be almost the same. The microstructure in the as-printed state consists of equiaxed prior beta grains and fine lamellar, alpha/beta colonies inside these grains. To compare, the machine chips are cut from rolled Ti-6Al-4V plates with elongated alpha grains. This microstructure evolution is not surprising as similar microstructures are widely found in the stir zone after FSP of Ti-6Al-4V

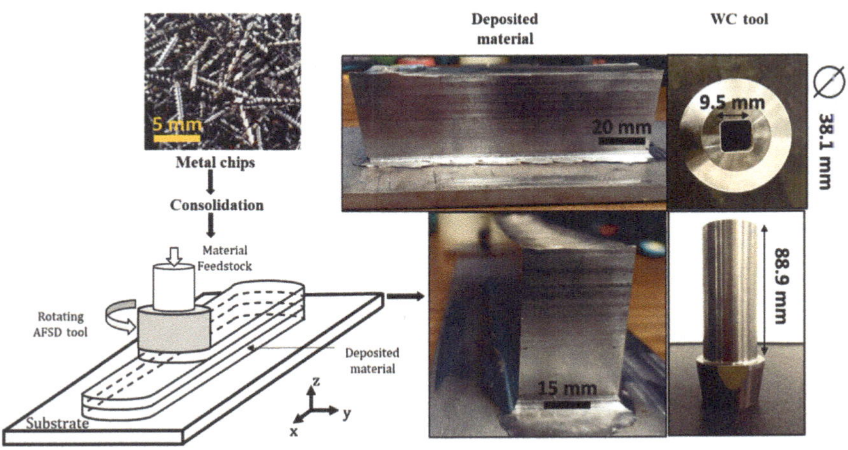

Figure 8.24 A summary of additive friction stir deposition of Ti-6Al-4V based on machine chips. *Reprinted with permission from Agrawal, P., Haridas, R.S., Yadav, S., Thapliyal, S., Gaddam, S., Verma, R., et al., 2021. Processing-structure-property correlation in additive friction stir deposited Ti-6Al-4V alloy from recycled metal chips. Additive Manufacturing 47, 102259.*

(Su et al., 2013). The processing parameters, which govern the thermal and deformation history, have a profound influence on the prior beta grain size, which is particularly sensitive to the pseudo heat index Ω^2/V (recall Chapter 2).

Fig. 8.25 shows a typical as-built structure as well as the detailed microstructures at three locations along the vertical direction. This structure is deposited at $\Omega = 340$ RPM, $V = 2.1$ mm/s, and $F = 0.75$ mm/s, consisting of 88.5% alpha phase and 11.5% beta phase. It is found that the bottom layer has a smaller prior beta grain size than the top layer. From backscattered-electron imaging, the colony size and lath size follow the same trend. The top region of the structure shows a colony size of 20 ± 0.8 μm and an average lath size of 0.72 ± 0.20 μm, whereas the bottom shows a colony size of 6 ± 0.5 μm and an average lath size of 0.60 ± 0.15 μm. The size differences in prior beta grains, colonies, and

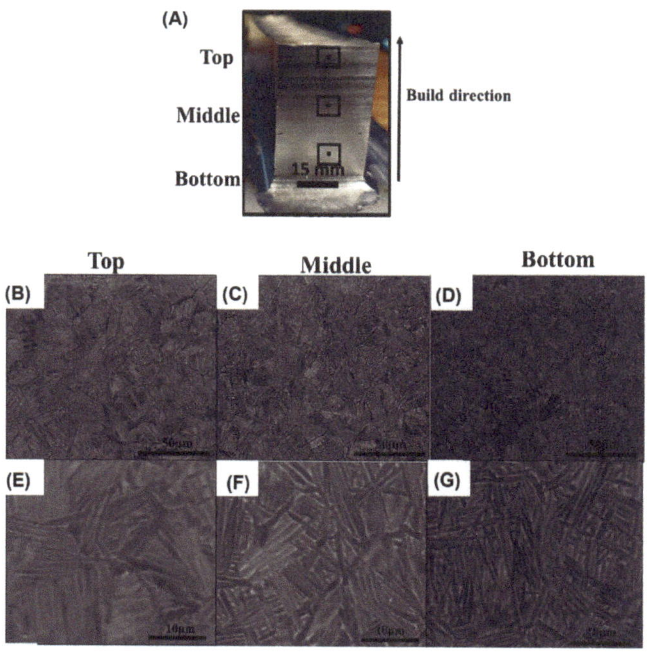

Figure 8.25 (A) A Ti-6Al-4V structure built by additive friction stir deposition using machine chips. Microstructure details are shown at low and high magnifications for the [(B) and (E)] top, the [(C) and (F)] middle, and the [(D) and (G)] bottom portion of the built structure. *Reprinted with permission from Agrawal, P., Haridas, R.S., Yadav, S., Thapliyal, S., Gaddam, S., Verma, R., et al., 2021. Processing-structure-property correlation in additive friction stir deposited Ti-6Al-4V alloy from recycled metal chips. Additive Manufacturing 47, 102259.*

laths between the top and bottom regions may be attributed to the different cooling rates as a consequence of the low thermal conductivity of Ti-6Al-4V (Boivineau et al., 2006). The bottom corresponds to layers that are deposited initially with a faster cooling rate, which likely results in finer grains, colonies, and laths. At higher cooling rates, the (0 0 0 1) texture also becomes stronger and the microstructure changes from the colony type to a mixture of colony and basket-weave structures.

Despite successful demonstrations for printing Al and Ti machine chips, scalability and cost are vital challenges to make additive friction stir deposition a mainstream recycling approach. Compared to alternative additive technologies, additive friction stir deposition is renowned for its high build rate, e.g., ~ 1000 cm^3/hour for Al. However, to match the industrial scale, the processing time per unit material still needs to be significantly reduced so that the environmental, energy, and cost benefits can be fully unlocked. The time spent on choosing, sorting, and preparing chips, establishing machine tools, and determining optimal process parameters must be greatly reduced as well.

8.4 Large-scale additive manufacturing

Additive friction stir deposition is a free-forming process allowing for printing at assigned locations. However, the current tooling size is large; the feedstock has a square cross-section of 9.525 mm × 9.525 mm. In addition, material flow and plastic deformation inevitably cause flash formation at the edges of the deposition tracks. Therefore, postprinting machining is essential; additive friction stir deposition is a near-net-shaping process rather than a net-shaping process. This makes additive friction stir deposition less attractive for making metallic complex geometries like powder bed fusion.

That being said, also thanks to the large tooling size, additive friction stir deposition is suitable for large-scale additive manufacturing of components, particularly on the order of meters or above. For example, two large ring structures made of Al alloys have been printed using additive friction stir deposition by the MELD Manufacturing Corporation (Editors, 2020). The larger ring has a diameter of 10 feet, which is equivalent to 3.048 m. The capability of quickly building large metallic structures is especially pressing for the defense sectors, which rely on supply chains for meter-scale, high-performance components like forged aluminum hulls for combat vehicles (Lenihan et al., 2019). Because of the unique capabilities of printing large structures with forging-like properties, additive friction stir deposition has been selected for the US Army's Jointless Hull program with the aim to print components measuring 6 m × 9 m × 3.5 m.

The challenges in large-scale additive manufacturing not only lie in the build rate and scalability, for which additive friction stir deposition excels, but also the ability to avoid significant warpage due to thermal stresses. Fusion-based additive manufacturing is characterized by large thermal gradients and residual stresses, whereas additive friction stir deposition is shown to be a thermally mild, solid-state process with low levels of thermal mismatch and thermal stresses. This view is supported by the findings in cladding on thin automotive sheet metals discussed earlier in this chapter.

8.5 Printing and repair under harsh conditions

Being able to print or repair under harsh conditions like in space or water will unlock unprecedented manufacturing opportunities for additive technologies. For example, the capability of repairing components directly in the water has great potential for submerged components in offshore wind energy facilities (Ren et al., 2021). Owing to constant exposure to corrosive seawater and particularly complex ocean current conditions, submerged steel components, such as platforms and anchors, suffer from accelerated damages by erosion, corrosion, and fatigue loading. Such damages can cause high risks of structural collapse and facility shutdown. This problem is typically addressed by overdesign, close monitoring, timed component replacement, or costly ex situ repair. However, the remote nature of offshore wind facilities makes these strategies prohibitively expensive and time-consuming. One solution to address the issue is through in situ underwater repair using arc welding by divers (Yusof and Jamaluddin, 2014). While this competing technology saves the total cost, it raises serious safety concerns: the diver must be isolated from the welding circuit during wet underwater welding.

Another example is the repair of hydropower turbine blades. Hydroelectric power is a reliable, renewable energy source but has more levelized costs associated with initial capital investment and operation and maintenance (O&M) compared to other nondispatchable sources of renewable energy (e.g., solar and wind) (Kaunda et al., 2012). This creates a strong initiative to extend the lifetime of legacy components through rehabilitation and to reduce costs associated with O&M. In hydropower plants, some of the most critical components are the submerged turbine runners, which over time become damaged by corrosion, erosion, cavitation, and fatigue cracking (Kougias et al., 2019). Under heavy demand, flow conditions around the turbines can produce cavitation bubbles, which collapse violently and create powerful erosive shockwaves. Abrasive sediments and corrosive chemical species also lead to damage on critical turbine surfaces, and cyclic fatigue can nucleate and grow internal cracks. Because replacing large turbine components is extremely expensive and requires long lead times for legacy components, periodic turbine inspection and repair is the typical solution. Current repair strategies involve dewatering the facility, manual inspection, building a scaffold up to the turbine, grinding down the damaged surface, using arc-welding variants to replace the lost material, and postprocess grinding to achieve appropriate dimensions.

To summarize, for both offshore wind energy and hydropower facilities, there are critical submerged components whose repair and maintenance play a critical role in energy efficiency and total cost. A safe, low-cost, high-quality, in situ underwater repair approach would significantly reduce the lead time, cost, and LCOE (levelized cost of electricity) (Aldersey-Williams and Rubert, 2019), while avoiding power operation suspension and lowering the risks of personnel accidents in repair operation. Additive friction stir deposition can be a strong candidate in this regard:

- Safe underwater repair. FSW is known to be operational in water (Wahid et al., 2018). Additive friction stir deposition is based on the same solid-state deformation principle. By avoiding the use of an electrical arc, this technology enables more robust underwater deposition. This addresses the safety concerns with the current in situ repair practice based on arc welding by divers.
- Prevention of biomass and particle entrapment. As a solid feed material is dispensed from the print head, it spreads out in the gap between the repaired structure and the print head. Anything previously in this gap, such as water, solid particulates, and biological matter due to biofouling, will be forced out by the stirring action. The chances of entrapped material from the seawater environment can be thus reduced.
- Good repair quality. During additive friction stir deposition, the extensive material flow leads to fully-dense material with fine, equiaxed microstructures and forging-like isotropic properties due to dynamic

recrystallization. The absence of melting leads to less residual stresses as compared to fusion welding. All these factors will improve the postrepair component lifetime and further reduce the maintenance cost.

- Semiautonomous capability. By coupling additive friction stir deposition with robotic manipulation, semiautonomous underwater repair may become possible to further reduce the safety concerns and human intervention. This capability can be implemented by incorporating the additive friction stir deposition facility with a Stewart platform architecture (Dasgupta and Mruthyunjaya, 2000)—a six-degree-of-freedom parallel robot—for multiaxis rotation and stable positioning.

For proof-of-concept purposes, a team at Virginia Tech has demonstrated the feasibility of underwater printing using additive friction stir deposition, winning an American Made Challenges prize (Griffiths et al., 2020). A wet enclosure is first designed and constructed to surround the print head. A MELD R2 system equipped with a nickel-superalloy tool is used to deposit a layer of 304 stainless steel onto a 304 stainless steel substrate. This experiment provides a realistic investigation of the underwater repair capabilities. The print is shown to be successful with good quality. Although the substrate is thin, 1/8 inch, there is no deflection or warpage after printing, indicative of low residual stress and low thermal exposure thanks to the presence of the water bath. This result demonstrates the potential of fully aqueous repair for offshore wind facilities and hydropower plants using additive friction stir deposition. In addition to printing underwater, additive friction stir deposition has the potential to be deployed in space. Since the process is based on compression and shear and the feed material is in the form of a solid bar, microgravity should have much less influence on additive friction stir deposition than alternative additive technologies.

8.6 Summary

Currently, additive friction stir deposition finds its most promising applications in the defense, aerospace, and automotive sectors. Compared to fusion-based additive technologies and cold spray, fully-dense material with forging-like mechanical properties can be obtained more straightforwardly by additive friction stir deposition. However, additive friction stir deposition results in wide deposition tracks along with flash formation; it is not an optimal solution for making complex geometries of small feature sizes. Therefore, the niche applications lie in where the load-bearing capacity is the top priority and postprocess machining is allowed.

The first notable niche application is structural repair, which means not only the original geometry is fully restored but the postrepair mechanical performance is also excellent. Additive friction stir deposition can repair surface divots, cracks, and in particular, can fill the entire volume of through-holes and large grooves. Under optimal conditions, the sidewall material and the deposited material are sufficiently mixed, resulting in an indistinguishable interface with a gradual change in microstructure. Additive friction stir deposition also enables selective reinforcement of thin components like automotive panels. Despite a large input of force and torque, high-quality cladding for automotive Al sheet metals can be achieved without local wrinkling. This also results in a good combination of the strength and ductility in the reinforced structure. With less thermal exposure than fusion processes, the thermal mismatch between the cladding and substrate is low, as is the thermal residual stress.

With a high tolerance of the feedstock quality, additive friction stir deposition holds great promise for solid-state metal recycling. Even if the feed material is porous, the compression, shear, and extensive material flow during printing will lead to fully-dense material with excellent mechanical properties in the as-printed state. This has been demonstrated for AA5083 and Ti-6Al-4V. Additive friction stir deposition has also been demonstrated for large-scale additive manufacturing of Al ring structures on the order of meters. It has been demonstrated for underwater printing as well, with the potential for repair and maintenance in the offshore wind and hydropower industries.

References

Agrawal, P., Haridas, R.S., Yadav, S., Thapliyal, S., Gaddam, S., Verma, R., et al., 2021. Processing-structure-property correlation in additive friction stir deposited Ti-6Al-4V alloy from recycled metal chips. Additive Manufacturing 47, 102259.

Aldersey-Williams, J., Rubert, T., 2019. Levelised cost of energy—A theoretical justification and critical assessment. Energy Policy 124, 169–179.

Baffari, D., Reynolds, A.P., Li, X., Fratini, L., 2018. Bonding prediction in friction stir consolidation of aluminum alloys: a preliminary study. In: AIP Conference Proceedings 1960, 050002.

Baffari, D., Reynolds, A.P., Masnata, A., Fratini, L., Ingarao, G., 2019. Friction stir extrusion to recycle aluminum alloys scraps: energy efficiency characterization. Journal of Manufacturing Processes 43, 63–69.

Bayliss, D.A., 2002. 34—Paint coatings for the plant engineer. In: Snow, D.A. (Ed.), Plant Engineer's Reference Book, second ed Butterworth-Heinemann, Oxford, pp. 34-31-34-14.
Becker, S., Bouzdine-Chameeva, T., Jaegler, A., 2020. The carbon neutrality principle: a case study in the French spirits sector. Journal of Cleaner Production 274, 122739.
Behnagh, R.A., Shen, N., Ansari, M.A., Narvan, M., Besharati Givi, M.K., Ding, H., 2015. Experimental analysis and microstructure modeling of friction stir extrusion of magnesium chips. Journal of Manufacturing Science and Engineering 138.
Bhadeshia, H., Honeycombe, R., 2017. Chapter 13—Weld microstructures. In: Bhadeshia, H., Honeycombe, R. (Eds.), Steels: Microstructure and Properties, fourth ed Butterworth-Heinemann, pp. 377−400.
Boivineau, M., Cagran, C., Doytier, D., Eyraud, V., Nadal, M.H., Wilthan, B., et al., 2006. Thermophysical properties of solid and liquid Ti-6Al-4V (TA6V) alloy. International Journal of Thermophysics 27, 507−529.
Buffa, G., Campanella, D., Fratini, L., Micari, F., 2016. AZ31 magnesium alloy recycling through friction stir extrusion process. International Journal of Material Forming 9, 613−618.
Carden, W.D., Geng, L.M., Matlock, D.K., Wagoner, R.H., 2002. Measurement of springback. International Journal of Mechanical Sciences 44, 79−101.
Cizek, J., Kovarik, O., Siegl, J., Khor, K.A., Dlouhy, I., 2013. Influence of plasma and cold spray deposited Ti layers on high-cycle fatigue properties of Ti6Al4V substrates. Surface and Coatings Technology 217, 23−33.
Cui, L., Yang, X., Wang, D., Cao, J., Xu, W., 2014. Experimental study of friction taper plug welding for low alloy structure steel: welding process, defects, microstructures and mechanical properties. Materials & Design (1980−2015) 62, 271−281.
Cunningham, C.R., Flynn, J.M., Shokrani, A., Dhokia, V., Newman, S.T., 2018. Invited review article: strategies and processes for high quality wire arc additive manufacturing. Additive Manufacturing 22, 672−686.
Dasgupta, B., Mruthyunjaya, T.S., 2000. The Stewart platform manipulator: a review. Mechanism and Machine Theory 35, 15−40.
Editors, D., 2020. MELD prints 10-foot Aluminum cylinder: digital engineering. <https://www.digitalengineering247.com/article/meld-prints-10-foot-aluminum-cylinder/>.
Energy, U., 2007. US Energy Requirements for Aluminum Production: Historical Perspective, Theoretical Limits and Current Practices. DOE Washington.
Finot, M., Suresh, S., 1996. Small and large deformation of thick and thin-film multi-layers: effects of layer geometry, plasticity and compositional gradients. Journal of the Mechanics and Physics of Solids 44, 683−721.
Fogagnolo, J.B., Ruiz-Navas, E.M., Simón, M.A., Martinez, M.A., 2003. Recycling of aluminium alloy and aluminium matrix composite chips by pressing and hot extrusion. Journal of Materials Processing Technology 143−144, 792−795.
Freund, L.B., Suresh, S., 2004. Thin Film Materials: Stress, Defect Formation and Surface Evolution. Cambridge University Press.
Fruehan, R., Fortini, O., Paxton, H., Brindle, R., 2000. Theoretical Minimum Energies to Produce Steel for Selected Conditions. Energetics, Inc., Columbia, Pittsburgh, PA.
George Nadim, M., 2018. Aerospace fasteners: use in structural applications. Encyclopedia of Aluminum and its Alloys. CRC Press.
Gottwald, R.B., Griffiths, R.J., Petersen, D.T., Perry, M.E.J., Yu, H.Z., 2021. Solid-state metal additive manufacturing for structural repair. Accounts of Materials Research 2, 780−792.
Griffiths, R.J.P., M.E.J., Hartley, W.D., Garcia, D., Yu, H.Z., 2020. Augmented repair via additive manufacturing <https://www.herox.com/iamhydro/round/656/entry/33593>.

Gronostajski, J., Marciniak, H., Matuszak, A., 2000. New methods of aluminium and aluminium-alloy chips recycling. Journal of Materials Processing Technology 106, 34–39.
Grujicic, M., Zhao, C.L., DeRosset, W.S., Helfritch, D., 2004. Adiabatic shear instability based mechanism for particles/substrate bonding in the cold-gas dynamic-spray process. Materials & Design 25, 681–688.
Güley, V., Güzel, A., Jäger, A., Ben Khalifa, N., Tekkaya, A.E., Misiolek, W.Z., 2013. Effect of die design on the welding quality during solid state recycling of AA6060 chips by hot extrusion. Materials Science and Engineering: A 574, 163–175.
Haase, M., Ben Khalifa, N., Tekkaya, A.E., Misiolek, W.Z., 2012. Improving mechanical properties of chip-based aluminum extrudates by integrated extrusion and equal channel angular pressing (iECAP). Materials Science and Engineering: A 539, 194–204.
Haase, M., Tekkaya, A.E., 2014. Recycling of aluminum chips by hot extrusion with subsequent cold extrusion. Procedia Engineering 81, 652–657.
Han, B., Huang, Y., Lv, S., Wan, L., Feng, J., Fu, G., 2013. AA7075 bit for repairing AA2219 keyhole by filling friction stir welding. Materials & Design 51, 25–33.
Hartley, W.D., Garcia, D., Yoder, J.K., Poczatek, E., Forsmark, J.H., Luckey, S.G., et al., 2021. Solid-state cladding on thin automotive sheet metals enabled by additive friction stir deposition. Journal of Materials Processing Technology 291, 117045.
Hu, M., Ji, Z., Chen, X., Zhang, Z., 2008. Effect of chip size on mechanical property and microstructure of AZ91D magnesium alloy prepared by solid state recycling. Materials Characterization 59, 385–389.
Jenney, C.L., O'Brien, A., American Welding, S., Welding Handbook, C., 2001. Welding Handbook: Welding Science and Technology, ninth ed. American Welding Society, Miami, FL.
Jordon, J.B., Allison, P.G., Phillips, B.J., Avery, D.Z., Kinser, R.P., Brewer, L.N., et al., 2020. Direct recycling of machine chips through a novel solid-state additive manufacturing process. Materials & Design 193, 108850.
Kaunda, C.S., Kimambo, C.Z., Nielsen, T.K., 2012. Hydropower in the context of sustainable energy supply: a review of technologies and challenges. ISRN Renewable Energy 2012, 730631.
Keicher, D., Smugeresky, J., Romero, J., Griffith, M., Harwell, L., 1997. Using the Laser Engineered Net Shaping (LENS) Process to Produce Complex Components from a CAD Solid Model. SPIE.
Kougias, I., Aggidis, G., Avellan, F., Deniz, S., Lundin, U., Moro, A., et al., 2019. Analysis of emerging technologies in the hydropower sector. Renewable and Sustainable Energy Reviews 113, 109257.
Lenihan, D., Ronan, W., O'Donoghue, P.E., Leen, S.B., 2019. A review of the integrity of metallic vehicle armour to projectile attack. In: Proceedings of the Institution of Mechanical Engineers, Part L: Journal of Materials: Design and Applications, 233, 73–94.
Li, X., Baffari, D., Reynolds, A.P., 2018. Friction stir consolidation of aluminum machining chips. The International Journal of Advanced Manufacturing Technology 94, 2031–2042.
Mishra, R.S., De, P.S., Kumar, N., 2014. Friction Stir Welding and Processing: Science and Engineering. Springer International Publishing.
Rahito, Wahab, D.A., Azman, A.H., 2019. Additive manufacturing for repair and restoration in remanufacturing: an overview from object design and systems perspectives. Processes 7, 802.
Ren, Z., Verma, A.S., Li, Y., Teuwen, J.J.E., Jiang, Z., 2021. Offshore wind turbine operations and maintenance: a state-of-the-art review. Renewable and Sustainable Energy Reviews 144, 110886.

Saboori, A., Aversa, A., Marchese, G., Biamino, S., Lombardi, M., Fino, P., 2019. Application of directed energy deposition-based additive manufacturing in repair. Applied Sciences 9, 3316.

Sankaran, K.K., Mishra, R.S., 2017. Chapter 5—Titanium alloys. In: Sankaran, K.K., Mishra, R.S. (Eds.), Metallurgy and Design of Alloys with Hierarchical Microstructures. Elsevier, pp. 177−288.

Shamsudin, S., Lajis, M., Zhong, Z., 2016. Solid-state recycling of light metals: a review. Advances in Mechanical Engineering 8, 1687814016661921.

Söderholm, P., Ekvall, T., 2020. Metal markets and recycling policies: impacts and challenges. Mineral Economics 33, 257−272.

Su, J., Wang, J., Mishra, R.S., Xu, R., Baumann, J.A., 2013. Microstructure and mechanical properties of a friction stir processed Ti−6Al−4V alloy. Materials Science and Engineering: A 573, 67−74.

Wahid, M.A., Khan, Z.A., Siddiquee, A.N., 2018. Review on underwater friction stir welding: a variant of friction stir welding with great potential of improving joint properties. Transactions of Nonferrous Metals Society of China 28, 193−219.

Wang, J., Yin, X., Murakawa, H., 2013. Experimental and computational analysis of residual buckling distortion of bead-on-plate welded joint. Journal of Materials Processing Technology 213, 1447−1458.

Woo, W., Choo, H., Brown, D.W., Feng, Z., Liaw, P.K., 2006. Angular distortion and through-thickness residual stress distribution in the friction-stir processed 6061-T6 aluminum alloy. Materials Science and Engineering: A 437, 64−69.

Yang, J., Yu, H., Wang, Z., Zeng, X., 2017. Effect of crystallographic orientation on mechanical anisotropy of selective laser melted Ti-6Al-4V alloy. Materials Characterization 127, 137−145.

Yusof, F., Jamaluddin, M.F., 2014. 6.07—Welding defects and implications on welded assemblies. In: Hashmi, S., Batalha, G.F., Van Tyne, C.J., Yilbas, B. (Eds.), Comprehensive Materials Processing. Elsevier, Oxford, pp. 125−134.

CHAPTER 9

Future perspectives

There is nothing like a dream to create the future.

—Victor Hugo

By now, the current understanding of additive friction stir deposition has been compiled in a few hundred pages in this book—through the struggle of the author. However, the research and application of this process are still in their early stages; this book should by no means be viewed as a conclusion. Instead, it is merely a preliminary summary for this burgeoning topic, with the hope that more folks in universities, industries, and research institutes will join the fascinating adventure to unlock greater possibilities for additive friction stir deposition in a variety of sectors.

Considering its salient advantages over alternative metal additive processes, the future potential of additive friction stir deposition cannot be over-emphasized. For example, quality control has historically been a major issue in fusion-based metal additive manufacturing. In contrast, additive friction stir deposition typically results in fully-dense material in the as-printed state, eliminating the need for postprocessing to reduce porosity. Cold spray, on the other hand, has long been seen as a promising contender for repair because of its ability to avoid hot cracking and high residual stresses. However, despite huge investments over the last decade, inadequate postrepair mechanical performance has limited cold spray to geometry restoration rather than structural repair. In comparison, additive friction stir deposition enables effective material mixing, strong interface bonding, as well as volumetric filling, and has demonstrated promising postrepair fatigue performance. It is seen, at least by some, as the next focus for structural repair in the defense and aerospace sectors. As more people become aware of the unique benefits of additive friction stir deposition, we anticipate unprecedented impact to be made in the coming years.

This chapter discusses potential future research directions for additive friction stir deposition. First, the physical process deserves an in-depth understanding. There is a lack of quantitative correlation between the

Additive Friction Stir Deposition
DOI: https://doi.org/10.1016/B978-0-12-824374-9.00006-X

process parameters, such as rotation rate and feed rate, and the physical variables, such as the strain rate, strain, and temperature. Second, there are numerous opportunities for material innovation that have not been explored, such as printing functionally graded materials, hierarchical materials, and heterostructured materials. Finally, artificial intelligence should be included for quality control and process optimization. This involves integration of physics-based and data-driven techniques while leveraging machine learning and statistical tools.

9.1 In-depth understanding of the underlying physics

Metallurgists have a good grasp of the process–microstructure–property linkage in thermomechanical processing—which includes hot rolling, hot extrusion, and forging. Here, the term "process" means the physical process variables, such as temperature and strain rate. Therefore, if we know the temperature and strain rate as a function of time during additive friction stir deposition, the metallurgy knowledge can be leveraged to optimize the final properties via process manipulation. However, the quantitative relationship between processing parameters and the resulting deformation and thermal history is currently poorly understood. The most fundamental question is: for a given set of the print head rotation rate, travel velocity, feed rate, and layer thickness, can we accurately and rapidly predict the temperature vs. time and strain rate vs. time plots? The answer remains no, unfortunately.

Experimentally, we have measured the temperature at the exterior surface of the deposited material by thermal imaging, but the internal temperature distribution remains elusive. Similarly, in situ monitoring provides the surface material flow information; however, less is known about the deformation history of each voxel in the feed material. For future research, detailed temperature monitoring can be performed to collect thermal data at various locations. For example, multiple thermocouples can be placed inside the print head, adjacent to the deposition track, and in the substrate. Thermal cameras can be installed from the top, side, and isometric views. To disclose the material deformation history, a more systematic design of the tracer experiments than those in Chapter 3, Material Flow Phenomena can be carried out. Time-resolved characterization will be performed throughout the entire printing process for tracers placed at various initial locations in the feed-rod. Ultrafast X-ray imaging capabilities, such as those present at the Advanced Photon Source, can also be leveraged for real-time characterization of tracer shape evolution (Cunningham et al., 2019). Another option is to print transparent feed material embedded with opaque particles and track the particle motion using high-speed optical imaging (Kumar et al., 2018). The suitable materials will be likely restricted to ultra-high-molecular-weight polyethylene or other polymers.

Experiments alone are insufficient to provide the entire thermal and deformation history due to limitations in temporal and spatial resolution. Physics modeling and simulation based on computational fluid dynamics or solid mechanics are complementary to the experiment and can be an important future research area. Since additive friction stir deposition follows the friction stir principle, the previous modeling methodologies in friction stir welding (He et al., 2014) can be leveraged by including material feeding and altering the boundary conditions. Experimentally validated and calibrated models will provide new physical insights into material flow and rotation, temperature distribution, and contact state, while being able to predict the temperature vs. time and strain rate vs. time plots for any set of processing parameters.

9.2 Material innovation

Additive friction stir deposition is an excellent technique for microstructure engineering and bulk-scale production. Depending on the cooling rate and material system, the grain size in the as-printed material may be in the submicron regime due to dynamic recrystallization. As a result, additive friction stir deposition could provide a bulk processing route for producing submicron materials (Valiev et al., 2000). In addition, while nanocrystalline alloys are routinely synthesized in powder form via ball milling, they are difficult to consolidate without grain growth (Chookajorn et al., 2012). Because of its short processing time and the presence of severe plastic deformation, additive friction stir deposition may enable bulk-scale production of nanocrystalline materials through powder consolidation. Furthermore, for materials undergoing continuous dynamic recrystallization, the deformation level may be tailored to manage the fractions of low and high angle grain boundaries, as well as special grain boundaries. The degree and type of shear textures can also be controlled.

Additive friction stir deposition is also a good tool for mesostructure engineering and fabricating functionally graded materials (Kieback et al., 2003).

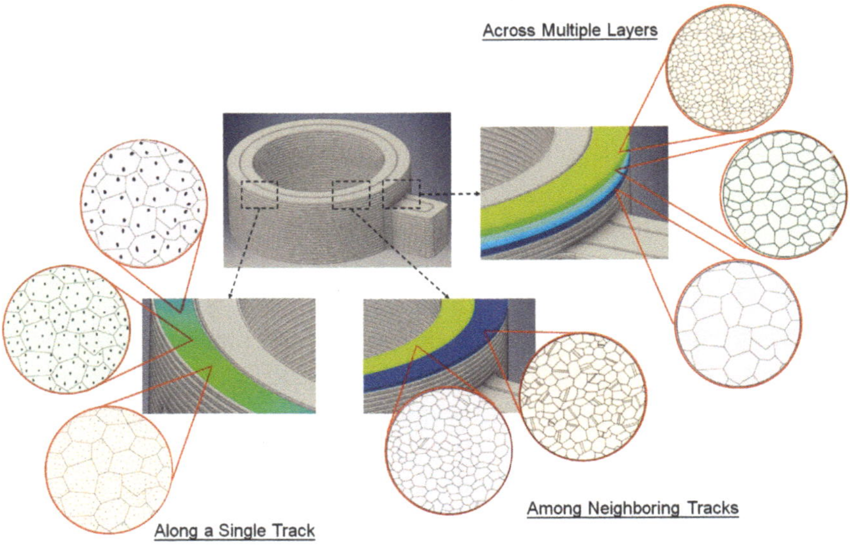

Figure 9.1 Illustration of a mesostructured object made by additive friction stir deposition. *Reprinted with permission from Yu, H.Z., Mishra, R.S., 2021. Additive friction stir deposition: a deformation processing route to metal additive manufacturing. Materials Research Letters 9, 71–83.*

As shown in Fig. 9.1, gradient microstructures can be made along a single deposition track, among neighboring tracks, or across multiple layers by programming the processing parameters. This is because the process conditions can sensitively influence the precipitate attributes, grain size, and grain boundary type. If we want to make compositionally graded materials, multiple feed-rods with graded compositions can be employed for different layers. By engineering the feed material, additive friction stir deposition can be utilized to construct hierarchical materials with complex internal structures (Fratzl and Weinkamer, 2007). For example, an Al–NiTi hierarchical material has been demonstrated with entangled NiTi cables embedded in the Al matrix. It is also possible to print heterostructured materials (Zhu et al., 2021), which enable long-range back-stress buildup and hetero-deformation driven strain hardening, showing promise for simultaneously achieving high strength and good ductility.

9.3 Incorporation of artificial intelligence

Exciting advances in artificial intelligence (AI) have occurred over the last decade, as evidenced by the widespread use of machine learning and deep learning tools in science, engineering, and business. For materials science and manufacturing, common AI applications include regression, classification, clustering, optimization, feature identification and extraction, design of experiments, reduction of model dimensionality, and transfer learning (Johnson et al., 2020). The central concept is to efficiently predict, optimize, and analyze using data-driven algorithms and methodologies such as principal component analysis, support vector regression, and k-nearest neighbors (Alpaydin, 2020), as well as convolutional neural networks (CNN) and general adversarial networks (GAN) (Goodfellow et al., 2016). AI-based solutions are capable of processing large amounts of data; it is well aligned with the trend of manufacturing digitalization and large-scale data acquisition platforms (Mozaffar et al., 2022). The incorporation of AI into additive friction stir deposition is a promising future direction, especially for quality control and property optimization.

That being said, data-driven approaches alone may not be the most efficient path. First, overfitting is a recurring problem that can mislead decision-making. Also, with purely empirical data, a large number of datasets are typically required. However, for key problems in additive friction stir deposition, collecting high-quality data by experiment is nontrivial and expensive. When introducing AI to additive friction stir deposition, therefore, we must integrate that with scientific ingredients that reflect the underlying physics and materials science principles. By including these principles, a relatively small dataset may be sufficient, while the established AI framework can be easily transferred to other problems characterized by different physical mechanisms. A relevant paradigm is termed mechanistic-AI (Mozaffar et al., 2022), which combines the raw mathematical power of AI methods with mechanism-driven principles and engineering insights (illustrated in Fig. 9.2). The physics principles are used to inform AI methods for more efficient and accurate predictions via physics-informed learning (Karniadakis et al., 2021), whereas data-driven discovery may give insights to new mechanisms that are not covered by the existing physics models.

For additive friction stir deposition, take the prediction of thermal and deformation history as an example. Computational fluid dynamics or solid mechanics-based process models must communicate with experimentally measured thermal and material flow data via AI tools. This will allow for more accurate prediction of the strain rate vs. time and temperature vs. time plots by calibration, while potentially capturing new physics that is not covered in existing process models. One notable AI framework of this

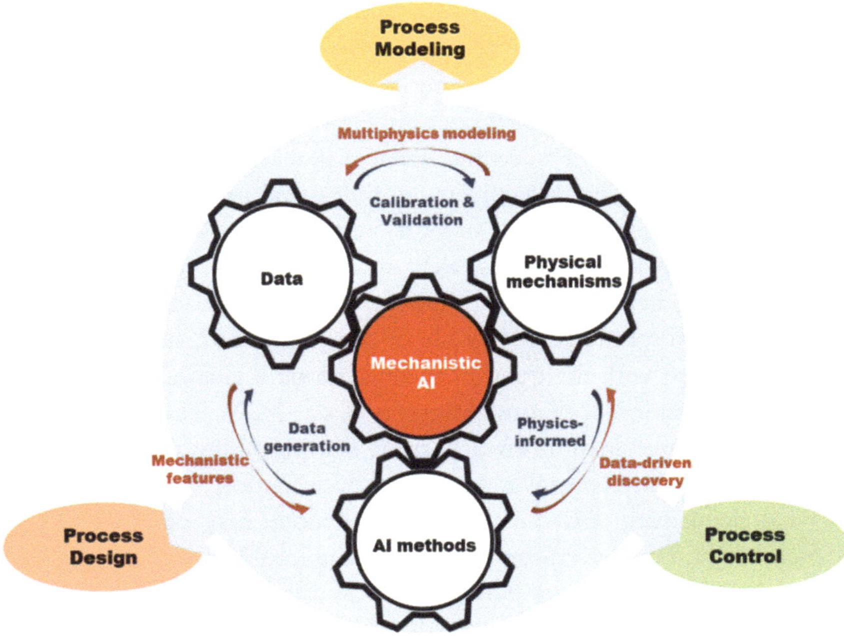

Figure 9.2 Illustration of mechanistic artificial intelligence for modeling, design, and control of advanced manufacturing. *Reprinted with permission from Mozaffar, M., Liao, S., Xie, X., Saha, S., Park, C., Cao, J., et al., 2022. Mechanistic artificial intelligence (mechanistic-AI) for modeling, design, and control of advanced manufacturing processes: current state and perspectives. Journal of Materials Processing Technology 302, 117485.*

type is based on Bayesian calibration and learning (Higdon et al., 2008), which formulates the reality as the sum of a surrogate model—usually based on a Gaussian process, a discrepancy term, and the measurement error. The surrogate model is based on physics simulation but significantly reduces the computational cost, while the experimental data is used to calibrate the surrogate model and identify the discrepancy. This framework optimally combines simulation and experiment to enable fast and accurate prediction, while fully accounting for all sources of uncertainty, which is not captured in conventional machine learning models. The discrepancy and uncertainty can be used to inform regions of deficient data sampling (Kim et al., 2022). This framework has proven effective at predicting the thermal field evolution in polymer additive manufacturing (Fig. 9.3), in which the thermal field evolution of different geometries is simulated based on heat transfer laws and is experimentally measured via thermal

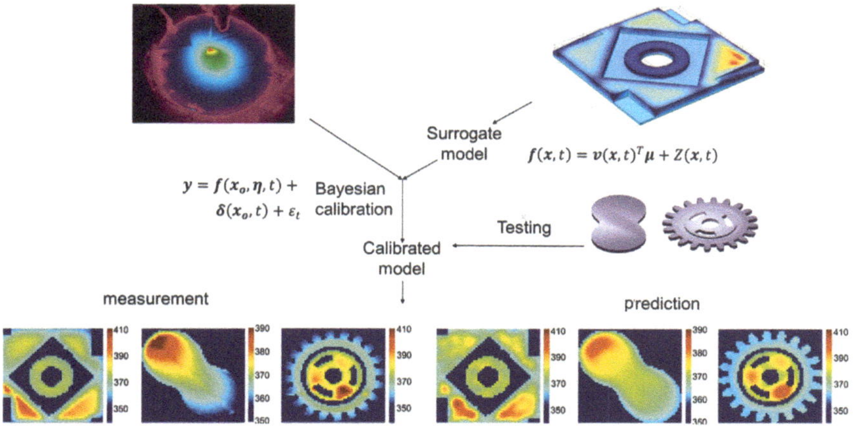

Figure 9.3 The Bayesian learning framework for predicting a thermal field in additive manufacturing. *Reprinted with permission from Li, J., Jin, R., Yu, H.Z., 2018. Integration of physically-based and data-driven approaches for thermal field prediction in additive manufacturing. Materials & Design 139, 473–485.*

imaging. The two sets of data are combined via a Bayesian learning framework for prediction at untried settings (Li et al., 2018).

9.4 Summary

Despite its salient advantages, including the capability of rendering fully-dense, as-printed materials and forging-like mechanical properties, as well as its promising applications, including structural repair, reinforcement, and recycling, the potential of additive friction stir deposition is far from fully unlocked. From a perspective of process fundamentals, the thermal and deformation history throughout the entire printing process needs to be quantified. This will likely involve creative in situ or quasi in situ experiments along with process modeling. A lot of material innovation opportunities exist as well; by engineering the feed material, solid-state additive manufacturing of functionally graded materials, hierarchical materials, and heterostructured materials will become possible. By introducing the power of AI to additive friction stir deposition, quality control and process optimization can be efficiently implemented by combining physics-based and data-driven strategies.

References

Alpaydin, E., 2020. Introduction to Machine Learning, fourth ed MIT Press.

Chookajorn, T., Murdoch, H.A., Schuh, C.A., 2012. Design of stable nanocrystalline alloys. Science (New York, N.Y.) 337, 951–954.

Cunningham, R., Zhao, C., Parab, N., Kantzos, C., Pauza, J., Fezzaa, K., et al., 2019. Keyhole threshold and morphology in laser melting revealed by ultrahigh-speed x-ray imaging. Science (New York, N.Y.) 363, 849–852.

Fratzl, P., Weinkamer, R., 2007. Nature's hierarchical materials. Progress in Materials Science 52, 1263–1334.

Goodfellow, I., Bengio, Y., Courville, A., 2016. Deep Learning. MIT Press.

He, X., Gu, F., Ball, A., 2014. A review of numerical analysis of friction stir welding. Progress in Materials Science 65, 1–66.

Higdon, D., Gattiker, J., Williams, B., Rightley, M., 2008. Computer model calibration using high-dimensional output. Journal of the American Statistical Association 103, 570–583.

Johnson, N.S., Vulimiri, P.S., To, A.C., Zhang, X., Brice, C.A., Kappes, B.B., et al., 2020. Invited review: machine learning for materials developments in metals additive manufacturing. Additive Manufacturing 36, 101641.

Karniadakis, G.E., Kevrekidis, I.G., Lu, L., Perdikaris, P., Wang, S., Yang, L., 2021. Physics-informed machine learning. Nature Reviews Physics 3, 422–440.

Kieback, B., Neubrand, A., Riedel, H., 2003. Processing techniques for functionally graded materials. Materials Science and Engineering: A 362, 81–106.

Kim, J.Y., Garcia, D., Zhu, Y., Higdon, D.M., Yu, H.Z., 2022. A Bayesian learning framework for fast prediction and uncertainty quantification of additively manufactured multi-material components. Journal of Materials Processing Technology 303, 117528.

Kumar, R., Pancholi, V., Bharti, R.P., 2018. Material flow visualization and determination of strain rate during friction stir welding. Journal of Materials Processing Technology 255, 470–476.

Li, J., Jin, R., Yu, H.Z., 2018. Integration of physically-based and data-driven approaches for thermal field prediction in additive manufacturing. Materials & Design 139, 473–485.

Mozaffar, M., Liao, S., Xie, X., Saha, S., Park, C., Cao, J., et al., 2022. Mechanistic artificial intelligence (mechanistic-AI) for modeling, design, and control of advanced manufacturing processes: current state and perspectives. Journal of Materials Processing Technology 302, 117485.

Valiev, R.Z., Islamgaliev, R.K., Alexandrov, I.V., 2000. Bulk nanostructured materials from severe plastic deformation. Progress in Materials Science 45, 103–189.

Zhu, Y., Ameyama, K., Anderson, P.M., Beyerlein, I.J., Gao, H., Kim, H.S., et al., 2021. Heterostructured materials: superior properties from hetero-zone interaction. Materials Research Letters 9, 1–31.

Index

Note: Page numbers followed by "*f*" and "*t*" refer to figures and tables, respectively.

C

D

F

G

I

N

P

T

V

FSC
www.fsc.org
MIX
Papier | Fördert gute Waldnutzung
FSC® C083411

Kolibri 360 GmbH
Ferdinand-Jühlke-Straße 7
99095 Erfurt, Deutschland
produktsicherheit@kolibri360.de